# William H. Webb: Shipbuilder

# William H. Webb: Shipbuilder

Edwin L. Dunbaugh
and
William duBarry Thomas

Webb Institute of Naval Architecture
Glen Cove, New York
1989

Frontispiece:
This portrait of William H. Webb hangs in the Main Lounge of Webb Institute of Naval Architecture,
Glen Cove, New York

Library of Congress Card No. 89-50473

Manufactured in the United States of America

This book is dedicated to all of those men and women, past and present, who, educated in the art of shipbuilding through the generosity of William Henry Webb, have contributed to the proud role of the United States on the seas.

# FOREWORD

"The very first naval architect in this country," the *New York Herald* called him, in a tribute reflecting the general opinion of mid-nineteenth century shipping people. This man of dedicated modesty, this William H. Webb, whose story is so well told here—this "affable and interesting talker," this "genial companion," as another newspaper reporter described him—earned that title of "first" by sheer performance.

And what a time to be first in! William Webb was born into pre-industrial America, and served out his youth in a New York familial apprenticeship a medieval shipbuilder would have been at home in. He went on to build his own flash packets, working for his father's friend Marshall of the famous Black Ball Line, and some of the most renowned of the American clippers, those ultimate sailing ships that flashed like meteors across the mid-century sky, outrunning the smoky steamers they competed with on long-haul oceanic trades. Then he went on to build great wooden steamers which, married to the Panama railroad, crowded the clippers out of the California trade, and which went on to carry the American flag across the Pacific to China.

These were the glory days of the American flag at sea, and to be first meant to be at the cutting edge of the young American republic's growing prosperity and power. American shipping opened new channels for American trade around the world. And American shipping populated the nation, bringing in the growing tides of immigrants from Europe, and carrying a flood of restless young Americans to settle the West Coast. Not inappropriately, Webb's most admired clipper was named *Young America*.

Though "congenial," and "affable," as the newspaperman said, Webb was a man who obviously preferred to have his work speak for him. When he did speak it was with compelling sincerity and force of conviction, reinforced by the fact that he spoke only of things he knew about and understood well. He was a superb spokesman for the America of his time. He was proud of being a mechanic, and announced himself as such wherever he went. Non-Americans must have found this fascinating, and European leaders, from Napoleon III to Bismarck, were glad to meet this man who crossed the ocean to call on them in such extraordinary ships—ranging from swift packets and steamers to the monster warship *Dunderberg*, an iron-encased wooden steamer that was soon to be as out of date as the armored knight on horseback.

The hectic pace of the advancing century did eventually overtake Webb's virtuosity, and he showed himself aware of this when he refused to go into building iron and steel ships. But. . . what an unmatched record he'd made in wood!

His ships were fast, very fast, but also capacious and above all durable. A good workman does not like to see things break or wear out through any fault in design or construction, and Webb's ships (except perhaps for the most radical of the immense fleet he was responsible for, the huge extreme clipper *Challenge*), were long-lasting and profitable. His clipper *Young Amer-*

*ica* kept on in the Cape Horn trade to California into the 1880s, into an age
when people came down to the waterfront when she came in, just to admire
her breathtaking grace of line. Webb's ships were, with rare exceptions,
beautiful—they had the integrity and purposeful flow of line that springs
from a happy resolution of forces to a clear purpose.

To be a leader in shipping was to be a leader in America—and certainly
in New York, America's leading seaport, where Webb grew up, launched his
history-making ships, and died. Before his life ended, both political parties
had asked him to stand for Mayor of New York. Unlike his colleague, the
shipping magnate W. R. Grace, he declined the honor. He had other things
to do, things he knew how to do, really, better than anyone else.

What does such a man leave behind him? Well, a large and devoted family,
for one thing. Webb in his long life became pater familias to an extended
family of cousins, nephews and nieces. And in an unassuming but distinctly
patriarchal way, he set up the Webb Institute to give shelter to the old and
infirm, and education to young people in his line of work—they were all,
one feels, "family."

He was a reticent man, personally, as Ed Dunbaugh notes in this distinguished biography. But if you follow Dunbaugh into the crowded streets of
Lower Manhattan, echoing with the ring of caulker's mallet and sledge upon
treenail, you will soon pick up the shape of Webb's beliefs and affections,
and the strength and positive ardor with which he pursued them. And in
Barry Thomas's life histories of the ships, which follow, you will soon pick
up another beat, the integrity that went into anything that Webb set his hand
to. You cannot understand the man, really, if you do not understand the
ships; and through their virtues you can read his—steady, on occasion inspired, always true to very evident principle. William H. Webb would have
it no other way.

Peter Stanford

President
National Maritime Historical Society

# INTRODUCTION

One hundred years ago, a noteworthy institution was created in the United States by a man of vision, a man remarkable in his time and in his profession. That man was William H. Webb. The institution he founded was Webb Institute of Naval Architecture.

Some have called William H. Webb the greatest shipbuilder of the nineteenth century, an era that saw the development of the clipper ship, the steam-propelled vessel and the ironclad warship, on all of which Webb imposed his genius. His reputation was based not only on the number, size and quality of the ships built at his New York City shipyard, but also on his innovative designs. This book is the biography of that extraordinary man.

The book came about because of unbroken affection for and gratitude to Mr. Webb that graduates of Webb Institute have felt through the century since his foresight brought forth the institution bearing his name.

In the waning days of shipbuilding at New York City toward the end of the nineteenth century, Webb concerned himself with the well being of his craftsmen and with the future well being of his profession. In the latter concern, he devoted much energy to the founding of The Society of Naval Architects and Marine Engineers, signing its Articles of Incorporation in 1893 and serving as a Vice President from that year until his death. At the Society's first meeting he pointed out that, "I have been engaged these last four years in preparing an institution that may educate young men to begin this fight, and I hope they will be able to continue it." The fight he referred to was the re-emergence of Yankee sea power on the oceans of the world. The institution was Webb's Academy and Home for Shipbuilders, now Webb Institute of Naval Architecture.

At first, the institution comprised two departments. One was the home for aged ship and marine engine builders and their wives. They were termed the "guests" and all had been in the employ of Mr. Webb while his yard was active.

The other department was the academy for boys whose parents were unable to provide an education in shipbuilding. The 1889 charter read, in part, that the object of the academy was to provide "free and gratituous education in the art, science and profession of shipbuilding—together with free board, lodging and the necessary implements and materials while obtaining such an education."

The century has seen much change. The aged guests are long gone. The name has changed. The Institute has moved from Mr. Webb's original site, then one of the finest within the New York City limits, to its present stately location on the shores of Long Island Sound in Glen Cove, New York. But much has not changed. All students are still selected through rigorous scholastic competition and all students are recipients of the tuition-free scholarship that Mr. Webb declared. Every student is a direct beneficiary of Mr. Webb's largesse. The school remains an independent, small college—perhaps the nation's smallest—with a total enrollment of between 70 and 90 students. With an annual graduating class of often less than twenty there is a strong bond among all alumni and an affectionate and lasting feeling of gratitude to Mr. Webb.

On the occasion of the Centennial of this remarkable school, a committee under the chairmanship of Eugene Schorsch, of the Class of 1952, determined to undertake a true biography of William H. Webb. Professor Edwin L. Dunbaugh and William duBarry Thomas, '51, volunteered to write this volume, drawing upon their considerable skills. Professor Dunbaugh, a member of the Webb faculty and an eminent marine historian, wrote not only of Mr. Webb's life and accomplishments, but also of the development of shipbuilding in New York harbor during the lifetimes of Isaac and William Webb. Mr. Thomas, whose great-grandfather and grandfather both served as Trustees of Webb Institute and whose family has been continuously involved with the school since 1891, contributed the extensively researched histories of Mr. Webb's ships.

Mrs. Barbara Stabile selflessly spent long hours typing the manuscript and preparing it for review. Robert G. Mende, '51, provided guidance in the printing of the finished work, and Henry J. Karsch, '44A, arranged the details of the special limited edition. Gailmarie Carlson assisted from the Institute's Development Office.

The project was ably served by a review committee composed of Messrs. Clifford E. Hoitt, '34; Victor W. Bethge, '53; and VADM B. L. Stabile, USCG (Ret.), President of the Institute.

Most importantly, this book has been made possible by the underwriting generosity of Clifford E. Hoitt, '34, and his wife, Flora. Their unstinting support and encouragement throughout the preparation of the volume reflected the appreciation felt by all alumni toward William H. Webb.

This has been a labor of love—a labor of great appreciation. It will be, we hope, a lasting tribute to one of America's finest men of vision.

Robert T. Young

Chairman Emeritus
Webb Institute of Naval Architecture

Past President
The Society of Naval Architects and
Marine Engineers

Chairman Emeritus and Past President
American Bureau of Shipping

# Table of Contents
# William H. Webb: Shipbuilder



# The Ships of William H. Webb

# List of Illustrations

# William H. Webb: Shipbuilder

by

Edwin L. Dunbaugh

# I

# William Webb's Predecessors: New York Shipbuilders, 1800-1815

THE years between 1840 and 1870 mark the great era of American wooden-hulled ships. At the start of this period the largest ships on the seas rarely reached more than about 200 feet in length or measured more than about 300 tons. By the late 1860s, just before iron and steel began to replace wood in hull construction, shipyards were producing wooden ships twice that length and measuring 4000 tons or more. This era saw the evolution of the ocean-going sailing ship from the bluff-bowed packets popular in the earlier part of the century to the tall, sharp-stemmed clippers, such as *Challenge* or *Flying Cloud*, whose record voyages across the oceans at mid-century were America's pride. The wooden-hulled steam-powered vessel, first introduced by Robert Fulton in 1807, evolved during this era in two different directions. On the one hand were the powerful black-hulled ocean steamships, which reached their apogee with the mammoth 4000-ton paddlers, like *China* or *Great Republic*, built in the late 1860s for the Pacific Mail Steamship Company. On the other hand were the myriad white-hulled steamboats that plied the rivers, sounds, and coastal waters of America, the crowning examples of which were *St. John* or *Drew* of the Hudson River Night Line or *Bristol* and *Providence* of the famed Fall River Line, all also built in the late 1860s and all measuring approximately 3000 tons.

In addition to these wooden titans of the sea there were, of course, hundreds, even thousands, of other smaller sailing craft, mostly schooners or sloops, built to carry coal or lumber or cotton; and there were also tugboats, small commuter steamers, cargo carriers, steam or sail, of every size or description, as well as yachts and small private sailboats.

This age of wooden-hulled ships was also the great age of American shipbuilding. Many fine ships took shape along the banks of the Clyde or elsewhere in Great Britain or on the Continent. But by far the greater number of the largest, the fastest, and the most innovative ships in this era came from American yards, largely because America possessed a wealth of every sort of timber needed for different aspects of shipbuilding that no European country could duplicate. Also, with their long coastline, Americans more than any other people, except possibly the British, were dependent for their sustenance on the sea. Something must also be said for the fact that American shipbuilders, very often first-generation entrepreneurs separated by an ocean from tradition-bound Europe, were rarely afraid to risk their reputations by trying something new or building something better, if they thought it would work.

In the early years of the century small ships were being built in the shipyards of many American coastal cities but mainly in Boston, New York, Philadelphia, or Baltimore. In the years following the War of 1812, however, and especially in the 1820s and 1830s, as orders for ships began to expand, New York, with its proliferating shipyards along the East River, soon outstripped the other ports and maintained its lead throughout the era of wooden ships.

Between the 1820s and 1870s, when iron hulls began to become popular and the demand for wooden-hulled ships dropped precipitously, there were a great many highly respected shipyards in New York, a few in Boston, such as that of Donald McKay, and also in Philadelphia or Baltimore. They turned out fine ships, some of which have earned enduring fame. But virtually all contemporary sources attest to the fact that the shipyard operated by William Henry Webb from 1840 to 1872, through the height of the great age of wooden ships, produced the largest number of hulls and won the soundest reputation for building strong and reliable ships in our country.

When William H. Webb, the son of a New York shipwright, was born on 19 June 1816, the American republic was just short of forty years old, and James Madison, its fourth President, was completing his last year in office. Just the year before, the United States had ended its second war with Great Britain—during which shipbuilding had been virtually paralyzed—and by 1816 was just beginning what was to become a spectacular recovery.

At the time of the Revolution New York had been America's second city, but in the census of 1810 it had passed Philadelphia and by the time of Webb's birth had reached nearly 120,000, almost double that of 1800. In 1816 the city proper did not extend above Canal Street. Beyond that were scattered farms and villages. Most of the businesses were located along the land side of South Street, which, then as now, ran along the East River. Successive land fills prior to that time had moved the waterfront from Pearl Street to Water Street, then to Front Street, and finally to South Street. Businesses congregated there because ships arriving in New York docked along this river. The East River, which is not in fact a river, but rather a narrow tidal strait connecting Long Island Sound with New York Bay and eventually the Atlantic Ocean, is known for its fast-flowing currents which often make docking or launching a ship difficult if not dangerous. But shippers preferred the East River nevertheless, because the more placid Hudson (or North River, as it was called in the city) on the opposite side of the island frequently froze over in winter.

Since businesses tended to locate along the East River, the West Side, from Broadway to the Hudson, remained mostly residential. There were still a few fine mansions north of the city fronting on the water along the East River, some dating from Colonial times, but these were soon to be engulfed by the unrelenting northward growth of the city. Those on the Hudson were to last longer. By 1816, however, many of the city's finer homes were located around the Battery, so that this area looked much as Charleston's Battery still appears today. Homes and churches also stretched along tree-lined Broadway from the Battery, past the new City Hall and its lovely park, almost to Canal Street, although by this time some of the houses near Wall Street were beginning to be converted to shops and counting houses, and business in general seemed inclined to expand westward. Across the river, Brooklyn, Williamsburg and Greenpoint were still small villages surrounded by farms, as were places such as Greenwich or Chelsea to the north on Manhattan.

The rows of wooden scaffoldings of New York's busy shipyards, including the one belonging to Henry Eckford, where William Webb's grandfather worked, could be seen along the marshy banks of the East River north of the city. By 1816 these yards extended from Montgomery Street, then in the northern part of the business district and just south of Canal Street, for about a mile up to Corlear's Hook, where the marshes gave way to a smooth

[From a Painting in the possession of the Long Island Historical Society.]

Shipbuilder Henry Eckford, born in Scotland in 1775, established his shipyard on the East River in 1796. He was Isaac Webb's mentor and his name was passed on to Isaac's sons, William Henry and Eckford.

sandy beach, a favorite spot for Sunday outings and church picnics.

One of the earliest yards was built by Thomas Cheeseman at the foot of Rutgers Street shortly after the War of Independence. By 1800 this yard had been taken over by his son, Forman Cheeseman, with a partner, Charles Brownne (1). In 1804, with business so prosperous they could no longer operate within the constraints of a small yard on the rim of the growing city, Cheeseman and Brownne purchased land along the river way north of the city, even farther north than Corlear's Hook. What they bought was a large tract of solid ground, then known, for reasons now forgotten, as "Manhattan Island," that rose amid the marshes in an area now largely occupied by the Houston Street entrance to the East River Drive. It was at this yard three years later, shortly after Forman Cheeseman had resigned, that Charles Brownne built the hull for Robert Fulton's *North River Steam Boat* (sometimes erroneously called *Clermont*), the world's first commercially successful steam-powered vessel.

Working in Brownne's yard at the time were two brothers, Noah and Adam Brown (no relation to their employer), who were soon to become respected shipbuilders in their own rights. By 1807, when they were helping Brownne build Fulton's steamboat, the brothers Brown had already experienced what some might consider a rather full life. Noah, born in 1770 (the same year as the Boston Massacre), and Adam, five years younger, had begun life in a small frontier village in what is now western New York. When their father and three older brothers were massacred in an Indian raid, the mother took her surviving sons back to her parents' home in Stamford, Connecticut (2).

Here, it would seem, they could have come to know Wilse Webb, who, born in 1767, was just three years older than Noah. Wilse Webb was the eldest son of Epenetus Webb, Jr., a respected mill owner in Stamford, a town where the Webb family had been among the original settlers in the 1640s.* Wilse Webb did not work in his father's mill, however. He loved sailing ships and preferred employment at the small shipyard out at Shippan Point.

Noah and Adam Brown meanwhile had become house carpenters in Stamford. However, in 1792, by which time Noah was twenty-two and Adam seventeen, the brothers decided to move on to New York, where this rapidly-growing city offered ample opportunities for their trade. Eight years later in 1800, Wilse Webb surprised his family by announcing that he also would move to New York. Shipbuilding in Stamford was not particularly prosperous at the time, and he wanted to try his luck at one of the big shipyards along the East River. So Wilse Webb, who was then already thirty-three, and also his wife, Sarah Jessup Webb, and their four children—Mary, aged eight; Isaac, aged six; Abigale, aged three; and the baby Harriet—packed all of their belongings in a wagon and left for the big city (3).

But Wilse did not have much luck finding shipyard work in New York either. As a result, although he had been the son of a fairly prosperous mill owner back in Connecticut, he had now to support his family as well as he could as a day laborer, mostly as a house carpenter, quite likely working with his friends Noah and Adam Brown. Since Wilse obviously lacked the cash to buy a house of his own, he had to rent, which meant that the

family was obliged to move often. Their first house in the city was on Hudson Street, a pleasant enough area, but far on the west side and some distance from the shipyards.

By 1804 shipbuilding in the city was becoming more active, probably because by that time the British were devoting their energies, and to some extent their merchant marine, to the war with Napoleon, and American entrepreneurs were taking advantage of this situation to establish commercial ties with South America or with other areas which had previously been purely British preserves. Thus in that year the Brown brothers and Wilse Webb all found employment in New York shipyards: Noah and Adam Brown with Cheeseman and Brownne, and Wilse Webb with Christian Bergh. When Webb finally got his job in a shipyard, which had been his ambition ever since moving to New York, he and his family were able to rent a house on Cherry Street, nearer the East River yards, where they remained for several years, and where Isaac Webb, now ten, passed his youth (4).

When Wilse Webb went to work at Bergh's yard, his foreman there was a Scotsman eight years Wilse's junior, named Henry Eckford. Born in 1775 in Kilwinning, Scotland, he had made his way alone at the age of sixteen to Canada. Here he apprenticed himself to an uncle who was a shipwright in Quebec. Apparently no great bond developed between uncle and nephew, for in 1796, as soon as he turned twenty-one and his apprenticeship ended, Eckford packed his bags again and headed for New York. For a while, young Henry Eckford took jobs in various shipyards, but by 1799 he had married, and by 1801 he had opened a small yard of his own across the river from Corlear's Hook near the Brooklyn Navy Yard, from which he obtained some of his earliest orders for ships.

Among the ships built by Eckford at this yard was a 400-ton packet for John Jacob Astor. Having already amassed America's largest fortune bringing beaver pelts and other goods back from the still largely uncharted regions of the Northwest, Astor was now ready to enter the China trade and needed a staunch ship for the purpose. The ship was named *Beaver*, for the species to which he owed his fortune. As a testimony to Eckford's talents, one might note that *Beaver* remained in active service over forty years, about twice the average age of a wooden sailing ship; and when she was broken up in the late 1840s, many of her timbers were used in a new ship then building.

After a short stint running his own yard, Eckford returned to Manhattan to accept a position as foreman at the yard of Christian Bergh. In the next few years, these two men, Christian Bergh in the office and Henry Eckford

---

* See Appendix A for a genealogy of the Webb family.

in the yard, not only built a reputation as shipbuilders but also developed a close personal friendship that was to last the rest of their lives (5).

It was during this period when Henry Eckford was a foreman at Bergh's yard that Wilse Webb came to work there in 1804. One year later, with the blessing and full cooperation of his friend Christian Bergh, Eckford decided again to open a shipyard of his own, this time, however, in partnership with Edward Beebe, whose role seems to have been limited to supplying capital. Their yard was at the foot of Jefferson Street, just south of Cheeseman and Brownne on a site now occupied by the LaGuardia Houses. When Henry Eckford left Bergh to set up his own yard in 1805, Wilse Webb, in whom he apparently placed great confidence, went with him as a carpenter.

Once on his own, Eckford soon earned a reputation as one of the most brilliant young ship designers in the country. According to John H. Morrison:

Henry Eckford, through his thorough knowledge of the trade, and skill in designing a vessel, built up a reputation in a few years as a first-class shipbuilder... Taking advantage of his practical experience he made many changes in the forms of his vessels that proved of value. (6)

In the Spring of 1807, Henry Eckford was invited by Charles Brownne in the yard next door to attend the launching of Robert Fulton's steamboat. That same summer saw the opening of what was to become another famous New York shipyard, for shortly after completing Fulton's boat, Charles Brownne sold his assistants a large tract of the land on the north side of "Manhattan Island" which now became the shipyard of Adam and Noah Brown.

This year of 1807, however, was not to be a propitious one for American shipbuilders. By 1807, following the Battle of Friedland in June, Napoleon had forced most of the monarchs of Europe, including those of the German states and of Russia, to recognize his dominant position. Only England remained completely independent. Napoleon and the British now faced each other for a final showdown, and since England was an island, each knew that its position depended on control of the seas. Each side therefore announced that it assumed the right to attack, capture, or sink any vessel attempting to trade with the other. These orders, needless to say, dampened the enthusiasm of American merchants ordering ships for trade across the Atlantic.

Matters worsened when the British, needing to maintain the strength of their navy, began a policy of kidnap-

ping men from American naval or merchant ships on the ground that they might be British. President Jefferson's cautious response to this outrage was to declare an embargo—i.e., a ban on all trade with belligerent powers—in December 1807, thus virtually putting New York shipbuilders out of business. Given this situation, Christian Bergh felt obliged to accept an invitation from the government to build a naval brig on Lake Ontario. This 85-foot brig was hardly a major war vessel and in better times would hardly have tempted a shipwright of the stature of Christian Bergh to leave his New York yard and make a long trip into what was still fairly primitive country. But with business in New York at a standstill, not only did Bergh go with several of his men, he was even able to persuade his friend Henry Eckford to join him. There is no evidence to show whether or not Wilse Webb accompanied Bergh and Eckford to Lake Ontario, but it seems likely that Webb remained in New York.

One of Thomas Jefferson's last acts before turning over the presidency to James Madison in March 1809 was to lift the unpopular trade embargo. By this time Bergh's brig (called *Oneida*) had been completed and launched, and both Bergh and Eckford as well as all their men were back in New York and ready to return to work. One advantage to the trip to Lake Ontario was that Eckford now had enough capital to buy out Beebe and to move to a new location of his own. As it happened Charles Brownne was now ready to retire and was willing to sell Eckford his southern half of "Manhattan Island." With Brownne now out of business, the two major yards in the city—other than Christian Bergh's—those of Adam and Noah Brown, and Henry Eckford, were located next to each other on "Manhattan Island" at the foot of North (now Houston) Street and Stanton Street (7).

In setting up his own yard in the new location, Henry Eckford reassembled much of the crew that had worked for him at Jefferson Street before the embargo, including, of course, Wilse Webb, who now appears on the roster for the first time not as "laborer" but as "shipbuilder" (8).

In 1809 Wilse Webb's only son Isaac was just fifteen and had begun begging his father to let him leave school and start learning the shipbuilder's trade. The father, whose own experience with shipbuilding had been neither particularly successful nor profitable, resisted encouraging his son's ambitions in this direction. But when Isaac persisted, Wilse Webb turned to his employer, Henry Eckford, for advice. Eckford's response was to tell Webb to have his boy come to see him.

Needless to say, an eager young Isaac Webb very soon presented himself in Henry Eckford's office. Their talk lasted a long time. When it was over, Eckford told young Isaac Webb to come back on his sixteenth birthday, and that he would be happy to take him on as an apprentice. Isaac was elated. His ambition had been merely to become a shipyard worker like his father, but to be told he would be taken on as an apprentice by one of New York's leading shipwrights was an honor he had never expected. Of course, in the process, the whole question of the boy's future had been taken out of his father's hands.

Isaac Webb turned sixteen on 8 September 1810, and on that day he was at Henry Eckford's shipyard ready to start his apprenticeship. This ritual required a contract, duly signed by Eckford and also by Wilse Webb and by his son, according to which Isaac Webb became all but the property of Henry Eckford. He had to promise always to do his master's bidding and never to betray his master's trade secrets. For his part, Eckford accepted the obligation to teach young Webb the craft and the "mystery" of the shipbuilder's trade. He also took from the father the responsibility for Isaac's behavior and moral education. The signing of this contract was a solemn rite, and it was solemnly regarded by all three parties concerned.

This must have been a proud day for Wilse Webb. Although he now carried the title of "shipbuilder," he was at best, at the age of forty-three, an employee in a shipyard. For his only son to be accepted as the legal ward and protege of one of New York's most respected shipwrights meant the promise of a future for Isaac far better than anything he had been able to attain for himself. Although Isaac continued to live at home with his parents and his sisters, his only real parent now, in a sense, was Henry Eckford, and during the five years of the apprenticeship, a very deep and lasting friendship, based on sincere mutual respect, developed between Isaac Webb and his mentor Henry Eckford.

Apprentices were apparently in demand in these years, for about this time we find an ad for shipyard apprentices that included the promise to teach "the art and the mystery of shipbuilding" (9). And indeed a mystery it often was, for such things as careful measurements or mathematical calculations did not become commonplace in the business before about the 1840s, and in fact, William H. Webb was one of the first to build ships in such a scientific fashion. Before that time the best caliper was a master shipbuilder's experienced eye.

In these days a young man did not attend a college or university to learn the abstract aspects of ship design. Rather, usually at the age of sixteen, he would seek a reputable shipbuilder and sign on as an apprentice, provided he found a shipbuilder who believed his potential worth the effort. An apprenticeship was a contract in which both parties accepted specific obligations. The shipbuilder had to promise his apprentice to teach him the trade, so that, theoretically, when the young man reached maturity at twenty-one and was released from the contract, he would have achieved a skill with which he could earn a proper living. As indicated above, the master shipbuilder also took on many of the responsibilities of a father for the term of the apprenticeship. He had to guarantee adequate food and board, and he become the guarantor of his ward's morals and personal behavior. Given this contractual relationship, some shipmasters took their apprentices into their homes where they were treated with some of the same affection as well as the same severity that would be accorded one of their own sons. More often, however, apprentices were boarded at the shipmaster's expense, at some local hostelry, though the nature of some of these "homes," according to contemporary accounts, must have made the master's job of guarding his ward's morals something of a challenge.

The apprentice, on his part, was expected to do any of the work in the yard that was asked of him. After all, his object was to learn all aspects of the trade. He was expected to work from sunup to sunset, which in summer could mean from 4:30 A.M. to 7:30 P.M., though Sundays and holidays were free days. In return the shipmaster usually granted his apprentices a munificent allowance of $2.50 a week.

Once an apprentice reached the age of twenty-one and was released from his contract, he was qualified to call himself a "mechanic," which was the generic name given to all experienced shipyard hands. The main difference between an apprentice and a mechanic was that a mechanic could earn all of $1.25 a day for his work, or even $1.50. He not only had to feed himself with this pay, he also by this time usually had a family and home to support as well. In fact, there was then little specialization of labor. "Mechanics," apprentices, and often even the shipmasters themselves could be found sawing or carrying wood, setting up ways and scaffolds, planking hulls with trunnels (tree nails or wooden dowels) or caulking a finished hull, in short doing whatever jobs were demanded depending on the stage of construction of the ship in the yard. For the mechanics and apprentices both, and also presumably for the shipmasters, the long sunup-to-sunset day was traditionally punctuated by an hour for breakfast during the morning, an hour or sometimes two for dinner at midday, and a break for grog—which mariners apparently preferred to the British tradition of tea—in the late afternoon. The timing of these breaks often depended on the whim of the shipmaster or yard foremen. And even the quitting time at sunset might be postponed if the shipmaster were in the middle of some pet project he believed could be finished by candlelight.

As shipbuilding in New York again reached its normal pace following the lifting of Jefferson's embargo, both Wilse Webb as a "shipbuilder," and his son Isaac, as an apprentice, continued their dawn-to-dusk service in Eckford's yard on "Manhattan Island," Isaac undoubtedly sometimes working alongside his father and the other carpenters, but probably more often receiving patient and detailed instructions from Henry Eckford himself. Other apprentices serving under Eckford in these years included Jacob Bell, John Dimon, and Stephen Smith, all of whom, as well as Isaac Webb, were later to become successful shipbuilders in their own rights. Since Jacob Bell and Stephen Smith had both come down to the city from Stamford, Connecticut, the home town both of Adam and Noah Brown and of Wilse Webb, it seems probable that they were family friends of the Webbs or the Browns or both. In any event, working together under the guidance of Henry Eckford, all four of these young men became and remained good friends during and after this experience.

The period of peaceful shipbuilding, however, was not to last. By 1812, less than two years after Isaac Webb began his apprenticeship, continued impressment of American seamen by the British as well as other issues forced President Madison and the Congress to declare war. And, with the British Navy now effectively blockading our harbors, once again American ships sat rotting at their piers, and the East River shipyards became idle. Soon after the war began, however, the American government became aware of the immediate danger of an attack from Canada across the virtually undefended Great Lakes. What was needed was the creation in record time of an entire Lakes navy, a job roughly analogous to the shipbuilding miracle performed by people like Henry Kaiser in the 1940s.

On 5 September 1812, Governor Daniel Tomkins, Mayor DeWitt Clinton, and Commodore Isaac Chauncey of the Brooklyn Navy Yard, met at City Hall to discuss the emergency. Commodore Chauncey had recently been reassigned to a small naval base at Sackett's Harbor, at the northeast corner of Lake Ontario, near the entrance to the St. Lawrence River. These men decided on the immediate necessity of establishing a shipyard at Sackett's Harbor in order to build a navy for the defence of the Great Lakes area. Remembering the remarkable speed with which Christian Bergh had assembled the brig *Oneida* in 1809, they summoned him to discuss the matter and asked him to bring other interested shipbuilders.

That afternoon the three men met again at the home of Colonel Henry Rutgers, north of the city on the East River. Joining them there was Christian Bergh and with him Henry Eckford, and both Noah and Adam Brown. Remembering too well the bone-chilling winter winds of Lake Ontario, Bergh, who was now nearly sixty, decided he did not wish to spend another winter on Lake Ontario, but he suggested the Navy assign his friend Henry Eckford to the job. Eckford accepted and, with business in the city at a standstill, the Brown brothers agreed to assist him. Eckford, with headquarters at Sackett's Harbor, was to supervise the entire project and provide most of the designs for the ships to be built. And, theoretically under Eckford's supervision, but, given the distances, probably operating fairly independently, Noah Brown was dispatched to Lake Erie and Adam Brown (somewhat later) to Lake Champlain to set up shipyards and start building. Christian Bergh would remain in New York to recruit carpenters and to arrange for the shipments of needed supplies.

Within twenty-four hours, Eckford had recruited forty ship's carpenters, most but not all from his own yard, and was ready to leave for Sackett's Harbor. Late in the afternoon of 6 September, Henry Eckford, his forty workers, and his apprentices, including Isaac Webb, boarded *Paragon*, one of Robert Fulton's pioneer Hudson River steamboats, for Albany. Also aboard was Governor Tomkins, returning to his duties in the capital.

With *Paragon*'s big paddles thrashing furiously upriver at what was then the very respectable speed of six miles an hour, the trip to Albany lasted thirty-six hours. Since this would have been the first time that Isaac Webb or the other apprentices had ever seen a marine steam engine in operation, and could even have been the first time Henry Eckford had seen one, we can assume that many of their waking hours must have been spent observing the engine's noisy mysteries.

*Paragon* paddled into Albany on 8 September, Isaac Webb's eighteenth birthday. From here Eckford and his crew made their way westward along the Mohawk River, alternating between small boats and hired carts, as far as Rome, New York. Working their way from Rome northward, however, as the party passed from the country roads connecting the villages of the upper Hudson valley into the completely unsettled and heavily wooded areas farther north, their progress became increasingly difficult. For much of the last leg of their expedition, Eckford's crew had to hew their own roads through the forests with their carpenter's tools (10).

Sackett's Harbor, once they got there, was a small village with several houses to offer welcome, and the small naval base not too far distant. Some of Eckford's men found bunks in the base barracks; some pitched tents near the shipyard; while a lucky few boarded in local homes. Eckford took over one of the houses as his headquarters, and his two apprentices, Isaac Webb and Stephen Smith, shared a bed upstairs in this house (11).

In the course of the next few months, in spite of the numbing cold of a Great Lakes winter, Henry Eckford and his fellow shipbuilders did indeed manage the kind of naval miracle that Americans have been known to produce in wartime emergencies. On all three lakes (Erie, Ontario, and Champlain) they set up shipyards with whatever materials they found at hand and started building. On Lake Ontario, Eckford's first ship, *Madison*, was completed in just eighty days. Later the crew learned to work even more efficiently, and some ships were completed in as little as thirty days.

For the men the pay was good, as it would have to have been to lure them away from their wives and families into this cold wilderness. As compensation for the Spartan conditions, the men were paid $1.75 a day plus board, as opposed to the $1.25 or $1.50 pay, without board, that was typical in the shipyards in New York.

In the course of the war, the navies these men produced on the Lakes performed a very valuable and necessary service in preventing a major British invasion. As it happened, the Lake Ontario theater, where Eckford was working, remained relatively quiet. But the ships built by Noah Brown for Commodore Oliver Hazard Perry on Lake Erie, and those by Adam Brown for Commodore Thomas Macdonough on Lake Champlain have been credited with maintaining American domination of the Lakes throughout the war.

On 11 February 1815 a ship arrived in New York with the welcome news that John Quincy Adams and Henry Clay had signed a treaty of peace with the British on Christmas Eve of 1814. The war was over at last and the whole city went wild with celebration and relief. Eventually the good news penetrated as far as Sackett's Harbor, and, leaving two partially completed hulls on the stocks, Henry Eckford and his shipbuilding family, happily packed up their tools and their belongings to make their way back to New York.

# II

# Isaac Webb: Shipwright, 1815-1831

SHORTLY after their return from Lake Ontario, Henry Eckford summoned Isaac Webb to his office to discuss the younger man's plans for his future. Isaac would turn twenty-one in September, which meant that his apprenticeship would be completed. Eckford pointed out that Isaac had three options open to him. First, as Eckford could see that he had genuine talents as a shipwright, Isaac should know that he could have a job in Eckford's yard, one that would offer security and promise of future advancement. Or, of course, Isaac would be free to find a position in another yard, as many new yards were opening in this post-war period. It was at the third option, however, that the young man's eyes lit up with enthusiasm: he could, now that he had finished an apprenticeship, open his own yard. Yes, this was what Isaac Webb wanted; this was what he had been waiting for. And his father, Isaac pointed out, had already put away $1000 to help him get started.*

Eckford was supportive of Isaac Webb's wish to start a yard of his own, but he urged caution. He suggested that, instead of taking such a risk by himself at so young an age, Isaac should start by joining in a partnership with his fellow apprentices, Stephen Smith and John Dimon. If his three apprentices were to open a yard together, Eckford promised to do what he could to help them until they could manage on their own. Isaac Webb was clear about the fact that this was not what he had in mind, but he allowed that he would be willing to try it "for now." Thus later in the year 1815, when he had finished his apprenticeship, Isaac Webb became a partner in the new yard of Webb, Smith, and Dimon, located south of Corlear's Hook (i.e., quite far downtown for a shipyard) on Water Street at the foot of Montgomery Street (2).

Even before the yard opened for business, however, Isaac Webb had an even more important matter to attend to. As an apprentice, of course, Isaac had not been allowed to marry. But apparently he and the young woman of his choice had already made plans, for as soon as the apprenticeship ended, Isaac and Phebe, the daughter of a Huguenot family which had settled first in New Rochelle, New York, but more recently in Manhattan, were married. Their first home was on Stanton Street, between Lewis and Cannon Streets, in a district where many of the shipyard workers made their homes, near the river between Corlear's Hook and "Manhattan Island." ** And it was here that the young couple's first child, William Henry Webb, was born on Wednesday 19 June 1816. The boy was named William for one of his father's uncles. The middle name, Henry, was chosen to honor Isaac's mentor, benefactor, and close friend, Henry Eckford.

By the time his first son was born, Isaac Webb, who was not yet twenty-two, and his partners were already hard at work on their first contract, the construction of a hull for the steamboat *Chancellor Livingston*, the fifth, last, and by far the grandest of the steamers built for the Hudson River line operated by Robert Fulton and his partner, Robert Livingston. It should not come as a surprise that this first job was actually a subcontract supplied by Henry Eckford (3).

All of Fulton's first steamboats had been built by Charles Brownne. But when Brownne retired, Fulton turned first to Brownne's successors, Adam and Noah Brown, for his next boat. Then in 1815 he asked Henry Eckford to build *Chancellor Livingston*, which presented something of a challenge, for there was then little precedent for designing a hull that had to carry a heavy engine

---

* Since Wilse Webb probably could not have been earning more than $1.50 or $1.75 a day at this time, $1000 would have represented about two years' income. Considering that these had been lean times in the shipbuilding business and that he and Sarah had three daughters living at home as well as Isaac, now that he was back, it seems unlikely that Wilse Webb could have saved such a large amount for his son without great sacrifice. Another source gives the more likely suggestion that the $1000 represented bonuses given Isaac by Henry Eckford during their stay on Lake Ontario, which Wilse Webb had been saving for his son (1).

** The area where this house was located, just two blocks north of the Manhattan entrance to the Williamsburgh Bridge, is now occupied by the Baruch Houses.

as well as the usual load of cargo and passengers. By the Spring of 1815, however, the plans had been completed and had been approved by Fulton. Unfortunately, in July, before construction began, Robert Fulton, still a relatively young man, caught a fever and died.

Whereas the earlier steamboats designed by Fulton himself, including *Paragon*, on which Eckford and his men had journeyed to Albany in 1812, had been little more than flat-bottomed barges fitted with engines and sidewheels, *Chancellor Livingston* was the first steamboat which was in every sense a "ship" with engines. Although by necessity of light draft (though not light enough apparently: alone of Fulton's steamers, she often had to transfer her passengers to coaches several miles below Albany when the river was running low, because she drew too much water), she otherwise had most of the characteristics of an ocean-going ship, such as a scrolled prow with a bowsprit and a slightly rounded hull, wide toward the bow and tapering to a high, richly decorated transom stern. With her engines placed roughly amidships, she carried beautifully furnished cabins fore and aft. An interesting innovation on *Chancellor Livingston* was another enclosed cabin, for women passengers only, placed on the after deck above the hull with an awning-covered sundeck above that (4).

*Chancellor Livingston*, with her two smokestacks (later three) athwartship, was a handsome as well as impressive ship and represented a major step forward in steamboat design. She was also at the time the largest and fastest steamboat afloat. In operation by 1817, she made the run from New York to Albany in eighteen hours, just half the time of *Paragon* only five years before. Hudson River steamboats were soon to cut this time by half again, but never by much more than that. After the Fulton-Livingston company closed down in 1825, *Chancellor Livingston* ran several years as a very successful opposition boat on Long Island Sound and ultimately ended her career as a pioneer steamboat out of Boston on an open-ocean run to the coast of Maine.

In 1817, two years after his three apprentices had left his tutelege, Henry Eckford succumbed to an offer by the Brooklyn Navy Yard to be their chief naval constructor. He seems not to have closed his shipyard, however, but rather to have divided his time between his own yard and the Navy Yard, though, obviously, he was not able to accept as many contracts as he had before. It seems also that during this period, in many cases, he accepted contracts, designed the ships, and then subcontracted their construction to Webb, Smith and Dimon (5).

In 1819 Henry Eckford received one of the most challenging contracts of his career when David Dunham, who had been operating sailing packets on various routes along the Atlantic coast, asked him to design and to build the steamship *Robert Fulton*—the first steamship to be designed specifically for an open-ocean service—for a route from New York to New Orleans by way of Charleston and Havana. Although, as most readers are aware, *Savannah*, which crossed the Atlantic that same year, is credited with being the first ocean-going steamship, *Savannah* was an experiment and as such she was not much of a success. She was essentially a sailing packet into which a steam engine had been fitted. For much of her voyage the collapsible paddles were folded on the deck while *Savannah* proceeded under sail, making as good time, by the way, as she did under steam. Also, since the wood carried for fuel took up a great deal of space, one of the conclusions from the *Savannah* experiment was that steam was clearly impracticable for transoceanic cargo carriers. Eckford was prepared to prove otherwise, and his *Robert Fulton*, the keel for which was laid at Eckford's yard before *Savannah*'s experimental voyage, was to become the first commercially successful American ocean-going steamship.

Although Eckford was eager to provide the design of this ship for Dunham, his responsibilities at the Brooklyn Navy Yard did not permit him to devote full time to supervising its construction. The one person Eckford trusted to act as his deputy in building a ship on which his own future reputation might rest, was the then twenty-four-year-old Isaac Webb. Yielding to Eckford's appeal, in 1819 Isaac Webb left his partners and returned to the yard of his former mentor to supervise the construction of *Robert Fulton*.†

Designing an ocean-going ship which had to carry heavy iron machinery amidships, in addition to the usual amount of cargo, presented problems never previously encountered by an American shipwright. All of the steam-driven vessels built in the dozen years since Fulton's first steamboat had been designed for service on the relatively

---

† In 1819, the year their son William turned three, Isaac and Phebe Webb's second child, Sarah Eckford Webb, named for Isaac's mother, Sarah Jessup Webb, was born in the house on Stanton Street (though later that year the family moved farther south to Water Street, near Walnut, closer to Isaac's shipyard). One might note that, of the five children born to Isaac and Phebe Webb between 1816 and 1823—William Henry Webb, Sarah Eckford Webb, Abigale Henrietta Webb, Samuel Wilse Webb, and Eckford Webb—all but Samuel bear one name honoring Henry Eckford.

placid waters of the Hudson River, Long Island Sound, or Delaware Bay. No one had yet provided any precedent for an engine-powered vessel strong enough and stable enough to operate exclusively and in all seasons on the open ocean.

*Robert Fulton* was not, like *Savannah*, a sailing ship with a steam engine. She was designed and built as a steamship with auxiliary sails. Her frames were much closer together than had been customary on sailing hulls, and she was planked with thicker and stronger timbers. She was also the first steamship designed specifically to burn coal, which used far less storage space on long voyages than wood.

*Robert Fulton* was launched at Henry Eckford's yard in May 1819, but building and installing her engine at the Allaire Works (successors to Robert Fulton's own engine-building operation) took almost a year. The eventual cost of this first ocean steamship, $130,000, was so great that part of the payment included bringing Henry Eckford in as part owner with David Dunham and other lesser investors (6).

Finally, on 25 April 1820, *Robert Fulton* sailed south out of New York on her maiden voyage to Charleston, Havana, and New Orleans. Whatever reservations shippers may have had about the viability of sending a steamship into the ocean were soon allayed by a statement issued by *Robert Fulton*'s passengers on their return to New York:

Having experienced on our passage, several heavy blows, we have had good opportunities to judge whether any good cause existed for the apprehension at first entertained as to her safety in heavy weather, and do not hesitate to declare them groundless. Being unencumbered with that heavy weight of spars and rigging, she consequently rolls and labours much less than a vessel rigged in the usual way, and propelled only by sails; and the manner in which she contended for several days with a heavy head sea, and wind, convinces us that. . . [she] will never be materially affected by the weather she has to encounter. (7)

Isaac Webb stayed on with Henry Eckford for his next assignment, the large warship *Ohio*, which was built at the Brooklyn Navy Yard. For this ship, Henry Eckford is listed as "Designer," and Isaac Webb as "Master Builder." Although *Ohio* was, in every sense, a successful vessel, since Eckford had made some modifications in her design of which the commandant of the yard had not approved, she was rarely taken to sea once completed. Frustrated by the restraints on his creativity by this and other incidents, and aware, particularly after the acclaim accorded *Robert Fulton*, of the lucrative opportunities open to him in private shipbuilding, Henry Eckford tendered his resignation to the Brooklyn Navy Yard as of 1 June 1820 and returned to work full time in his own yard at "Manhattan Island." Before leaving the Navy, Eckford recommended Isaac Webb as a possible successor as Chief Naval Constructor. Webb, however, wisely declined and instead stayed on as an assistant at Eckford's yard (8).

The next decade marked the high point of Henry Eckford's distinguished career. Though some other yards certainly built more ships than Eckford's in this period, no other ship designer was more highly regarded. And, needless to say, no small part of Eckford's reputation reflected, quite deservedly, on the trusted associate Isaac Webb, who managed his yard and supervised the construction of his ships.

One reason for the success of Henry Eckford, as well as of all of the shipyards now proliferating along the East River waterfront as far north as Tenth Street, was the phenomenal growth in American commerce in the period following the War of 1812. Immigrants arriving on our shores by the thousands contributed to the doubling of the populations of our coastal cities virtually every decade. At the same time, these people plus the offspring of earlier settlers, were rapidly filling in both the Northwest Territory and the vast lands of the Louisiana Purchase, in the process producing a bumper agricultural surplus for export and providing a nearly insatiable market for imports. Also in this era, southern plantations were sending literally hundreds of bales of cotton daily up to the port of New York via sailing packets. Here they were transferred to cargo sloops or to one of the Long Island Sound steamboats for delivery to the mills and fabric factories then being set up in eastern Massachusetts, or to a large packet for delivery to English factories.

These were also the years in which Latin American countries, taking their cue from the United States, were fighting for their independence from Spain or Portugal, for which they needed their own navies. In the process they also opened most of Central and South America to enterprising American traders, who in turn needed ships to conduct their business. And too, for a country not yet completely converted to coffee, American packets were now competing with British ships sailing all the way across the Pacific to bring home not only tea, but also the exquisite silks and ceramics that some sophisticated Americans were beginning to appreciate.

In this era many powerful merchant houses appeared in New York to manage this prosperous trade, among which Goodhue and Co.; Grinnell, Minturn and Co.; G.

The illustrious naval architect John W. Griffiths. The original print of this portrait was given to the late Cedric Ridgely-Nevitt, '39, in 1938 by Griffiths' granddaughter.

G. and S. Howland (after 1832, Howland and Aspinwall); A. A. Low and Bros.; or N. L. and G. Griswold were the best-known. These companies generally ordered their own ships for trading down the coast, with Europe, or even with China.

It was also in this period, on the other hand, that many shipping companies were formed. Instead of individual owners or captains sailing their ships wherever a good lading led them, these companies advertised sailings at regular times for specific ports. These lines might order several ships at a time, usually from different shipbuilders. Such a company was the famed Black Ball Line, founded in 1817 by a consortium of American merchants. With their four initial packets of roughly 400 tons each, the Black Ball Line, operating between New York and Liverpool, was the first packet company to schedule regular sailings between two ports, in the process establishing an American dominance of the North Atlantic route that was to last for many years. Their ships were advertised to leave on a specific day whether or not they had loaded a full cargo or had a full complement of passengers. Within another decade, by which time the line was operating eight or ten vessels, a Black Ball ship was scheduled to leave each port on the first and fifteenth of every month. The Black Ball Line was to last as a scheduled carrier on the Liverpool route until the late 1870s.

Two local events at this time also helped bring business to the New York shipyards. Before 1824 the company formed by Robert Fulton, the inventor, and his politically-influential backer Robert Livingston held an exclusive charter for operating steam-powered vessels in the waters of the state of New York. This restriction, needless to say, limited the number of steamboats that could be built in the yards of New York, or indeed anywhere else, since much American commerce depended on the use of the port of New York. Then in 1824, when Daniel Webster brought the matter to the Supreme Court in the case of Gibbons v. Ogden, Chief Justice John Marshall found that, as the regulation of interstate commerce had not been specifically assigned to the states by the Constitution, it was therefore a function of the federal government. Thus, while the state of New York could pass laws concerning commercial vessels operating completely within its own waters, regulating ships operating between the state of New York and other states was unconstitutional. This landmark decision had the immediate effect of rendering the Fulton-Livingston monopoly void and, in consequence, putting the company out of business. Following the decision, many merchants and packet companies now began ordering steamships for their services, most of which were built in New York shipyards.

Another event was the opening of the Erie Canal the following year. This had the effect of making the port of New York, rather than New Orleans, as previously, the port of entry and of export for any part of the new mid-West within travelling distance of the Great Lakes. Now there was not only a need for ships to handle the significant increase in the commerce out of New York into the Atlantic, there was also a tremendous demand for cargo sloops, barges, and steamboats for service up the Hudson River.

All of these factors plus the continued expansion of the country meant that the era between 1815 and 1860 was an unusually prosperous time for the New York shipyards. Among the older yards established before the War of 1812, both Christian Bergh and Adam and Noah Brown reopened their former yards as soon as the war was over, though now Christian Bergh had taken on Jacob A. Westervelt as his assistant. Adam Brown died in 1817, after which Noah continued the business alone for a while at their yard on the northern section of "Manhattan Island." Then in 1822, he turned most of the yard over to a nephew, David Brown, a young man about Isaac Webb's age, whom he had adopted and raised as his son. The younger Brown entered into a partnership with Jacob A. Bell, who had meanwhile completed his apprenticeship with Henry Eckford, so that this yard was henceforth known as Brown and Bell. Noah Brown, however, maintained an office in the yard and still occasionally built a few ships there under his own name until his retirement in 1833, the same year that old Christian Bergh retired and turned his yard over to Jacob Westervelt.

North of Brown and Bell, as we have already noted, was the yard of Smith and Dimon. In the period of the 1820s and 1830s these two yards unquestionably built more ships than any other, but their ships were not considered to be of the quality of those built at the yard of Henry Eckford and Isaac Webb. Henry Eckford's "thorough knowledge...and skill in designing a vessel" made him, according to Morrison, "the ruling spirit in the shipbuilding line in New York at this time" (9).

In 1822 Isaac Webb's former partners, John Dimon and Stephen Smith, purchased a large riverside lot quite far uptown, even farther north than Eckford's yard at "Manhattan Island," and established a new business there. In time the yard of Smith and Dimon became one of the most successful in the city. With the yard he once shared with his former fellow apprentices now available, Isaac Webb decided that the time had come at last for him to have his own shipyard. So in 1822, after three years with Eckford, Webb again left to set up his own shipbuilding establishment, Isaac Webb and Company, at the downtown Montgomery Street location.

Although Isaac Webb remained only three years at his Montgomery Street Yard, the reputation earned working with Eckford on *Robert Fulton* and *Ohio*, plus several subcontracts supplied by Eckford himself, brought him orders for several interesting and even innovative ships. Notable among these were the China packets *Superior* of 575 tons in 1822 and *Splendid* of 642 tons in 1823. One

source credits this pair with being the largest packets built in America up to that time. But, although they were indeed large ships for their day, other reliable evidence, alas, does not allow us to sustain this claim. In this same period Isaac Webb also completed the coastal packet *Silas Richards* and the hull of the steamboat *Oliver Ellsworth* (named for a Connecticut man, formerly a member of the Constitutional Convention of 1787 but then serving on the Supreme Court of the United States) for a steamer line operating between New York and Hartford, Connecticut.

On 14 March 1824, during the period when Isaac Webb was located downtown, a fire broke out in the yard of Brown and Bell, just north of Henry Eckford's yard on "Manhattan Island." The fire started at about five in the morning, and before anyone could get there, it had nearly destroyed the Brown and Bell yard, including two large steamboats almost ready for launching.

Although a small building used as a saw-pit in Henry Eckford's yard next door was destroyed, as well as a valuable stock of choice lumber, the Eckford yard, fortunately, was not seriously affected by the fire. Nevertheless, Henry Eckford, Isaac Webb, and other neighboring shipbuilders, decided to take steps to prevent such a disaster from taking them by surprise ever again.

There was no such thing as a municipal fire department in those days, and such equipment as then existed depended on steam-operated pumps that had to be fired up before they became operative. Thus a typical piece of fire-fighting equipment would consist of a small steam-driven pump and a large wooden water tank placed on a cart. This cart, of course, had to be hauled by human beings, since horses could never get up much speed in the crowded narrow streets of the city. Though there were many private fire-fighting companies, made up mostly of volunteers, around the city, with the many fires that frequently flared up in or near the shipyards, a group of local shipbuilders held an emergency meeting and decided that to prevent another major fire like the one that had nearly destroyed Brown and Bell, they would organize their own fire-fighting company, manned by their own workers, to be ready for any unexpected emergency.

The steam engine they built, which was kept on North Street (later Houston) near Lewis (which would be about across the street from Brown and Bell's yard), was dubbed "Live Oak, Number 44," in reference to a type of wood (actually beech) brought up from Georgia that was favored for hull planking. Isaac Webb himself was foreman of the newly-organized fire company. From this time, whenever the gong sounded, certain men, among them Isaac Webb and Jacob Bell, would leave their work, or even jump from their beds if it were during the night, to pull "Live Oak, Number 44" to the scene of the fire. Many years later, some of the older shipyard hands remembered seeing William Webb, when still a small boy, eagerly running alongside his father, when he was pulling the engine. Later, as a teenager, William Webb became a member of the fire company and was himself usually one of the men pulling the engine whenever the gong sounded (10).

Meanwhile, events were taking place in far-off South America which were destined to effect a major change in the shipyard of Isaac Webb and Company. In 1822, the year Isaac Webb started his own yard, Henry Eckford had drawn up plans for the fast corvette *Hercules*, which he then subcontracted to Isaac Webb to build. According to Howard Chapelle, probably America's foremost authority on sailing ships, the small 121-foot *Hercules*, which displayed many of the innovative characteristics later associated with ships built by Isaac Webb, was an excellent early example of what he calls a "pre-clipper" design.

[Hercules] had very little sheer, a straight keel with some drag, a curved raking stem rabbet, and a slightly raking post. She had a round tuck with upper and lower transoms. Her midsection was formed with very rising, straight floor, high round bilge, and marked tumble home. The entrance was convex and moderately sharp; the run was very long and fine; the buttock at the quarterbeam became straight a short distance forward of its intersection with the load line. (11)

Soon after completion, *Hercules* was purchased by the government of New Granada (Colombia), where she was renamed *Bolivar* in honor of the popular South American revolutionary general.

The success of *Bolivar* prompted other recently-independent Latin America countries to turn to Eckford, now recognized as America's leading ship designer, for warships, and in 1825 he accepted simultaneous orders for four identical forty-four gun frigates from the governments of Brazil, Colombia, Chile, and Peru. By this time, however, Eckford was so occupied with contracts for ship designs as well as with other businesses spawned by his investments that he decided it was no longer profitable for him to devote valuable time to managing a shipyard as well. And once again, the only person he would trust to manage his shipyard was Isaac Webb.

Thus in 1825 Isaac Webb and Henry Eckford worked out a new arrangement. The yard on "Manhattan Island" would still be owned by Henry Eckford, but Eckford himself now restricted his activities to designing ships, either for that yard or for others, and to managing his other business affairs. For the actual operation of the yard, a new company was formed consisting of Isaac Webb, who now gave up his own yard; John Allen, Eckford's half-brother; and a Mr. Blossom, who was apparently mainly an investor, since he seems not to have played much of a role in the business.†† Although John Allen was a shipbuilder, he too seems to have been a fairly silent partner, as it is clear that from this time Isaac Webb managed the yard by himself. This new company was first known as Webb, Allen, and Blossom, but when Blossom died the following year it became simply Webb and Allen.

The new company's first contract was to build one of the four corvettes for Latin America. Another was subcontracted to Christian Bergh, and one each was built at yards in Philadelphia and Baltimore. The success of these vessels begot yet another order, this one from the revolutionary Greek government, then still in the throes of its war for independence from the Turks, for two similar frigates. One, named *Hope*, was again subcontracted to Christian Bergh, and the second, *Liberator*, was built at Eckford's yard by Webb and Allen. But, soon after the delivery of *Hope*, with Greek independence virtually assured by the great naval victory at Navarino by their British allies in 1827, the order for *Liberator* was cancelled. She did not languish long on the stocks, however, but was soon purchased by the American government, which she served many years as *Hudson*.

By 1825, the year Isaac Webb gave up his own shipyard at Montgomery Street and moved back to Henry Eckford's yard, his eldest son William was nine years old. William was a quiet boy, but full of energy and devoted to his father, who was himself then only just past thirty. Since William was also obviously an exceptionally intelligent boy, his father wanted to start planning early for his future. Though Isaac Webb was a talented and highly-respected shipwright, by now he was beginning to view his own profession as a great deal of hard work with relatively small remuneration. He wanted to help his son find a career that was both more respectable and more financially rewarding, perhaps one of the professions,

††  Blossom had also, for a short time, been associated with Smith and Dimon when they first opened their yard.

# III

# William H. Webb: Apprentice Shipwright, 1831-1840

IN 1831, the year Isaac Webb assumed the management of Henry Eckford's yard, his son William turned fifteen, the age at which he might be allowed to start an apprenticeship in a shipyard and begin the life's work he knew he wanted. But his parents, as we have seen, had other ambitions for William. They wanted him to finish his studies at the Columbia Grammar School, to attend a university, and ultimately to find a career in one of the professions. William, however, was adamant. The happiest hours of his young life had been those spent in his father's shipyard. That was where he wanted to be. That was where he belonged.

Since his twelfth year when he had built that first small skiff completely by himself, William had returned to his father's yard during his summer vacations, each year building himself another larger boat, launching it, and then testing its merits in the crowded East River. In the summer of his fifteenth year, his father, fearing for his safety—and also with the hope of dampening his son's interest in becoming a shipbuilder—sent William to work in the mold loft,* where the work could be tedious and the hours long. But William, as it turned out, was just as happy, and just as adept, in the mold loft as he had been in the yard. Many years later William Webb himself explained how he finally managed to persuade his father to allow him to begin his apprenticeship as a shipwright:

...when I was fifteen years old he sent me to work in the molding loft during my vacation to keep me out of danger. When Co-lumbia College Grammar School opened in the fall, I did not dare to ask my father to let me give up my studies, but instead of going there I went to the molding loft....My father thought I would soon tire of the work and let me have my own way, but I never tired of it. (1)

Since Isaac Webb had been through much the same experience with his own father just twenty-one years earlier, he probably understood well enough both the intensity of William's ambitions and the futility of opposing them. So in the fall of 1831, the fifteen-year-old William H. Webb officially became an apprentice in the yard of Webb and Allen. This time, however, no indenture papers were necessary. The indenture papers associated with starting an apprenticeship served to transfer the responsibility for a young man who had not yet reached the age of maturity from his father to his master. Since William, in this case, was apprenticed to his own father (as had been the case with David Brown), such papers, obviously, were unnecessary. Nevertheless, William was now legally an apprentice, and Isaac Webb was legally obliged to teach his son the art of shipbuilding and initiate him into the "mysteries" of the profession. Given what we know about William Webb's serious and stolid demeanor plus his dedication to hard work and the number of hours a day he devoted to the shipyard, his father's obligation to guard his moral behavior probably required very little attention. For this period William, like the other apprentices, was to receive an allowance of $2.50 per week. The $40 per annum "in lieu of meat, drink, washing, lodging, clothing...." he did not receive, for this apprentice Isaac Webb was prepared to maintain in his own home.

One might expect that these arrangements were mere legal technicalities, and that, as the master's son, William Webb could expect to have a fairly easy time of it. But this was definitely not the case. If Isaac Webb was not demanding of his son during his apprenticeship, and, given what we know of his character we can assume that he was not, William Webb was demanding of himself. It was his determination not only to be a shipbuilder but to be a good one. For the next six years he was at the yard every morning at six A.M. Through the day he did all

---

* The mold loft was a large room where sections of a ship's lines were first drawn on the floor. From these lines, molds or templates were created so that pieces of timber, specially chosen for shapes that roughly conformed with these lines, could be marked and then sawed to exact specifications. The mold loft continued to be an important aspect of shipbuilding until recent times, when most of its work was superseded by computer-related procedures, leaving the loft floor to cope only with the knotty problems of ship construction, such as hawsepipe penetrations and shaft bossings. Apprenticeship in a mold loft will be missed, for it was an ideal way to learn the complex spatial relationships of a ship's hull and the difficulty of making recalcitrant materials such as wood (and later iron or steel) conform to its warped surfaces.

of the work that other apprentices were doing, alternating among all of the various jobs in the yard: sawing heavy planks, steaming timber to make it more pliable, carrying heavy planks from the sawpit to the hull taking shape on the ways, climbing high on the scaffolding to help hold planks while experienced mechanics bored holes before hammering the planks into place with trunnels. He took his breaks with the others when the great Mechanics' Bell pealed, and stayed at work until its final gong at six P.M. and probably often long after that.

At night William, often working by candlelight, studied such books as were then available on ships and shipbuilding (although naval architecture as a science did not yet exist). In short, he spent the six years, from age fifteen to twenty-one, both pursuing intensive study and gaining practical experience in the career he had chosen. Like most of us who in our youth had to work and study hard in order to qualify for a chosen profession, Webb apparently had days when he became discouraged and even questioned his ability to reach his goal. As he was to note in an interview many years later,

... after about two years I began to discover what a difficult occupation I had chosen, and I feared I should never become a master shipbuilder, although I worked and studied very hard. The only thing that sustained me was the thought, 'others have learned it, why can't I?' (2)

There was apparently one fairly long period early in William Webb's apprenticeship during which he was away from the shipyard for some time as a result of illness. Our sources are frustratingly vague about the nature of this illness, but we do learn both that he was taken sick in the summer of 1832 and that he came very close to death. (Isaac Webb was frequently concerned for his son's health after this time.) Combining this information with the known fact that in the summer of 1832 a cholera epidemic swept the city of New York, there seems fair reason to assume that William Webb was stricken by cholera during that summer. If so, considering that the majority of those afflicted died during the epidemic, Webb was fortunate to have survived it.**

Other than the occasion when he was out as a result of this illness, only once during his apprenticeship did William Webb take time away from his work, and even

that was to be able to learn something about shipwork he could not have found in his father's yard:

During six years I had only one vacation. Then I asked permission to go to Boston to see the new stone drydock there, the first of the kind in this country. I traveled by steamboat and stage coach, and put up at the new Tremont House. When I came to settle my bills, I found them larger than I had anticipated. The result was that I had to walk much of the way home. (3)

Some time during the first year of William's apprenticeship, Isaac Webb, aware that the yard he had inherited from Henry Eckford was not large enough for his needs, sold this yard to Brown and Bell, who now occupied all of what had been "Manhattan Island," and purchased a large waterfront tract much farther uptown where the land was less expensive, along Lewis Street between Fifth Street and Seventh Street, just north of Smith and Dimon.

By this time there were no less than thirty-three shipyards along the East River waterfront extending from just below Christian Bergh's yard at Corlear's Hook all the way north to Thirteenth Street.

In 1834 Bishop and Simonson opened a new yard at the foot of Seventh Street, just north of Webb and Allen. Jeremiah Simonson was a nephew of Cornelius Vanderbilt, and for much of the history of this yard, Vanderbilt's orders alone would keep them in business, though by 1834 Vanderbilt had not yet become a particularly wealthy man.

Another new yard established at about the same time was that of William H. Brown (no relation to the Browns of Brown and Bell), located north of Bishop and Simonson. Like its neighbor to the south, the William H. Brown yard tended to specialize in the construction of local steamboats.

As noted earlier, it was about this time also that Christian Bergh, who had been Henry Eckford's mentor and closest friend, decided to retire, leaving his shipyard to his two sons and his two assistants, Jacob A. Westervelt (then 35) and Robert Carnley. Westervelt and Carnley took a year off to travel in Europe to learn what they could about new techniques of shipbuilding practiced there. Later Jacob Westervelt returned and purchased control of the yard from the other heirs.

In the areas north of Bishop and Simonson's new establishment were several yards, such as that of the Novelty Iron Works, which built engines rather than ships; or yards specializing in repair or joiner work. Most of the larger yards, such as Brown and Bell or Webb and Allen contracted to build only a hull, not an entire ship.

---

** Cholera is an infectious disease spread by bacteria, usually through contaminated drinking water. The severe attendant vomiting and diarrhea cause dangerous dehydration, which in turn is the usual cause of death.

Once one of these yards had launched a hull, including all of the heavy interior framing necessary for strength, stability, and structure, if it was a steamship, it was sent on to another yard for the installation of the engine. In the case of sailing ships, some of the yards furnished the interiors as well as the hulls, but as a rule the ship went to a different yard for the joiner work, the finely-painted and often very elegant interiors of ships carrying passengers, and, usually, also for the installation of the tall spars to carry the sails. William H. Webb, however, did most of the work in his own yard. As a rule Webb's ships were launched with their masts in place and the finished yards on the deck. Webb also employed special carpenters to finish fine joiner work right at the yard.

Still farther north, at the foot of Tenth Street, was the yard of Thomas and Steers, established in 1824. Henry Steers had begun his career as a shipwright in Plymouth, England. In 1819 his friend John Thomas, then working for the American government, persuaded Steers to come to this country and to take a position in Washington. Two successful naval vessels were subsequently built on plans submitted by Steers. However, after only five years of battling governmental bureaucracy, both Thomas and Steers, sensing the possibilities of success in building commercial vessels, opened their yard in New York. In later years, as we shall see, after the yard had moved across the river to Williamsburgh and had come under the management of Henry Steers' two sons, James and George, this yard often worked closely with that of William Webb and earned a reputation second only to Webb's.

There appears to have been a fair amount of co-operation, as well as friendly rivalry, between the yard of Webb and Allen and the yard of Smith and Dimon, their neighbors to the south. Isaac Webb, Steven Smith, and John Dimon were, after all, old friends who had served their apprenticeships together under Henry Eckford. They had also, we remember, been partners in an earlier venture. In their yard Stephen Smith concerned himself mainly with the designing and building of ships, while John Dimon specialized in the repair work which this yard also accepted and which was apparently more profitable. In an interview Dimon once declared: "Smith builds the ships and I make the money" (4).

During much of the period of William Webb's apprenticeship, two young men, Donald McKay and John W. Griffiths, both former apprentices of Isaac Webb, were working at Smith and Dimon's yard. Considering that these two men—as well as William H. Webb—were later to produce many innovative concepts in ship design, it would be surprising indeed, if the much younger William Webb, known already for his inexhaustible curiosity about ships, had not frequently wandered next door to Smith and Dimon's yard to watch Griffiths or McKay at work, to ask them questions, or to engage them in lively discussions on matters of mutual interest.

The 1830s must have been an exciting time for a young man to have been an apprentice shipwright. For just as this was an era of social and political revolution, of scientific experimentation, of technological innovation, and of rapid industrial development throughout the western world, it was also a period when experimentation and innovation in ship design and construction were challenging the accepted practices of centuries.

By the 1830s, New York shipbuilders were routinely using half-models (which had been used in New England yards since about the 1790s) and lines drawings as guides to hull shapes during construction, though more often at first only the half-model was used. With a half-model a builder could experiment with hull shapes and configurations and then, when he was satisfied that he had the best design, simply copy the form of the model when he built the full ship. These models were usually created by screwing together horizontal lifts of wood—sometimes of different kinds, so that the result had a striated effect. The screws (or dowels) could be removed and the lifts taken apart to be studied separately. In this way, it was easier to see what shape the hull should take at any given water line. Though various sources credit different New York shipbuilders with first using the half-model, one source specifically attributes its introduction to Isaac Webb (5), though John W. Griffiths, who is probably more reliable, claims that Stephen Smith built one when he was still an apprentice under Henry Eckford (6). In any event, Isaac Webb consistently used half-models to design hulls in advance, as did his son William H. Webb, many of whose half-models now hang on the walls of Webb Institute of Naval Architecture.

Although Fulton's first steamboat had appeared in 1807, the developments which made the steamship generally practicable—the use of coal rather than wood for fuel, or of iron rather than copper boilers, for instance—first appeared in the 1830s. During this decade Robert Stevens, who began running steamers on the Hudson River once the Fulton-Livingston monopoly was broken and who is credited with many of these improvements, also experimented with hull designs, often fitting a series

The Mechanics' Bell announced the hours of work for East River shipbuilders for many years. The bell was later displayed at Webb Institute's campus in the Bronx until it was sacrificed to the scrap metal drive when the United States entered World War II.

of false bows on his steamers and measuring their effects on speed and stability. In the process Stevens was able to conclude that sharper bows did reduce drag and thus increase speed. But while his suggestions were soon put to practice on river steamers, it was still generally held that bluff bows were necessary for ocean-going ships, and that a sharper bow, though it might increase speed, would tend to cut through a large wave rather than ride over it.

There were some shipbuilders in this era, however, who were willing to try bold experiments with hull designs, and it was these experiments of the 1830s and 1840s that led to the evolution of one of America's proudest accomplishments: the clipper ships of the 1850s. In the forefront among those shipbuilders willing to experiment with new designs was Isaac Webb. According to Howard Chapelle, "The packet period had produced many skilled shipbuilders, but one was to have especially great influence on the development of the clipper ship. This was Isaac Webb . . ." (7). This courage to depart from established practice and to experiment with what could prove a superior design, Isaac Webb had learned from Henry Eckford. And, as the shipping world would soon discover, he was also able to pass it on to his son, William Henry Webb.

One of Isaac Webb's most innovative and most successful designs was *Natchez*, built in 1831 for a packet line running between New York and New Orleans. As she was expected occasionally to pick up cargoes farther up the Mississippi, Webb designed her with a somewhat flatter-bottomed hull than the V-shaped hull then more commonly used for ocean-going ships. Perhaps to compensate for the carrying space thus lost, he also designed her with relatively less sheer. With an eye to speed, her bow was not quite so rounded as was customary, and her entrance at the waterline was somewhat sharper than usual. Almost everyone, probably including Isaac Webb, had been surprised when *Natchez* was regularly making the fastest runs between New York and New Orleans. It was because of her legendary speed, in fact, that *Natchez*, though only 132 feet in length, was later purchased by the firm of Howland and Aspinwall for the trade with China. Since one of their major imports was tea, which tended to bring far better prices when it was fresh, the firm kept an eye out for ships known to be fast. Some years after she had been on this line, *Natchez* made a run from Macao (on the China coast) to New York in only seventy-seven days, at the time a record for the route (8).

*Oxford*, built by Isaac Webb in 1836 for the Black Ball Line's New York to Liverpool packet run, was another ship with innovative hull lines. Her design was also apparently the first in which William Webb, now nearing the end of his apprenticeship, took a major role. *Oxford* was ordered by Charles H. Marshall, the general manager of the Black Ball Line and also a close personal friend of Isaac Webb. It is interesting to note that the first large ship designed by William Henry Webb in 1836, as well as the last, in 1869, were both built for Charles H. Marshall's Black Ball Line.

Charles H. Marshall, though born on a farm in upstate New York, went to sea on a whaler in 1807 at the age of fifteen. A few years later he was serving as a deckhand on a ship smuggling American goods into Russia, past Napoleon's blockade. By 1814 he was back in America, where he joined an uncle in setting up a temporary trading post at Sackett's Harbor to serve the naval base there. It would seem certain that young Marshall, then twenty-two, would have met the twenty-year-old Isaac Webb here and that their friendship dated from this time.

When the war ended, Marshall went back to sea and received his first command in 1816 at the age of twenty-four. In 1822 he was made captain of the packet *James Cropper* of the Black Ball Line and continued to command Black Ball ships for another twelve years. In 1834, when the American merchant firm of Goodhue and Company bought a controlling interest in the line, they appointed Captain Marshall as general manager with offices in New York.

The first time after becoming general manager that Marshall found he needed a new ship for the line, he turned to his friend Isaac Webb to build it for him. When Marshall also accepted Webb's suggestion that he allow William to assist in the design and construction of the new ship, he was apparently not disappointed, for every ·succeeding ship built for Marshall's Black Ball Line was designed and constructed by William H. Webb.

At 495 tons and 147 feet 6 inches in length *Oxford*, believe it or not, was considered a large ship in her time. Whether designed by father or son or both, *Oxford*, according to Howard Chapelle, displayed the usual char-

acteristics now becoming associated with Isaac Webb's designs: "little sheer, straight keel with no drag, moderately raking and slightly curved stem rabbet,. . . . The entrance was short but rather sharp for a full-model ship; the run was about 1/3 the waterline length, with a straight quarterbeam buttock near the load line crossing" (9). The straight quarterbeam buttock was critically important to high-speed sailing, avoiding the excessive drag produced by a convex line, or the tendency to squat that usually accompanies a concave one.

As *Oxford* was building, she attracted sufficient attention for the Havre Line, the primary American packet line between New York and France, to order an exact sistership, which slid off the ways at the Webb yard four months later as *Burgundy*. The twenty-year-old apprentice shipbuilder was already developing a personal reputation.

Just as William Webb's career was getting started, however, the New York shipbuilding world was shaken by the great financial panic of 1837. It would be impossible to find any single cause or even a simple series of causes, for the panic. But the most apparent immediate cause, most would agree, was the conflict between the Bank of the United States and the "people's president," Andrew Jackson. Between 1791, when it was chartered, and 1833, when Jackson started his second term, the Bank of the United States had played a valuable role in the rapid economic development of the young republic. As the depository for government revenues, it was able to lend money to new businesses just getting started. It gave financial backing to the local state banks that were lending smaller amounts to the homesteaders seeking lands in the new West. It helped expand available currency by issuing notes its firm reputation could sustain. And, with these notes, it supplied a sound paper currency to American entrepreneurs doing business abroad.

Alas, the Bank had many enemies. The agricultural areas of the South and West, constituencies that had supported Jackson, saw the bank as the tool of pompous businessmen and avaricious moneylenders in the North. Jackson himself called the Bank "the Monster." Taking his election to a second term as a mandate to end the special relationship between the government and the Bank, Jackson withdrew government deposits, greatly reducing the Bank's resources.

At the same time, newly-chartered local banks were lending money to homesteaders for the purchase of government land in the West. Devised as a sensible way of assisting people with little or no capital to settle and thus help develop the new territories, the system was being sadly abused by speculators who used credit to buy land cheaply from the government only to resell it at a profit to themselves. Again the western settlers saw this, not without reason, as a scheme of northern businessmen to exploit the poor farmers, and turned to Jackson for relief. Then in 1836, Jackson, who had never been comfortable with paper payments, announced that all land must be purchased with gold or silver, thus in one swift blow placing an impossible drain on the specie available in the country. At about the same time, British business houses, noting that American credit on imports from Britain was perilously overextended, ceased extending credit and began to demand payments in specie. Thus, just at the time when specie was scarce, eastern business houses dealing with Britain were obliged to go to their banks and withdraw their deposited funds in gold. It was not long before they were told that their banks did not have the gold to give them, which meant that American merchants did not have the specie to pay for foreign imports already ordered.

Jackson left office in March 1837, leaving his hand-picked successor, Martin van Buren, to deal with the situation as well as he could. But by then there was little Van Buren could do to head off the inevitable. On 10 May 1837 all New York banks announced they could no longer honor their customers' withdrawals in specie. Within a week all banks in America had followed suit. With creditors to satisfy and no cash available, many American merchants were ruined.

Not only were many American shipping firms themselves caught short in the panic, but also the amount of goods being shipped was considerably reduced in the depression that followed. This depression was even further aggravated by a poor cotton crop that year. As a result, orders for new ships, particularly from the American coastal lines which supplied much of Isaac Webb's business, fell off sharply.

The shipbuilding business suffered another blow in 1837, when in October one of the coastal steamships foundered with a great loss of life, and the subsequent newspaper accounts suggested strongly that ships being built for these runs were not of sufficiently sound construction to guarantee the safety of their passengers. The ship in question was the 198-foot steamer *Home*, built the previous year by Brown and Bell for the Southern Steamship Company's line between New York and Charleston. *Home* had sailed southbound from New York on 7 October. Leaving the harbor, she stranded on an underwater shoal but floated off with the rising tide. The captain, assured that his ship had sustained no serious damage, continued on his way. The following evening, however, steaming off the Virginia coast just north of Cape Hatteras, *Home* encountered gale winds and high seas. As the small ship was being tossed about by gigantic waves, it became apparent that the hull had indeed been strained by the stranding and that the ship was taking water rapidly. Though both passengers and crew were set to bailing through the night, the rising water eventually doused the fires, rendering the steamer powerless in the raging storm.

Raising the sails maintained for such an emergency, *Home* headed shoreward. She sailed along safely for awhile and had almost reached the coast when she suddenly ground to a halt on a sandbar six miles offshore. Held in this grip and pounded by gale-force winds and crashing seas, *Home* broke up in less than half an hour. Boats were launched but some capsized as soon as they hit the water. Less than a third of the 135 people aboard managed to reach shore. Ninety-five people were killed (10).

Since this was the first major accident causing loss of life on one of the steamships which were now beginning to replace sailing ships on the coastal lines, it was a costly lesson to the New York shipbuilders that vessels carrying heavy machinery in their hulls had to be built with much greater support than the sailing ships they had been accustomed to producing. Isaac Webb and his son, William, had not as yet been building steamships, so that their reputations were not tarnished by this incident. But the lesson was not lost on William Webb who was later to build many steamships for the coastal lines.

The year 1837 was not only the year of the Panic, the year Martin van Buren became President of the United States, or the year the young Victoria became Queen of England, it was also the year that William H. Webb on 19 June turned twenty-one and officially ended his apprenticeship under his father. Not that this event made much change in William's position at the yard. As we have seen he was already designing ships on his own and supervising their construction. Since he was in essence already a partner of his father's before he had completed his apprenticeship, he simply continued in this role. During this period, William Webb, working now as a subcontractor for Isaac Webb, built *Pennsylvania* for Fish, Grinnell & Company (later Grinnell, Minturn & Company) and *Ville de Lyon* and *Duchesse d'Orleans* for Fox and Livingston's packet line between New York and Le Havre.

In 1838 William Webb received an order from Marshall's Black Ball Line to build a ship similar to *Oxford* and *Pennsylvania*, but even larger, for their line between New York and Liverpool. Launched on 24 October 1839, this vessel, named *New York*, measured 152 feet 6 inches in length and 862 tons, more than double the capacity of the first North Atlantic packets. *New York* displayed most of the same characteristics, already described as typical of Webb-built vessels, as *Oxford* and *Burgundy*. With her somewhat sharper entrance, and her minimal and unbroken, clean sheer, *New York* emerged as an exceptionally handsome ship.

In November 1839 *New York* was ready for her maiden voyage. Since William Webb had designed this ship himself, he wanted to make the trip with her to see just how she performed at sea and to take notes on her responses under a variety of circumstances. And, as always, interested in the shipbuilder's trade, he thought he would like to do what Jacob Westervelt had done a few years before: to tour some of the shipbuilding facilities of Europe, particularly those in Scotland along the Clyde.

Probably the most compelling reason for this trip, however, was William Webb's health. Occasional weakness had apparently plagued him ever since his brush with death in 1832. In 1839, again in poor health, according to his own report as a result of overwork, he had been told to take time abroad to rest and regain his strength. This advice was based on the persistent but insupportable myth that an ocean voyage and travel in Europe in winter is both healthy and restful!

But William Webb was not able to complete his mission either to study British shipbuilding or to rest and regain his strength. Only two months after his arrival in Europe, he received news of his father's unexpected early death on 14 January 1840 at the age of forty-six. William, himself then only twenty-three and still less than three years from his apprenticeship, hurried home on the next available sailing to take his father's place at the helm of Webb and Allen.

# IV

# William H. Webb: Shipbuilder, 1840-1847

IN March 1840, William Webb returned from Europe to assume the management of his father's shipyard. Though he was still but twenty-three years old, no one—not his mechanics, not his customers, not his family, and certainly not himself—questioned either his right or his ability to do so. William Webb was by now thoroughly familiar with every aspect of his father's shipbuilding business. Though privately deeply bereaved by his father's death, William lost no time in taking full charge of the yard, in getting his father's business affairs in order, and in personally attending to settling family matters specified in Isaac Webb's will.

In checking over his father's books, however, William discovered something his father had apparently been able to keep from his family: Isaac Webb was, in essence, bankrupt. The company of Webb and Allen, it seems, had been another victim of the Panic of 1837. Orders for new ships had been slow since 1837. Also, it seems that some of the merchant houses which had ordered new ships had, on their completion, either been unable to make full payment, or, in some cases, simply gone out of business. Another son, on finding a business inherited from his father in such serious financial difficulty, might have closed out accounts and, however reluctantly, sought a living elsewhere. But here, at the outset of his professional career, we get a sense of the stamp of William H. Webb. With a confidence in his own abilities born of a combination of natural talent and hard work, he never lacked the courage to take a calculated risk when he believed in what he was doing.

After a thorough perusal of his father's books, William Webb worked out reasonable arrangements with his father's creditors and then closed out the affairs of the firm of Webb and Allen. At the same time, William Webb and his father's former partner, John Allen, formed a new company, also called Webb and Allen, which took over the yard, consolidated such other assets as existed, and inherited the good will of the former company. Since the new concern had the same name and operated at the same location, neither the public nor the customers were ever aware that there had been a problem.

Once the affairs of the company had been reorganized, William Webb proved himself not only a master shipwright on a level with any then in business but also a far better businessman than his father had been. One might remember that, even as a schoolboy, he had always excelled in mathematics; throughout his life, Webb continued to have a quick sense of figures. Thus, in addition to understanding how to build ships, Webb also understood how to order materials, how to court solvent customers, and how to make sensible estimates. By scouting immediately for new orders and by producing hulls which showed the shipping world from the start that, in spite of his youth, he knew how to build ships, William Webb soon had his yard running on a sound financial basis. When David Brown of Brown and Bell wandered over to have a look at *Malek Adhel*, the first ship completed by William Webb after his father's death, he was obliged to note that "young Webb's future [is] not going to be a problem" (1).

By May 1843, just three years after William Webb had taken over his father's business, eleven hulls had been launched at the yard, and three more were on the stocks. William Webb, regardless of his youth, was now clearly established as one of New York's primary shipbuilders.

In this year, John Allen, who was considerably older than William Webb, announced his wish to retire. So prosperous had the business become in three years that William Webb was able to purchase Allen's share by himself. From 1843 until it stopped building ships in 1869, the yard at the foot of Sixth Street was never again Webb "and" anything, not even "Webb and Company." It was simply "William H. Webb." And never again, though the prosperity of the country was to undergo many fluctuations, were the finances of this yard in question.

There was another yard, however, known as Webb "and" something, which opened that same year. When William Webb took over his father's shipyard in 1840, one of the apprentices learning the trade from Isaac Webb was Eckford Webb, William's youngest brother, then seventeen. Eckford had continued his apprenticeship when his brother took over the yard, but by 1843, when the term of his apprenticeship ended, the two brothers agreed, apparently amicably, that Eckford Webb should have his own business at another location. Nathaniel Bell, the son of Jacob A. Bell of the Brown and Bell yard, was then also completing an apprenticeship under his father, so it was arranged for the younger Bell and Eckford Webb to join in a partnership. (Remember that Jacob Bell and Isaac Webb had both come to New York from Stamford at about the same time, and that both had served as apprentices to Henry Eckford, so that the family connections had apparently been close for some years.) Thus in 1843 the yard of Webb and Bell was established over on Long Island at Greenpoint (not yet a part of Brooklyn, as it is today), roughly across the river from the yard of William H. Webb. Although the details of the funding of the new yard are no longer available, it can be assumed that Jacob Bell and William Webb probably provided the capital to get the yard started. Also, over the years, both Brown and Bell and William H. Webb were to subcontract many ships to Webb and Bell when their own yards were too busy to accept them.

It was also in 1843, in the month he turned twenty-seven, that William H. Webb was married to Henrietta Amelia Hidden, the twenty-two-year-old daughter of a prominent Manhattan merchant. Soon after the marriage, the now sound finances of the shipyard, assisted undoubtedly by another major contract from Charles H. Marshall that spring, allowed the couple to build a fine townhouse at 415 Fifth Avenue, between Thirty-Seventh and Thirty-Eighth streets. By this time the elegant downtown houses, especially along or near Broadway north of City Hall, were giving way to shops and other commercial buildings, and the prosperous merchants of this fast-growing city were now building homes along the new numbered streets stretching north from Houston Street, quite far from the commercial district. Even so, the fine three-story house built by William and Henrietta Webb, which was to remain their home as long as either lived, while in an area that was to become highly fashionable, was then still considered very far out of town. It was certainly very far from William Webb's shipyard on the East River at the foot of Sixth Street.

The years following William H. Webb's establishment of his own business and his marriage in 1843 were to be particularly prosperous years for his shipyard. There were

Portrait of William H. Webb, aged thirty-three, painted in 1849.

undoubtedly many factors contributing to the early and continuing success of William H. Webb as a shipbuilder, though probably none played a greater role than the extraordinary talent, industry, and business acumen of Webb himself. One must also note that by the early forties the financial depression of 1837 had run its course and American business, and even America itself, was expanding beyond all expectations. American commerce extended not only along our coast, but also to Europe, and now to South America and even to China, creating an almost inexhaustible demand for ships. What is remarkable, however, is that, with the many yards building ships in this prosperous era, none was more productive than that of William H. Webb, and none had a better reputation for producing strong and reliable ships. According to Robert Greenhalgh Albion, the doyen of historians of New York Port and North Atlantic sail, William Webb produced "the greatest aggregate of tonnage of any American shipbuilder up to that time" (2). The same author also notes that, while some shipwrights specialized in constructing certain types of vessels and thus faced hard times if the demand for those vessels fell off even temporarily, Webb's yard, throughout its history, produced ships of a variety of types and sizes, from small fishing craft to packets or clippers that crossed the oceans, large steam-powered liners, or even big sidewheel steamboats for service on nearby Long Island Sound. Since the Webb yard was frequently adapted to deliver whatever kind of ship the market demanded, it managed to prosper even in periods when others did not.

As the head of the shipyard, in which he was both the naval architect and the shipwright, it was never William Webb's way to sit in an office and tell others what to do. When asked many years later how he would account for his success, his terse answer was, "Attention to detail." (3) And this was true. Throughout his career Webb tended to take personal charge of virtually every operation in the yard. Though most shipbuilders did not begin rendering lines drawings of proposed ships much before the mid-fifties, it was William Webb's practice to do so from the time he started designing ships, not only making the drawings himself, but also writing out every specification for every ship by hand. Webb once observed that he had all the information needed to reproduce exactly any ship he had ever built.

Webb, at least in his early years, also crafted his own half-models, carefully shaping the exact configuration of each proposed hull on a half-model of one-eighth inch to the foot, or about four to five feet in length. With this model in hand, Webb and his senior ship carpenters would climb to the molding loft, where they would meticulously draw each section of a proposed hull on the floor, checking each against the lines in a section of the now-disassembled half-model. Once the lines had been studied and carefully corrected, and here again the experienced eye was the most reliable caliper, the molds were shaped and the planks were laid over them and sawed to specification.

Next, from the great stockpiles of lumber stacked in the yard, carefully selected in advance for type of wood, shape, and strength, Webb would himself choose which pieces were most appropriate for a given section of a particular ship. In addition to the planks for the outer hull, he also had to choose timber for the keel, the keelsons, the frames, and the knees, all of which then had to be carefully shaped to provide the greatest possible support. Once the keel was laid and enough frames were in place to give some sense of the shape of the hull under

Portrait of Henrietta Hidden Webb, aged about twenty-eight, painted in 1849.

construction, Webb would check each detail again and again to be sure that every part of the structure was tight. At this stage Webb himself usually "marked the place for every stick of timber in every vessel" (4).

It was also the shipbuilder's job to establish relations with shipowners to solicit orders, to make careful estimates, and to draw up contracts. He also had to hire and fire his workers and train his apprentices. When a ship was completed, Webb himself had to make all the arrangements for the launching, which often turned into something of a public spectacle. A launching was usually announced in the papers and, in the case of a ship of any size or note, would draw a large crowd of spectators. For the day of the launch Webb would have to arrange a fairly formal reception for the official party, which usually included the officers of the company that owned the ship and their families, the officers of the ship, and, of course, a sprinkling of local dignitaries to offer toasts and speeches. There was food and some sort of drink, and also a subsidiary party with less interesting food and drink in some other part of the yard for the men who had actually built the ship (5).

Then came the launch. The ways were tallowed in advance so the hull could glide down with ease. Then men came along with mallets and knocked out the wooden chocks which were holding the hull in place. As the last chocks were being removed, someone, usually a young woman relative of the owner, would swing a bottle of something—not usually foreign champagne in this era of American pride—and the great ship would slide slowly down the ways toward the sea. Or at least one hoped it would.

There is something compellingly beautiful, but also exciting, overwhelming, and even somewhat frightening in watching an enormous object larger than a building, suddenly begin to move on its own. A launch could certainly be an anxious event for the shipbuilder as well. His whole career and reputation hung on its success, for the launch was the most dangerous stage of the operation. Once the stern of the ship had entered the water and was afloat, while the bow was still supported by the fore poppet, or cradle, the midship portion of the vessel might be strained as much as it ever would be, even in a tossing sea. A ship that moved too slowly might stick on the ways, leaving the cheering crowd with little to cheer for. A ship that moved too fast could glide beyond the reach of the steam towboats that were to return her to the shore leaving a fortune's worth of destruction in her wake. The greatest concern of the shipbuilder, however, was whether he had achieved the proper center of gravity. A poorly designed hull, especially since it was still completely unloaded, was known more than once to slide down to the water only to capsize, the shipbuilder's heart and soul slowly sinking with it (6).

Though there are many hair-raising stories of launch disasters along the New York waterfront, the one most often recounted was the launch of the clipper *Sweepstakes* at the yard of Westervelt and Mackay in June 1853. When the bottle cracked over her bow, *Sweepstakes* slid gracefully down the ways, as expected, for about a hundred feet, and stopped. Then, instead of proceeding into the river, she rolled over onto the scaffolding of another ship under construction next to her, with a group of spectators standing atop the scaffolding thus being tossed unexpectedly into the river, though all were dragged out, wet and somewhat shaken, but safe. The cause of this accident, apparently, was that the ground was too soft. The weight of the moving hull had pushed the ways so deep into the mud that the hull itself hit the ground. Thus lifted from the poppets that had been supporting her on the ways, the ship heeled over before she hit the water (7).

Although William Webb produced more hulls than any New York shipbuilder of his time, he never had to preside over an unsuccessful launch. And one can be safe in asserting that this record was not entirely due to luck.

Considering the number of responsibilities the shipbuilder had to assume in order to ensure that the ships he produced not only suited the requirements of the owner for carrying capacity, speed, or draft but also proved strong enough to survive any sort of turbulent sea anywhere in the world, one can understand that, during the first twenty years or so of business at any rate, William Webb frequently had to be the first person to enter the yard before six in the morning, summer or winter, and the last to leave at night.

William Webb's career as an independent shipbuilder came at the start of a decade during which ship design was passing through a period of transition. During the 1840s, the traditional bluff-bowed sailing packet was beginning to give way to ships with the sharper and sleeker hulls which would emerge as the American clipper ship. The same period also saw the beginning of the use of steam-powered vessels, already long in general use on American sounds and rivers, for ocean travel.

The prime reason for these changes was that speed was becoming a greater factor in ship design than it had been earlier, in part at least because of an increased interest in trade with the Far East. In 1839, the year that William Webb was abroad, England had started the infamous Opium War with China. At that time the British had been enjoying the enormous profits of growing opium in India and selling it in China. When certain conscientious Chinese officials, noting the detrimental effect of opium smoking, not merely on their addicted countrymen but even on the morale of their whole nation, attempted to put an end to the nefarious trade, the British countered by declaring war. With their superior ships and munitions, they soon brought the Chinese to heel. Not only was the opium trade allowed to continue unhindered by any conscience, and Hong Kong ceded outright to the British, but China was now also obliged to open five of its seaports to unrestricted foreign trade.

American merchant houses such as N. L. and G. Griswold; Grinell, Minturn and Company; Howland and Aspinwall; or A. A. Low and Brothers, were quick to take advantage of this new opportunity for profit, and the many new ship orders for this trade also obviously worked to the benefit of the New York shipbuilders, including William H. Webb. These American firms not only brought tea and other Chinese exports, including beautiful Chinese porcelain, back to America, they even began competing with British shippers by carrying tea to England itself. Since the subtler tastes of teas disappear with time, the freshness of the shipment became a factor in the profits. So now merchant houses were coming to the shipbuilders seeking ships designed especially for speed.

The traditional sailing packet generally had a fairly high freeboard, a V-shaped hull, and a full bluff bow. The hull reached its extreme beam within the first third of its length and then tapered toward a fairly high and narrow transom stern. Experience over the centuries had shown that such a design created the least disturbance in the water, and therefore it was assumed, also created the least amount of resistance.

As we have seen, first Henry Eckford and later both Isaac and William Webb, experimented with hull designs, most particularly with sharper bows and longer runs, to achieve better speed. Those favoring the more traditional shape, however, feared—despite considerable evidence to the contrary—that ships with concave rather than convex bows would not be carried over a large wave, but would slice right into it, dangerously dousing both crew and cargo and perhaps even sinking the ship.

One of the first ocean-going ships to attempt an innovative design was *Ann McKim*, launched in Baltimore in 1822. Ordered for Isaac McKim, a local merchant engaged in trade with South America, she was a smart and speedy ship, long and narrow, with an exceptionally low freeboard. The inspiration for this design was a much smaller type of ship, known as the Baltimore clipper, built especially for speed to run the British blockade of Chesapeake Bay during the War of 1812. Although *Ann McKim* did not carry the more concave bow found on some of the ships built by Henry Eckford or Isaac Webb, her unusually narrow beam helped prove the contention that a low length-to-beam ratio improved speed, for she was a fast ship (8).

As we have already noted, when Isaac Webb built *Natchez* with minimal deadrise—that is, with a flat-bottomed rather than V-shaped hull—so that she could, when necessary, sail further up the Mississippi than New Orleans, she turned out to be much faster under sail than had been expected. With the opening of the China trade and the consequent change in priorities from capacity to speed, the New York trading firm of Howland and As-

pinwall, then just beginning to move into the China market, arranged to purchase both *Ann McKim* and *Natchez*, because of their reputations for speed, for this service. Both ships made some impressive runs across the Pacific, and both proved themselves seaworthy, which helped dispel doubts about their novel designs.

Through the 1840s a fairly friendly controversy prevailed between shipbuilders who held tenaciously to traditional designs and those who felt that, while space might be sacrificed, safety would not, if hulls were modified for speed with a concave bow, a long sharp run, a narrow beam, and a fairly flat hull with little or no deadrise. William H. Webb, throughout his career, was never afraid to experiment with a design in which he himself felt confidence, but in almost all of his ships, Webb, by conviction, held to a compromise in design that sacrificed space as little as practicable and safety not at all.

Almost all of Webb's ships through the 1840s followed the pattern he had established when still designing ships such as *Oxford* or *New York* as a subcontractor for his father. Webb ships in almost every instance had a moderately concave bow and a short but sharp run. He favored a full mid-section of long straight sides with little or no sheer and little or no tumblehome, and almost perpendicular to a fairly flat-bottomed hull. With this design Webb believed he could achieve—and in fact did achieve—respectable speed while compensating for cargo space lost with the narrower beams. Webb also tended to believe that a rounder counter stern rather than a transom stern allowed a smoother flow and less resistance.

In 1842 the firm of N. L. and G. Griswold, rivals of Howland and Aspinwall in the China trade, asked William Webb to build them a fast ship that could compete favorably with *Ann McKim* or *Natchez* in speed. This ship, named *Helena*, appeared with an even more concave bow and sharper entrance than Webb's earlier ships, evoking considerable comment, mostly negative, along the waterfront. In designing so controversial a hull in his second year of operation, when he was still only twenty-five years old, Webb was again showing not only his independence from tradition but also his complete confidence in his own abilities. It hardly needs mentioning that *Helena* was successful from the start. She was not only fast on the run but also belied all doubts by proving a good sea boat as well.

Ships in the China trade generally followed prevailing winds and made China by sailing eastward across the South Atlantic and across the Indian Ocean. On one occasion, however, when *Helena* had a delivery to make in San Francisco, she took the westward route instead. When she arrived in China, it was discovered that she had made what was at the time a record westward crossing of the Pacific (9).

The controversy over hull design received considerable publicity when in 1843, at a commercial fair held in New York, John W. Griffiths exhibited a half-model of a sharper hull and subsequently offered a series of lectures explaining the advantages of the newer concepts. Griffiths, who at the time was in the employ of Smith and Dimon, had, as we have seen, started his career, like William Webb and Donald McKay, both also known for their experiments with sharper hulls, as an apprentice under Isaac Webb, and had been with Webb at the time of the building of *Natchez.*

In tests conducted in his own towing tank (one of the first designed for this purpose), Griffiths was able to show that ships with traditional high convex bows and V-shaped hulls, while they stirred less wake and therefore appeared to be creating a smoother flow of water around them, were, in fact, creating an invisible underwater drag which seriously impeded their movement. He was also able to show that a sharp-bowed ship would not, as was commonly believed, plow into a wave, but would actually ride a sea more smoothly, and thus also more speedily, than a bluff-bowed ship (10).

These ideas were not entirely new. Robert Stevens had experimented with sharp-bowed hulls on the Hudson River many years earlier and had come to similar conclusions. And, as we have seen, many of Griffiths' concepts had already been incorporated in ships built by Isaac or William Webb or others. The contribution of John W. Griffiths was the more-or-less scientific way in which he conducted and recorded his experiments, as well as the publicity gained for these ideas in his writings and his public lectures.

William H. Webb's belief in a compromise between capacity and speed was well illustrated in his beautiful *Yorkshire,* ordered by his father's good friend, now William's primary patron, Charles H. Marshall, managing agent of the Black Ball Line of New York-Liverpool packets. Though at 163 feet 6 inches in length and 996 tons she was somewhat larger than *Oxford,* which Webb had earlier built for the line, *Yorkshire* generally followed a similar pattern, with a relatively sharp entrance, a straight keel, and minimal sheer. Compromise or not, *Yorkshire,* which was launched in 1843, shortly after William Webb was married, proved fast under sail and held

the record for trans-Atlantic runs until the appearance of the true clipper ships in the 1850s. Though the average time for packets crossing to Liverpool at this time was about twenty-three days, *Yorkshire* regularly made the run, under favorable conditions, in about fifteen days (11).

It was about this time that William H. Aspinwall, the relatively young head of the firm of Howland and Aspinwall, impressed with the speed of his ships *Ann McKim* and *Natchez,* asked Smith and Dimon to build him another larger ship of a similar type, specifying that John W. Griffiths provide the design. The ship that Griffiths produced was *Rainbow,* which, though her keel was laid in late in 1843, was not launched until 22 January 1845. The reason *Rainbow* spent this long time on the ways was that Aspinwall, possibly given pause by negative comments from traditionalists along the waterfront, held up *Rainbow*'s construction midway while he consulted some British shipbuilders about the viability of her narrow beam and of the enormously tall spars and extra spread of sail which Griffiths had designed for her. (12).

Meanwhile, also in 1844 (the year that William E. Webb, William and Henrietta Webb's first son, was born), William Webb laid keels and even completed not one but two more ships for firms engaged in the China trade: *Panama,* for N. L. and G. Griswold, the owners of *Helena;* and *Montauk,* for A. A. Low and Brothers. Both had sharper bows than *Rainbow,* as indeed had several of Webb's earlier packets, but, following Webb's rule of compromise, they did not sacrifice capacity for speed with as narrow a beam as Griffiths had designed for *Rainbow.* While neither proved a record-breaker (they were not intended to break records) both were fast. *Montauk* once made a run from the China coast to New York in seventy-seven days, not much over the existing record of seventy-two days (13).

At the same time that Webb was building these two ships and Griffiths' *Rainbow* was still languishing on the ways, yet another ship with sharper lines was taking shape at the yard of Brown and Bell farther downriver. This ship, commissioned by A. A. Low and Bros., was named *Houqua* for a highly esteemed Chinese merchant with whom they had been doing business. The idea for this ship had begun when William Low, a member of the firm, returned from a business trip to China aboard *Paul Jones,* captained by Nathaniel B. Palmer, who had once been captain of *Natchez* when she was still running to New Orleans. From his experiences with Natchez as well as with other ships he had commanded, Palmer had inde-

The Black Ball Line's 996-ton packet ship *Yorkshire*, built at the Webb yard in 1843, was the subject of artist Richard Schlecht's painting used on the United States Postal Service's 28-cent international surface postal card, first issued at Mystic, Connecticut, on 29 June 1988. (Courtesy of the United States Postal Service)

pendently arrived at conclusions similar to those of Webb and Griffiths, though in addition to a sharper bow and a greater length-to-beam ratio, he also agreed with William Webb that a flatter hull with little deadrise, like that of *Natchez*, would give a ship greater speed. A. A. Low and Bros., eager for theirs to be the first tea ships arriving from the East, took the risk with fewer qualms than Aspinwall and told Brown and Bell to build *Houqua* just as Palmer suggested, without seeking consultants from Britain or anywhere else (14).*

William Low's faith was rewarded, for *Houqua*, commanded, of course, by Nathaniel Palmer, making her maiden voyage in May 1844, proved one of the fastest ships on the China run. Nor did her narrower beam

appear to affect her stability. *Rainbow*, finally launched the following year, was not so fortunate. Griffiths' plan for higher masts and added canvas, was simply not successful. Not many days out of New York on her first voyage, *Rainbow* was hit by a gale which snapped all three of her topgallants. The crew managed to get her rerigged, with less sail, of course, while remaining underway. Even with her sail reduced, however, *Rainbow* proved a fast ship, logging on one occasion up to fourteen knots on the return trip (15).

*Rainbow*'s success on her maiden voyage (she was to disappear at sea on the outward leg of her second voyage) persuaded Howland and Aspinwall to order another ship of similar design from Brown and Bell. She was launched in 1846 as *Sea Witch*, one of the most beautiful, and unquestionably the fastest ship of her day. When *Sea Witch* achieved new records on the run to China, once again we find that Isaac Webb's *Natchez* had played a

---

* No drawings or plans of *Houqua* survive, but it is Howard Chapelle's belief that she probably closely resembled William Webb's *Helena*. (Speed Under Sail, p. 327)

role. The captain of *Natchez* on some of her fast trips across the Pacific in the 1840s was Robert Waterman (known as "Bully" Waterman to his detractors, who claimed that his placing speed as a priority over any human considerations led him to a treatment of his crews for which they said he probably should have been hanged). His appetite for speed whetted by his fast crossings with *Natchez*, Waterman had specifically requested command of *Sea Witch*. And it was his controversial though certainly effective talent for getting the most from his ship, as well as Griffiths' innovative design, that won records for *Sea Witch* (16).

All of the ships mentioned thus far, from *Ann McKim* or *Natchez*, to *Helena* or *Montauk*, or even *Houqua* or *Rainbow*, although each evidenced many bold new concepts in design, have to be classed as proto-clippers. Given the definitions that have emerged on the subject, *Sea Witch*, which was considered John Griffiths' masterpiece (17), and which embodies virtually all of his design recommendations, was the first ship that could honestly be classed as a "clipper."

With *Sea Witch* of 1846, the age of the American clipper ship had begun, though most of the great clippers did not appear until the 1850s. When they did appear, many of the finest and fastest were to be built by William H. Webb. But before this could happen, another matter horned its way into the world of American shipbuilders, for in 1846, the year of the launch of *Sea Witch*, the United States was edging toward a war with Mexico, a war which would in time dramatically change the nature of our nation and in the process create a whole new and unexpected demand for ships.

# V

# Busy Years in the New York Shipyards, 1847-1848

THE annexation of Oregon in 1846 and of California in 1848 meant that Americans now occupied two coasts three thousand miles apart. With a vital need for regular and dependable communication between the two, and with overland transportation at the time difficult almost to the point of impossible, the obvious answer was to establish connection between the East and West coasts by sea. With the consequent demand for ships strong and dependable enough for this long and arduous voyage, the New York shipbuilders were swamped with orders. One of the shipbuilders who responded to this demand was William H. Webb, who in these years greatly expanded the productivity of his yard and who, by completing a record number of ships, both sail and steam, all of the highest standards, became recognized by mid-century, when he was still in his early thirties, as the foremost shipwright in America.

The inexorable American push to the West had begun even before we achieved our independence. And, although the Louisiana Purchase of 1803 had nearly doubled the size of the country, settlers hungry for land continued to push westward even beyond our new boundaries. Americans crossing into the vast unsettled areas of Mexico south and west of Louisiana declared themselves the independent Republic of Texas in 1836. By the 1840's, so many American settlers, as well as commercial adventurers like John Jacob Astor, had found their way to the areas along the Columbia River, in the Oregon Territory still disputed between the United States and Great Britain, that the annexation of both of these areas became a major campaign issue in the presidential election of 1844. In that year, when the more conservative Whigs put up Henry Clay, the Jacksonian Democrats countered with the previously little-known James K. Polk. When Polk's campaign of blatant expansionism won him the presidency, he considered his election a mandate to push for the annexation of both Texas and Oregon.

One of the first acts of Polk's Congress early in 1845 was a vote to annex Texas to the United States, whereupon General Zachary Taylor was dispatched with an army to the Rio Grande to defend what we considered our new territory. In May the Mexicans, who viewed Taylor's presence more as an invasion of their lands than a defense of ours, declared war. As the war progressed, with the United States clearly maintaining the upper hand, Americans began eyeing the rich and sparsely-settled lands of California as ripe for annexation as well.

In June 1846 a peacefully negotiated treaty with Great Britain recognized the American claim to lands along the Pacific coast from the Mexican border at 42 degrees (the present northern border of California) to 49 degrees (the present northern border of the state of Washington). Our claim to Oregon settled, Congress now had to take up the issue of establishing communication with our newly-acquired territories on the Pacific coast. Not only did the government wish to provide means of encouraging settlers, it also needed to dispatch officials and troops, to send supplies, and especially to establish a regular and reliable mail service. Given these objectives, Congress in 1847 was prepared to award substantial mail contracts to private parties who could provide transportation to these new territories. The next question was whether this transportation should be overland or by sea, and if by sea, whether by sail or steam. The overland route was early eliminated. Not only were mountain passes difficult and Indian raids dangerous, but also all routes would be virtually closed during the long winters. The answer was clearly in favor of a sea route. Congress also determined that, although steamships had not yet come into common use on the oceans, steam-powered vessels would be more dependable for regularly-scheduled services. Congress further decided that the fastest service could be provided by steamship to some relatively narrow point in Central America, overland transportation across to the Pacific, and another steamship from there to Oregon. With this in mind, they arranged an agreement with the republic of New Granada (Colombia) allowing the United States the right of passage across the isthmus of Panama, the narrowest part of Central America, from Chagres, at the mouth of the Chagres River, to the city of Panama on the Pacific.

In 1845, therefore, Congress empowered the Postmaster General to subsidize mail services via this route. At the same time they also authorized the Post Office Department to award mail contracts to operators of steamships who would establish regularly-scheduled mail routes to Europe and South America. Unfortunately, Congress did not allot sufficient funds to the Post Office Department to make the services to the Pacific feasible. But, as we were then at war with Mexico, the Congress, aware of our shortage of naval vessels, gave the Department of the Navy virtual carte blanche to award contracts for two steamer services, one from New York or New Orleans to Chagres (on the Atlantic side of Panama), the other from Panama City (on the Pacific side of Panama) to Oregon, provided their ships met Navy specifications and could, in time of war, be easily converted to naval vessels. With these acts, Congress not only doubled the number of contracts to be handled by New York shipyards over the next few years, it also forced them into adapting without delay to the construction of steam-powered vessels (1).

When these contracts came, the yard of William H. Webb was prepared to accept them. In the two years of 1847 and 1848, no less than a dozen keels were laid in the Webb yard, half of them large steamships.

The first to take advantage of the new mail contracts was the Ocean Steam Navigation Company, organized in May 1846 by Edward Mills to supply mail service between New York and Bremen via Cowes. By this time the British Cunard Line had been operating steamships across the Atlantic for six years, but this was the first American line to offer regular steamer service between the United States and Europe. Thus also, the two steamships built for the line by Westervelt and Mackay, *Washington*, launched on 30 January 1847, and *Hermann* (named for the Teutonic hero, known as Arminius to the Romans, who defeated a Roman army in 9 A.D.), launched just eight months later, were the first large ocean-going steam vessels built in the United States. Given the lack of previous experience in building such ships, one is not surprised that neither was particularly successful. *Washington* was decidedly unstable until she was rebuilt after two voyages, and both were heavy on coal. But, once modifications had been made, the pair continued in the New York-Cowes-Bremen service for nearly a decade (2).

A third ocean-going steamship, *United States*, was built at the yard of William H. Webb. In the fall of 1846, Charles H. Marshall came to Webb and asked him to build a large ocean-going steamship. Though this ship was ostensibly intended for a service Marshall planned to start between New York and New Orleans, later events suggest that he actually had his eye on one of the lucrative European mail contracts. At that time Webb's experience with steamships was limited to a few small harbor craft, but he was willing. To finance this venture, Marshall organized a consortium which included Charles Morgan, who operated a shipping line in the Gulf of Mexico, and Henry Chauncey, of the merchant firm of Alsop and Chauncey, specializing in trade with South America. It also included T. F. Secor, who supplied the engines for *United States* and William H. Webb. It was from this time, it should be noted, that William Webb began to include a share of the ship in his contracts whenever feasible. As we shall see later this turned out to be a significant practice.

The keel for *United States* was laid at the Webb yard on 30 January 1846, the same day *Washington* was launched down at Corlear's Hook at Westervelt and Mackay's. Although Webb had built thirteen packets for ocean runs, the largest to date had been 174 feet in length. *United States*, whose wooden hull had to be staunch enough to carry several tons of iron machinery, was 256 feet overall, even forty feet longer than *Washington*. She also carried three full decks, compared with two on *Washington* and *Hermann*. A fuller description of this superb pioneer steamship can be found in Cedric Ridgely-Nevitt's fascinating American Steamships on the Atlantic (pp. 140-148), which also includes a two-page lines drawing, a two-page outboard profile, and deck plans. For the present, however, let a short quotation from this book suffice.

The United States presented a major advance over the Washington and Hermann in that she had a true steamship hull rather than that of a sailing packet with paddle boxes hung over the sides. The transom stern was replaced with a more seaworthy counter. For the first time on a steamship, the deadrise was reduced almost to zero and the bottom made practically flat. The bow was considerably sharpened, at the load water line, but the full deck line, fine stern, and maximum section well forward, elements of a sailing ship, were all retained. The hull was sturdy enough to resist the impact of the sea. The flat bottom was particularly important, for it put the engines low in the ship; as a result, there were no stability problems. (3)

*United States* was completed in only seven months and launched on 20 August 1847, a month before *Hermann*, the second of the Bremen boats. She was then towed from the Webb yard up to T. F. Secor's pier at the foot of Ninth Street to have her engines installed. Building a steamship involved a somewhat different process from building a sailing packet. The packets were launched in

The sidewheeler *United States*, completed in 1848, was the first ocean-going steamship built by William H. Webb. This handsome vessel, shown in an Endicott lithograph, set the style for many of William Webb's later steamers. (Courtesy of The Mariners' Museum, Newport News, Virginia)

a nearly completed state, but the hull of a steamer had to be launched with parts of the supportive structure of the interior of the hull still incomplete so that the engine could be installed. Only after the engine was in place did the shipyard's workers go over to the engine builder's yard to complete the internal structure of the hull. While *United States* was still fitting out, Webb received orders for four more steamships, all of which were built in his yard almost simultaneously, with incredible dispatch, and all of which were ultimately successful.

In 1847 the New York and Savannah Steam Navigation Company, better known as Mitchill's Line for its president Samuel Mitchill, was organized to operate between these two cities. William H. Webb was asked to build their first two steamers, to be known as *Cherokee* and *Tennessee*. By that time the Navy Department had finally offered contracts to the companies that were to run steamships from New York and New Orleans to Chagres and from Panama to Oregon, and it was the owners of the latter of these companies who also approached Webb late in 1847 to build two steamships for them (4).

The awarding of these mail contracts by the Department of the Navy turned out to be a rather complicated affair. The Congress seems not to have appreciated—or did they?—that these potentially very lucrative contracts would appeal as much to speculators as to serious steamship operators. Thus when the contracts were announced, the one for the line to Chagres was awarded to A. G. Sloo of Cincinnati, Ohio, and the one for the line from Panama to Oregon to Arnold Harris of Nashville, Tennessee. While both of these men had considerable influence in Congressional circles, neither was known to the shipping world, neither had any resources for carrying out such a contract, and neither appeared to be located in a place from which he could have done so. In short, both contracts were up for grabs—for the right price (5).

One man who was particularly eager to get these contracts was William H. Aspinwall, head of the New York merchant firm of Howland and Aspinwall, which already owned a fleet of packets and which did business in Europe, South America, and more recently also China. Aspinwall reasoned that controlling the route between the East and West coasts of the United States could, in time if not at first, become a very good business, and he was one of the few who could afford to wait for the investment to pay. Aspinwall was beaten to the East Coast contract, however, by a group of investors led by George Law and Marshall Owen Roberts but including also a variety of less active partners. Known, for some reason, as "Live Oak George" among his associates, Law had made money operating horse-drawn streetcars in New York. His interest in steam vessels had begun more recently when he joined Cornelius Vanderbilt, then just starting his entrepreneurial career, in a variety of water and rail lines between New York and Boston. The association of Law and Vanderbilt in these ventures suggests that the "Commodore" may have been behind the scenes in this venture as well, or at least at the start of it (6).

The government's contract with Law and Roberts stipulated that regular sailings between New York and Chagres be scheduled at least twice a month; that they employ five steamships on the run, four of which should measure no less than 1500 tons; that their plans be approved by the Navy Department to assure the easy conversion of the steamers to naval vessels; and that the ships be ready to begin service by 1 October 1848 (7).

Frustrated in his effort to secure the Atlantic part of the project, Aspinwall was still determined to get control of the rest of the route—the passage across Panama and the steamer line from Panama to Oregon—in his own hands. And in this he was successful. Harris signed his contract with the government on 16 November 1847, and he in turn sold the contract to Aspinwall on 19 November. This contract stipulated that there be three steamships operating between Panama and some as yet undetermined point in Oregon, with a sailing from each terminus once a month. Since by this time the war with Mexico was nearing conclusion and it was now apparent that California would also soon be part of the union, a requirement for stops at Monterey and San Francisco was added (8).

Some of Aspinwall's associates were less enthusiastic than he about operating steamships on the Pacific. The number of settlers there did not promise much passenger traffic, nor was there enough business to guarantee revenues from freight. The small villages along the coast, either in Mexico or Oregon, could hardly be counted on to furnish coal or supplies. Also, steamships tended to have mechanical problems, and the remote Pacific coast as yet offered no facilities for repair (9).

But Aspinwall was adamant and immediately ordered the construction of the three vessels he needed to start the line. For one, to be named *Oregon*, he turned to Smith and Dimon, the yard which had built most of his sailing packets. The contract for the other two, *California* and *Panama*, went to William H. Webb, whose big steamship *United States* was still on the ways, and who had also recently accepted orders for the two Mitchill Line steamers *Cherokee* and *Tennessee*. The cost of each of these three steamships for Aspinwall was to be $200,000 (10).

Sometime after ordering construction of these steamers, Aspinwall and his associates obtained a charter for their new company, now called the Pacific Mail Steamship Company. Among those on the board of directors were William H. Aspinwall, John L. Stephens, Henry Chauncey, and William H. Webb. Webb, apparently appreciating the future of American westward expansion, had become a major stockholder in Pacific Mail, very likely by accepting stock in the company as part payment for California and Panama. William H. Aspinwall was elected president, and the firm of Howland and Aspinwall designated as agents for the line (11).

Similarly, George Law, Marshall O. Roberts, and their associates, operators of the Atlantic side of the route, incorporated as the United States Mail Steamship Company. Orders were placed for two steamships, not five as the contract had required, though each of the two ordered was considerably larger than called for in the contract. Each was also somewhat larger than the three steamers being built for Pacific Mail. One, to be named *Ohio*, was built by Bishop and Simonson. As this yard built few ships for owners other than Cornelius Vanderbilt, the choice of this yard again hints at the Commodore's presence in the shadows of the organization. The second ship, *Georgia*, larger by a few feet, was ordered from Smith and Dimon, who were also building *Oregon* for Pacific Mail (12).

The work on the four steamships built at the Webb yard during 1848 progressed with amazing speed. Since Webb had already taken orders for the Savannah steamers, he was obliged to build them. But the contract for the two West Coast steamers for Aspinwall stipulated a completion date which had to be met. The keel of *California* was laid first on 4 January 1848. *Cherokee* followed on 14 February and *Panama* one week later. *Tennessee* was not started until *Cherokee*'s ways were cleared for her in June. By this time, understandably, there were nearly a thousand men employed in Webb's yard. Yet William Webb himself designed all four of these ships and took personal responsibility for every stage of their construction.

Webb designed *California* and *Panama*, but not *Oregon*, which was built next door at Smith and Dimon and was probably designed by John Griffiths. But, as all three were built to certain specifications, those of both Aspinwall and the Navy, they emerged virtually sisterships. All were within a foot or two of two hundred feet in length, and the tonnage ranged from 1047 for *California* to 1099 for *Oregon*, no appreciable difference. Again, for a detailed description of these pioneer steamships, one should turn to Ridgely-Nevitt who again includes an outboard profile and lines drawings.

Given the route for which they were designed, these ships were strong but simple. They were the first major American ocean steamships to give up the elaborate scrolled bowsprit, typical of sailing vessels of the time, for the straighter bow then typical on inland steamboats. These were also the first American-built ocean steamers constructed with diagonal iron strapping: that is, with their interior framing strengthened by crossed iron bars to counter the stresses that heavy engines placed on wooden hulls of such length.

All three had side-lever engines. A side-lever engine requires more moving parts and is therefore somewhat less efficient and more expensive to operate than the more customary walking-beam engine. But a heavy cast-iron walking beam, working high above the superstucture, can render a ship unnecessarily topheavy in a rolling sea, whereas the side-levers, located in the hull, lower a ship's center of gravity and tend, therefore, in the case of steamers designed for use in the open ocean, to make them more stable. The engines for *California* and *Oregon* were supplied by Novelty Iron Works, *Panama*'s by Allaire.

Since not many passengers were expected to patronize steamers operating on the distant shores of the Pacific, the three were designed to carry only about fifty people in cabins and about 150 in a steerage area which could be converted for cargo if steerage passengers failed to materialize.

By April 1848 the hulls of three large steamships were taking shape in the yard of William H. Webb. On 8 April, Webb's first steamship, the 256-foot triple-decked *United States*, finished fitting out at the joiner yard and sailed on her maiden voyage to Liverpool. Judging from Ridgely-Nevitt's drawings, she was a beautiful ship, far more graceful and better balanced than the rather ungainly *Washington* and *Hermann*. William Webb must have been proud indeed when he watched his first steamship sail down the river and out into the open sea. Though stately and stable, she was not intended as a fast ship. bOn this first trip across the ocean, which she managed without mishap, *United States* averaged about nine knots, taking fourteen days to reach Liverpool. Some of Webb's sailing packets had done as well.

*United States* was not to last long as a trans-Atlantic steamer in any event. By the time she was ready to sail for Charles H. Marshall, the mail contract for Liverpool had already been awarded to Edward K. Collins, operator of the Dramatic Line (so-called because his ships were named for famous actors) of packets. Without this contract, there was no way Marshall could compete on the Liverpool run.

On the night of the departure of *United States*, William Webb was at home and about to retire when there was an insistent knock at his door. It was a messenger with the news that his yard was on fire. Webb hastily called for a carriage and hurried to the yard as fast as he could get there.

The fire had started about ten P.M. in a stable across from the yard at Lewis and Sixth Street. Before any firemen had arrived, the flames had leapt across the street to Webb's yard. By the time Webb got there, his small office building facing on Lewis Street was in flames, and with it many of his drawings and half-models, though someone had already run in and rescued most of Webb's papers, which included notes on ships he had built as well as financial records concerning ships then building.

The firemen, seeing the owner arriving, asked what part of the yard deserved their first consideration. Webb asked them please to save the steam chest, a large building where timber was steamed to make it pliable enough to shape. But soon another object caught their attention.

The steamship *Cherokee*, built for Samuel L. Mitchill's New York & Savannah Steam Navigation Company in 1848 and pictured in this lithograph, was in service only five years. A fire at her North River pier ended her career in August 1853. Her sister ship, *Tennessee*, sold to the Pacific Mail, was even less fortunate—she stranded while attempting to enter the Golden Gate in March of the same year. (Courtesy of The Mariners' Museum, Newport News, Virginia)

The fire was spreading to the ways, and *Panama* was on fire. The firemen, dragging their pumps and hoses, hurried to the ship and directed their hoses on her with as much force as they could produce. Fortunately they were able to put out the fire before too much damage had been done.

Just as this emergency was being dealt with, there was a crash in the center of the yard followed by screams. Several people who had come to observe the fire had climbed up on the shed which housed the saw-pit. But the roof of the shed, not built to hold so much weight, collapsed, dumping its human burden into the saw-pit. Several people were seriously injured with broken bones; one man later died.

By morning the fire had been put out. William Webb's office and the mold loft were gone as was most of the valuable stockpiled lumber. But Webb, who undoubtedly remembered that two nearly-completed steamers had been destroyed in the fire at the Brown and Bell yard in 1824, was grateful that, aside from superficial damage to *Panama*, which could be repaired in time, none of the ships then under construction had been lost. In spite of the cost and time lost in rebuilding the offices, all three of the steamships on the ways were finished on schedule.

*California* was first and, as she was the first Pacific Mail steamer to be launched, the event was celebrated with the requisite number of banners, toasts, and speeches, as well as a fair amount of champagne. *Cherokee* was next a few weeks later on 12 June, with the keel of *Tennessee* laid in the same berth within a few days. *Oregon* was launched at Smith and Dimon on 5 August and joined *California* at Novelty Iron Works to have her engine installed and her final joiner work completed. When *Panama* was launched at Webb's yard on 29 July, all three Pacific Mail steamers were in the water just a few weeks more than six months after their keels had been laid. With the launch of *Tennessee* on 25 October, just four months after construction began, one can begin to see at what level of efficiency the William H. Webb yard must have been functioning in the year 1848, especially considering that the yard had to contend with a major fire and much consequent reconstruction in the same period.

Late in September a group of naval officers came aboard *California*, still docked at Novelty Iron Works, and spent several days going over every part of her hull and engines. Finally, on 5 October they pronounced her acceptable to the United States Navy and thus serviceable as one of the three steamships the Pacific Mail had to have in operation to qualify for the government subsidy.

The following day, 6 October 1848, *California*, commanded by Captain Cleveland Forbes, and well bedecked with banners and bunting, sailed out of New York on her way to the West Coast to inaugurate the service of the Pacific Mail Steamship Company. She was not going all the way to Oregon, however. When the Treaty of Guadaloupe-Hidalgo was signed in February, ending the war with Mexico and annexing all of what is now the American Southwest, including California, to the United States, *California* began to loom larger than Oregon for potential development. Thus Aspinwall was able to arrange a modification of his agreement with the government to the effect that his steamships would run from Panama only to Monterey and San Francisco. From San Francisco north, the company would be permitted to use sailing vessels until they had time to procure some smaller steamers for the purpose.

As *California* paddled south down New York harbor, she had about one hundred invited guests aboard, including both William H. Aspinwall and William H. Webb. On board these guests were served a banquet to celebrate the event, and once more there were toasts, speeches, and champagne. Somewhere off Sandy Hook, the smaller steamer *Orus* came alongside to take the guests back to the city, while *California*, relieved of her revellers, continued quietly down the coast. Since few people were ready yet to emigrate to remote California, she carried only seven paying passengers, these all bound for destinations in South America. Headed for a long journey to

unknown parts, she also carried a year's worth of stores and a full set of spare parts for her engine. *California*'s cruising speed was an unspectacular nine knots (anything faster would have been unduly expensive and consumed far more coal than she could carry), though with a fair following wind, she could raise sail and reach as much as twelve knots.

By late October, with *California* cruising somewhere south of the Equator, Captain Forbes, whose lungs had been hemorrhaging, found he was too ill to retain responsibility for the ship, which he turned over temporarily to Mr. Duryee, his mate. On 31 October *California* arrived in Rio de Janeiro, where she remained three weeks for repairs and recoaling, giving the captain a chance to recuperate.

*California* steamed out of Rio on 24 November, Captain Forbes again in charge, and headed south. She began her passage through the Straits of Magellan, the first steamship flying the American flag to do so, on 5 December. But, as she had to stop and anchor on several occasions because of fog, she did not emerge on the Pacific side until the sixteenth.

By the time *California* was in the Pacific and paddling up the coast of Chile, Captain Forbes realized that he was much too ill to retain command. On arrival in Valparaiso on 18 December, he reported his wish to be relieved of command to the office of Alsop and Chauncey, the local agents for Howland and Aspinwall. As it happened, Howland and Aspinwall's fast packet *Natchez*, which we have already encountered often before, was then at anchor in Valparaiso. It was therefore arranged for her captain, John J. Marshall, to take over *California*, and for *California*'s mate Duryee to become captain of *Natchez*. Since Marshall had no previous experience with steamships, it was agreed that Forbes would remain in command as far as Panama, with Marshall observing, and that Marshall would take command at Panama, where Forbes hoped he could find medical care. Once these arrangements had been worked out, *California* sailed again on 22 December. Five days later, having passed Christmas at sea in what there was early summer, she put in at Callao, the port for Lima, Peru. Here fifty-two Peruvians taking passage to California joined the ship. Then on 29 December, *California* headed north again toward Panama.*

---

* For a fuller account of *California*'s maiden voyage, see Victor M. Bertold, *The Pioneer Steamer California, 1848-1849* (Boston and New York, 1982).

Meanwhile, both of her sisterships had been completed and were ready for their trips to the West Coast. Webb's *Panama* sailed on 1 December, amid celebrations only slightly less enthusiastic than those accorded *California*. But on 6 December she returned to New York rather ignominiously under sail. Someone at the Allaire works, incredibly, had left a piece of wood in the main steam pipe before sealing it, and not until *Panama* was well out to sea did the wood find its way to the cylinder, creating considerable damage to the engine when it got there. *Panama* went back to Allaire for repair and was not ready to sail again until 17 February 1849. Meanwhile, *Oregon* had sailed on 9 December and was following *California*'s path down the coast.

While all three of the Pacific Mail steamers were ready, or in the case of *Panama*, nearly ready, for service in compliance with the terms of their contract, the United States Mail's steamers were nowhere near completion. *Ohio* had been launched in August but was not finished fitting out for over a year. *Georgia*, launched in September, was not ready to sail until January 1850 (13). In order to supply at least some semblance of the service for which they were contracted, United States Mail purchased the smaller steamship *Falcon* to start the New York to Chagres route. *Falcon*, under Captain Notestein, sailed from New York on 1 December, the same day as *Panama*'s first attempt, and headed for Chagres by way of New Orleans.

As the year 1848 came to a close, *Panama* was still at the Allaire yard having her battered engine repaired, and her two sisterships were both steaming peacefully toward California: *Oregon* in the Atlantic about to cross the Equator and *California* about to cross the Equator in the Pacific. If their owners, their officers, and their crews believed, as they undoubtedly did, that these steamships were on their way toward a routine and rather tedious service in a wilderness, they were very soon to get a very big surprise.

# VI

# Pacific Mail and the Gold Rush, 1848-1850

**B**ETWEEN the time *California* and *Oregon* steamed out of their piers in New York and their arrival at Panama, a new element—gold—had entered the world of the Pacific Mail Steamship Company. Gold had been discovered by James Marshall near Sutter's Mill on the Sacramento River as early as January 1848, even before the treaty of Guadaloupe-Hidalgo had assured California's addition to the United States. But the news was very slow in travelling, and even those who heard of the strike were inclined not to take it too seriously. It was not until 5 December, when President Polk, in a message to Congress justifying the annexation of this remote area, announced that the governor of California had assured him of "the abundance of gold in that territory," that the country began to take notice. Even then several days passed before the President's announcement set off the unexpected avalanche of human migration westward known as the Gold Rush, an event that almost overnight would change William H. Aspinwall's "risky" venture into a golden bonanza and contribute also to making William H. Webb a wealthy man.

On 5 December, *California* was still paddling her lonely way through the foggy Straits of Magellan; *Falcon* of the U. S. Mail was five days out of New York on her way to Panama via New Orleans; and *Panama*, which had left port the same day, was making her way back under sail. When *Oregon* left New York five days later, there was still not enough interest in passage to California to give her more than a handful of paying passengers, most of whom had transferred from *Panama*.

By the time *Falcon*, with ninety-five passengers from New York aboard, reached New Orleans on 18 December, however, the rush had started. Literally hundreds of people were on the docks seeking passage to California by any means available. Captain Notestein, torn between the opportunity for revenue and the safety of his ship, leaned toward the latter and allowed only another ninety-eight aboard, for a total of 193. Most of these passengers held through tickets to California and therefore expected to be transferred at Panama to *California*.

*Falcon* dropped anchor in Chagres on 27 December, more than three weeks before *California*'s arrival in Panama. Passengers, rowed ashore in small boats, found only chaos waiting for them. At the small village of Chagres, hardly equipped to accommodate even the 193 people coming in from *Falcon*, hundreds had been arriving every day, with more still to come, on sailing ships, coastal sloops, fishing boats, anything that would float, and all expected to board a steamer at Panama to take them to California and instant fortune. Not only were accommodations meagre and food in short supply but also the trip across the isthmus was not exactly inviting. Passengers were taken in native canoes up the Chagres River, much of which was a mosquito-infested mangrove swamp with alligators splashing about; then by foot or burro down again through a jungle that also hosted various forms of wild life. Finally, on the Pacific side, they found Panama, another small village ill-prepared to welcome over a thousand unexpected travellers (1).

When *California* steamed unsuspectingly into Panama harbor on 17 January 1849, there were over 1500 anxious souls waiting to board her. Thus the first official assignment of her replacement captain, John J. Marshall, was to resolve this problem. When it became apparent that *California* could take no more than two hundred at best, both the American consul and Captain Marshall were besieged with protests. Meetings were held to insist that the Peruvian passengers be put ashore and that as an American ship *California* should take aboard only American passengers. This recommendation Marshall, to his credit, refused to accept. His first decision, a sensible one, was to give priority to passengers who had bought through tickets to California when they boarded *Falcon* in New York or in New Orleans. These alone would overcrowd *California* and tax her limited food supply. But, as she was carrying little cargo other than her own extra engine parts, Marshall cleared the freight deck and had wooden bunk beds hastily built in the space (2).

The Peruvians, pleased enough not to have been dumped ashore, happily accepted berths in this area, while the regular steerage section, as well as corridors, public rooms, and even deck space, were taken by as many Americans as Marshall deemed safe. *California* finally steamed out of Panama with 365 passengers, which, with a crew of thirty-six meant that there were over four hundred people aboard a steamer designed to carry half that number. They included virtually the entire range of humanity, from dignified representatives of eastern business firms to swashbuckling adventurers to plain down-and-out vagrants. In these crowded conditions, with food always in short supply, all were apparently reduced in time to the level of beasts, so that Captain Marshall eventually gave up any attempt to maintain order. On the way north, stops were made at Acapulco, San Blas, Mazatlan, and Monterey. After *California* left San Blas, part of the crew mutinied but were delivered to a local jail at Mazatlan. Out of Mazatlan, the ship ran out of coal, so that parts of the ship itself had to be dismantled to keep the fires burning, though in the process of tearing up some lower deck planking, more coal was found stored beneath it (3).

Finally on 28 February 1849, *California* steamed into San Francisco, her passengers cheering themselves hoarse with excitement (or with relief). As it happened, the Webb-built *California* in the process became the first steamship to sail through the Golden Gate. She had just travelled 14,000 miles in 144 days, though only 76 of these had been in transit (4).

In theory, *California* was now supposed to prepare for a return trip to Panama, thus inaugurating the regular San Francisco-Panama service for which she had been designed. But as soon as she arrived, her entire crew, including Captain Marshall, took off in search of gold, leaving only the ailing Captain Forbes and one boy in the engine room aboard. But she could not have sailed in any event. The collier due from England had not arrived, and there was no coal available anywhere on the bay.

By the time *Oregon* reached Panama on 23 February, far more people were still arriving at Chagres than available ships could possibly pick up at Panama. But Captain Pearson was a rules man, so *Oregon* sailed on 13 March with just 250 passengers, the number allowed by her license. Also, on arrival in San Francisco, Pearson, summoning the aid of the local naval commander, kept his entire crew aboard literally at gunpoint. By this time the coal ship had arrived, so *Oregon* rather than *California* became the first to start the service between San Francisco and Panama with her departure on 12 April 1849 (5).

Cleveland Forbes, again, by necessity, in command of *California*, managed to assemble a new crew only by offering each man $120.00 a month, or ten times what his former crew had received. *California* then sailed for Panama on 1 May. On arrival there, Forbes asked to be relieved permanently and took ship back to New York (6).

*Panama*, on arrival at her namesake city, encountered problems similar to those of her two sisterships. She finally arrived at San Francisco on 4 June with 290 passengers aboard and was ready to sail south again on 20 June. *Oregon* was back in time for a 1 July sailing from San Francisco, and from this time Pacific Mail was able to maintain sailings once a month from each port, usually on the first day of the month.

While these events were taking place in California, Webb's much larger *United States* arrived in New York on 5 February after two unsuccessful trips to Europe. Without the mail contract, Charles Marshall could not hope to compete with the Cunard ships, which were supported with a substantial subsidy from the British government. At just this time a group of German states was attempting to form a federation—which was not to last—and this short-lived federal government, seeking to assemble its own navy, made Marshall a healthy offer for *United States* which financial realities obliged him to accept. As a result, in the spring of 1849 William Webb was given the unpleasant task of converting his recently-completed masterpiece, designed to help establish an American steamship service between New York and Liverpool proud enough to compete with Cunard, into a warship for a foreign nation. With the upper (of three) decks completely removed, and gunports cut into her sides, *United States* sailed on 31 May for England, where, renamed *Hansa*, she was transferred to the Germans. Not long in this service, she was later used briefly as a trans-Atlantic emigrant ship by a German company. Then in 1854 she was sold to English parties and renamed *Indian Empire* for service between Great Britain and India, though her first assignment was a charter to the government to carry troops to Sevastopol during the Crimean War. After a useful, if not very glamorous, career, she ultimately burned in the Thames on 23 July 1862 (7).

Meanwhile, the mail contract for a New York-to-Liverpool line had been awarded to Edward K. Collins. In his effort to provide a service that could establish the supremacy of an American line over the Cunard Line on this route, Collins began by ordering the construction of four large steamships, designed to be unsurpassed in com-

This painting of the Taylor & Merrill packet ship *Guy Mannering*, built in 1849, shows her outward bound from New York, while another packet (under the tip of her jibboom) makes her way up the harbor. The owner's Black Star Line house flag— a black star on a red field—flies from *Guy Mannering*'s mainmast.

fort and elegance, as well as in speed. For some reason, none of these orders went to William H. Webb, possibly because Webb's yard at that time was already backlogged with orders from Pacific Mail and from Mitchill's Line. Two of the Collins liners, *Pacific* and *Baltic*, were built by Jacob A. Bell, who carried on alone for a few years after the death of his partner, David Brown. The other pair, *Atlantic* and *Arctic*, was built by William H. Brown. With overall lengths averaging about 280 feet and tonnages of about 2850, all were considerably larger than any of the ships Webb had built to date. Once the ships were completed during 1850 and the Collins Line started running, it was soon apparent that all four were fast as well. *Baltic*, fastest of the four, soon established a record

by being the first to cross the Atlantic in less than ten days, which certainly eclipsed the thirteen-day passage of Webb's *United States* the previous year.

During the Spring of 1849, now that the four steamships for Mitchill's Line and Pacific Mail had been completed, the yard of William H. Webb began construction of no less than five merchant packets for as many owners, both in the coastal and the trans-Atlantic trade. The grandest of these was *Guy Mannering** for Taylor and Merrill, a New York trading firm. In this service, Guy

* An excellent painting of *Guy Mannering*, once in the private collection of William H. Webb, now hangs in the main lounge of Webb Institute of Naval Architecture.

Mannering joined *Marmion* and *Ivanhoe*, which William H. Webb had built earlier for the same company and which, like *Guy Mannering*, had been named for heroes in novels by Walter Scott.

The following year Webb received an order for another large packet for Charles H. Marshall's Black Ball Line. It was Mr. Marshall's special request that this ship be named *Isaac Webb* for his good friend, the builder's father. According to one writer, "the packet ship had developed into the highest class of passenger vessel in the year 1850, when....Isaac Webb, of 1800 tons, was launched from William H. Webb's shipyard" (9).

In the same year Webb also built the large steam towboat *Goliah*. *Goliah* and other similar vessels were used to tow large sailing ships through the tricky channels of the East River and New York harbor, until they reached the open sea, or, similarly, to tow them into the harbor and safely up to their piers after an ocean crossing. They were also frequently sent out to do salvage work on ships which had come to grief along the coast.

Also of interest to William Webb, construction was getting underway that spring on the railroad in Panama which would connect the steamship landing on the Atlantic coast with the city of Panama on the Pacific. Even before he realized how much improved the future of Pacific Mail would become with the discovery of gold, William H. Aspinwall had been working to get control of the second of the three links in the route between New York and California, namely the railroad across Panama. As early as 1 December 1848, several days before President Polk's speech to Congress (although not necessarily before Aspinwall could have known its message), Aspinwall and his partners, John L. Stephens and Henry Chauncey, had petitioned Congress for permission to build a railroad across the isthmus of Panama. It seems that Stephens had already been to Panama with a group of engineers to survey the feasibility of such a route and that Aspinwall, after passing an impressive amount of gold into the hands of the government of New Granada through the agency of its minister in Washington, had secured a grant of the land, for a term of ninety-nine years, on which the railroad would be built (10).

Once Congressional permission had been secured, the group incorporated the Panama Railroad Company, a company independent legally of either Howland and Aspinwall or the Pacific Mail Steamship Company. Among those who owned a large amount of stock in the new railroad, the future of which, for the present at least, appeared to be "golden" indeed, was William H. Webb,

who had probably been brought in from the beginning by his association with Aspinwall and the Pacific Mail Steamship Company. Among those who were not stockholders, however, was George Law, president of the United States Mail Steamship Company, operating between New York and Chagres. Law had been eager to have a share in the proposed railroad, but Aspinwall, who apparently never trusted him, had made a point of excluding him (11).

Aspinwall's next step was to dispatch another group of engineers to Panama to choose a site for the Atlantic terminus of the railroad, presumably somewhere near the steamship landing at Chagres. But when the engineers arrived there, they found that George Law had purchased—or obtained options to purchase—every available site along the coast (12). Although this move might seem shortsighted of Law, considering that passengers landing from his steamers would be the ones who needed the railroad, it was not Law's intent to block the building of a terminus at Chagres, merely to blackmail Aspinwall into letting him buy into the railroad company.

But Aspinwall would not rise to the bait. He would build his terminal elsewhere and, if necessary, run his own line of steamers to meet it. Aspinwall's directions to his engineers were to find some available spot along the coast not already controlled by Law. And they found one. It was Manzanillo Island, which was not so much an island as an alligator-infested mangrove swamp several hundred feet off the coast and several miles south of Chagres. The coast itself, once reached, was also a mangrove swamp, similarly inhabited by alligators, snakes and mosquitos, among identifiable creatures. At best, the area was not designed for railroads. Nevertheless, over the next several years, Aspinwall and his partners persisted, pouring a great deal of money as well as untold tons of rock and sand into the area over which they eventually began laying the tracks of the railroad to Panama (13).

Although only two years earlier Aspinwall's associates had questioned the wisdom of spending sound money to send steamships to the Pacific, by 1849 the original trio, *California*, *Oregon*, and *Panama*, could not come close to handling the traffic. More and larger steamers were needed. Also, with Pacific Mail and United States Mail maintaining positions of independent hostility toward each other, each determined in 1849 to begin operating steamers on the other's route. Even without the government mail subsidy, the amount of business (with the number of prospectors now returning home, even the West-to-East traffic was becoming profitable) would seem to guarantee success. The reason for the decision of Aspin-

The 1169-ton packet ship *Vanguard*, a typical three-decker built in 1850 for service between New York and the Gulf Coast, was painted by S. Walters soon after she was completed. She was constructed for James W. Phillips, and was lost in 1877 while preparing to load lumber at Matane, Quebec. (Courtesy of a Private Collector)

wall and associates to operate on the Atlantic as well as the Pacific was apparently not necessarily to get more business but rather to provide a complete East-to-West service of uniformly high quality and dependability. United States Mail had been slow in starting. Also its lower quality of service, the reputation of its steamships as rollers, and the fact that its arrivals in Chagres did not always dovetail as well as they might have with the carefully scheduled departures of Pacific Mail steamers at Panama, were damaging the reputation of Pacific Mail. Only by providing its own service between New York and Chagres could Pacific Mail maintain the reputation it had so carefully cultivated on the West Coast.

To provide the necessary new steamships, Pacific Mail again turned to William H. Webb. Probably through Webb's agency, Pacific Mail was able to purchase both *Cherokee* and *Tennessee* from Mitchill's Line. These two

steamships, as mentioned, had been built at the Webb yard at the same time as *California* and *Panama*, and, like these two ships, both of which had proved successful in operation, they had been designed by Webb himself. They were similar to the Pacific Mail's steamers, though at 210 feet somewhat larger, and, capable of eleven knots speed, also somewhat faster.

*Tennessee* was sent immediately to join the Pacific fleet of the company, and here she soon established herself as a favorite with the passengers. Before starting service, some of her cargo area was converted to passenger use, giving her a capacity for 200 passengers in cabins and 350 in steerage, almost double that of the three earlier steamers. With four steamships available, Pacific Mail was now able to schedule two sailings a month from each port, one on the first and one on the fifteenth. *Tennessee* had not been inspected by the Navy when she was built.

But the contract stipulated that only the steamer sailing on the first be accepted by the Navy, so *Tennessee* was simply scheduled to sail on the fifteenth. Later, however, when Pacific Mail asked for an increase in the subsidy for providing two guaranteed mail sailings a month, the Navy did inspect *Tennessee* and pronounced her acceptable with no modifications.

*Cherokee*'s assignment was to inaugurate Pacific Mail's competing service on the Atlantic side, running between New York and Chagres, as the Manzanillo terminus was not yet even habitable let alone serviceable. Also on this route was *Philadelphia*, borrowed from another line.

Now that their first and only steamers had been sold to Pacific Mail, at a pleasant enough profit, Mitchill's Line to Savannah placed orders with William H. Webb for two even larger steamships, *Alabama*, the keel for which was laid almost immediately, and *Florida*, which was started later.

Meanwhile, George Law's United States Mail Line was not in such good shape. The two steamships they had ordered, while larger and grander than the three built for Pacific Mail, took far longer than expected to be completed, leaving *Falcon* to cover the route alone, with occasional assists from chartered steamers. *Ohio* finally came from her builders on 20 September 1849, nearly a year after *California* had been delivered by Webb. *Georgia* was not ready until January 1850. Once in service, only *Georgia* ran from New York to Chagres, stopping en route at Havana. *Ohio* filled the New Orleans part of the contract, running from New York to Havana and then to New Orleans. The new assignment for *Falcon* was to run out of New Orleans to Havana and Chagres, picking up *Ohio*'s passengers for Chagres or *Georgia*'s passengers for New Orleans, at Havana.

Neither *Ohio* nor *Georgia* proved very satisfactory in service. Both burned prodigious amounts of coal. According to Ridgely-Nevitt, with whom one should not argue on such matters, their excessive beam led to both severe rolling and erratic steering. Ridgely-Nevitt also points out that these steamers, unlike the Webb-built *California* and *Panama*, did not have the diagonal iron strapping, which he considers essential for an ocean-going wooden-hulled ship of that length carrying so much iron machinery (14).

Nevertheless, even before both of their Atlantic steamships were in operation, United States Mail had started its own line, known as the "Law Line," on the Pacific to compete with Pacific Mail. Although their fares were lower, their steamers were inferior, so the competition did not seriously hurt Pacific Mail. If anything, the Law Line had the effect of siphoning off some of the more unsavory passengers, a type for whom the voyage to California seemed at the time to have a special attraction, thus reserving the cream of the trade, such as it was, for Pacific Mail.

Before the year 1849 was over, however, two more competitors had appeared on this apparently lucrative field. Charles Morgan, already operating steamers out of New Orleans on the Gulf of Mexico, began a service from New Orleans to Chagres by chartering two fine new steamships, *Empire City* and *Crescent City*, recently built by William H. Brown for a line between New York and New Orleans. Morgan's line, known as the Empire City Line, also chartered two lesser steamers to run from Panama to San Francisco.

By this time Cornelius Vanderbilt, who had by then obviously dissolved whatever earlier connection he may have had with United States Mail, decided that he too wanted a share of the action, and also that he wanted a New York-to-San Francisco route all to himself. Since Aspinwall already had the rights to the railroad across Panama, Vanderbilt decided on Nicaragua. The overland route here from San Juan del Norte to the Pacific was considerably longer than the one across Panama, but it was also less arduous. Much of it could be accomplished comfortably aboard a sternwheeler up the San Juan River to Lake Nicaragua. From there it would be but a short train ride to San Juan del Sur on the Pacific. Furthermore, the total distance between New York and San Francisco via Nicaragua was nearly four hundred miles shorter than via Panama and therefore could be travelled in less time. To create his river and rail route across Nicaragua, Vanderbilt obtained the necessary permits from Nicaraguan officials and set up the Accessory Transit Company, with himself as its major stockholder.

During the last week of 1850 Vanderbilt started his new line, or at least part of it. There were as yet no facilities for crossing Nicaragua nor any steamers on the Pacific. So Vanderbilt's start amounted only to running a chartered steamer from New York to Chagres where passengers could cross to Panama and board a steamer of one of the other two lines to San Francisco.

In preparing to mount its simultaneous assault on three competitors during the summer of 1850, Pacific Mail increased its capital stock from the original $500,000 to $2,000,000. Even with this large increase the stock had become so popular that the entire amount was subscribed by friends or business associates (including William H.

With her two stacks and clipper bow, *Golden Gate* was not typical of Webb's steamers. Built in 1848, she was the third of the Pacific Mail liners constructed at the yard. Bad luck dogged her throughout her career, which ended when she burned in July 1862. (Courtesy of The Mariners' Museum, Newport News, Virginia)

Webb, who accepted more stock at this time) and never went on the market. Armed with extra capital, Pacific Mail was able to make two moves against the competition. First, the Empire City Line was eliminated in October 1850 with the simple expedient of purchasing all of Morgan's chartered steamships behind his back. Thus the Empire City Line was left with no ships to run, and Pacific Mail found itself with some fine additions to its fleet, with which they started their own line between New Orleans and Chagres, thus throwing yet another gauntlet at the feet of George Law.

The second move was to order two new steamships, each larger and far grander than any ships then on their own line or on any of the other lines. One, named *Golden Gate* and intended for the New York-Chagres route, was to be built by William H. Webb. The other, to be called *Louisiana* and apparently therefore intended for the New Orleans-Chagres route, was assigned to Smith and Dimon.

*Golden Gate* was certainly the finest steamship built by William Webb to date and at the time the largest and finest American steamship afloat, including even the vaunted quartet of Collins Line steamers. At 265 feet *Golden Gate* was longer than the original ships by nearly a third. With almost twice the tonnage (2200), she had accommodations for 800 passengers: 300 in cabin and 500 in steerage. Moreover, with two oscillating engines supplied by the Morgan Iron Works, she was the fastest steamship in American coastal waters.

By late 1850 the construction of the Panama Railroad was also progressing, so that arriving passengers could now take a train at least part way to Panama. But the construction to date had taken a far greater toll than anyone had anticipated. Hundreds of lives had been lost to disease, and the railroad company, unlike the prosperous Pacific Mail, was by then close to bankruptcy (15). At the same time, United States Mail discovered that they were losing business to Pacific Mail because their passengers were being denied the use of the railroad. It was therefore at this juncture that George Law and William Aspinwall discovered it would be in the interests of both to come to some agreement. Had they been willing to do so two years earlier, both companies could have saved their stockholders a great deal of money. In essence, the settlement signed in January 1851 stipulated that United States Mail would take over all of the Atlantic routes and Pacific Mail all of the Pacific routes of both lines. Another aspect of the agreement, and one essential to both parties, stated that United States Mail would put enough cash into the Panama Railroad to make it possible to continue construction; in return, needless to say, Aspinwall had to accept George Law as a member of the railroad's board. In the process of working out the new arrangement, Pacific Mail purchased all of the ships operated by United States Mail on the Pacific. Since none of these ships was the equivalent of Pacific Mail's own ships, they were not added to the San Francisco-Panama service but were placed on some of the company's subsidiary services such as those between San Francisco and various Mexican ports or ports to the north in the Oregon territory. United States Mail got the better of the bargain in purchasing Pacific Mail's Atlantic fleet. Now both *Georgia* and *Ohio* ran from New York to Havana and Chagres, with sailings increased to twice a month; *Empire City*, *Crescent City*, and *Cherokee* made up a separate route from New Orleans to Chagres via Havana, and *Falcon* was retired.

*Louisiana*, being built by Smith and Dimon, was purchased by Law, who changed her name to *Illinois* and planned to use her on his New York-Chagres run. *Golden Gate*, being built by Webb, was retained by Pacific Mail, which now planned to send her to the West Coast. An inspection of *Golden Gate* after her completion in November moved the San Francisco Herald to high praise:

She is the finest specimen of naval architecture in the Pacific, being the largest and swiftest steamer in our waters, and equalled by few anywhere. Her long deck forms a noble promenade, while her beautifully fitted up accommodations and well ventilated cabins, remind one more of a drawing room than the interior of a ship. (16)

To this Kemble adds: "With the Golden Gate it may be said that the Pacific Mail had struck its stride in the construction and operation of steamers adequate for the service . . ." (17).

Another byproduct of the 1851 agreement was that United States Mail, now that it was in an alliance, however uneasy, with Pacific Mail, placed an order with William H. Webb for another new steamship to be as large, as powerful, and as elegant as *Golden Gate*. When this fine steamship, named *George Law*, was delivered in 1853, she met these requirements and more. Now United States Mail, with two fine new steamships, *Illinois*, built by Smith and Dimon, and *George Law*, built by William H. Webb, was able to retire both *Georgia* and *Ohio*, which, though less than five years old, had never proved very satisfactory.

The alliance was arranged just in time for both companies, for by July 1851 advertisements began to appear on both coasts for "Vanderbilt's Independent Line." As the Accessory Transit Company had not yet finished creating the new route across Nicaragua, the Vanderbilt ships still operated via Panama. Nevertheless, a new competitor, known both for his tenacity and his penchant for foul play, had entered the ranks, and another battle was about to begin.

Shortly after the Pacific Mail steamship *San Francisco* was lost in late December 1853 while on her delivery voyage to the West Coast, N. Currier produced this famous lithograph of artist F. E. Butterworth's impression of the rescue of survivors by the ships *Antarctic* (left) and *Three Bells*. Despite having been pounded for days by heavy seas, *San Franscisco* simply refused to go under—a testimonial to the strength of Mr. Webb's staunch hull. (Courtesy of The Mariners' Museum, Newport News, Virginia)

# VII

# The Clipper Ships, 1850-1856

NO vessels in history have excited so romantic a response as the American clipper ships. These long, lean ships, with their tall spars and billowing white sails have been called, with good reason, the most beautiful products in the history of American ingenuity. Though their European copies started later and lasted longer, almost all of the great clipper ships, which were indeed purely American products, were built between 1850 and 1856. A few more came off the ways as late as 1859, but, after that, none at all in this country.

The clipper was designed for speed, and some American clippers made extraordinarily fast runs across the Atlantic or the Pacific, racing at times up to an astounding eighteen knots, which even today can be bettered only by very powerful motor-driven vessels. But the long fine lines that gave the clipper its speed were often achieved at the expense of revenue-producing cargo space, so that the functions of these beautiful craft were limited. Clippers were rarely used on the busy trans-Atlantic runs, for instance, where the traditional wide-bellied packets served better for bulk freight and where steamships, which could maintain more regular schedules, were by 1850 already replacing sailing ships for express freight and passengers.

There were two areas, however, where the demand for fast sailing ships with limited cargo space peaked during this short period and gave the clipper ship its moment of glory: the China trade and the Gold Rush. We have already noted that time was a major factor in the tea trade and that for American importers fast profits from a fresh shipment via clipper more than compensated for the limited space. In 1849, when the British abandoned protectionism for free trade and removed the requirement that British imports arrive in British bottoms, there was great surprise and considerable consternation when American clippers began delivering tea from China at the London docks several days before the British ships arrived.

The clippers also played a vital role in the early development of California. The enormous human migration to the West Coast in the early 1850s included not only prospectors for gold, but settlers, businessmen, or, in fact, anyone who wanted to try a new life in the West. In the few years between 1850 and 1853, virtually any vessel leaving the East Coast, including many which should never have been allowed into the ocean and some which were never seen again after they cleared New York, could count on more passengers and more cargo than they could safely carry. And the faster the ship, the higher the passage. The highest rates were on the steamships via Panama, which probably carried the bulk of the California trade. Since prevailing winds on the Pacific made a direct run from Panama difficult, and required a sailing ship to veer nearly a thousand miles westward to catch the winds into San Francisco, sailing ships were rarely used on this route. Sailing ships were employed almost exclusively, however, on the all-sea voyages from the East Coast to California around South America. Not only did sailing vessels not have to put into port for frequent refueling on these long ocean voyages, but it was also not necessary for owners to ensure that coal would be readily available at the remote bunkering ports. On this run, where fast passages meant high profits, the clipper ships were in great demand.

The period between the onset of the demand for fast wooden-hulled clippers on these routes and their replacement by steam-powered iron-hulled vessels, which, while not necessarily faster, were certainly more dependable, was short, lasting only a few years. But they were years which contributed far more than their share to the romance of the sea. The speed of these majestic and lofty clippers on their runs to California and back captured ·the interest of America and produced almost as much excitement as the Gold Rush itself. Newspapers featured sailings and arrivals of famous fast clippers, regularly reporting statistics and record runs, and exhorting their readers' support for one favored vessel or another. Many bets were placed along Manhattan's waterfront, a few days' pay perhaps in the bars or cafes close by the docks on Water Street, but also large fortunes among the merchants in their counting houses. Reporters often kept watch atop the bluffs at Atlantic Highlands, either to clock the exact time a California clipper caught her first full breeze after crossing the bar, or to be first to spot the sails of some expected vessel as she approached the finish line.

In the short era between 1850 and 1856 there were many clippers built in America at a variety of yards, mostly in New York or New England. But the two yards which produced by far the greater number and whose names are most closely associated with the clipper era, are those of Donald McKay in East Boston and of William H. Webb in New York.

As noted in an earlier chapter, Donald McKay, who was Webb's senior by ten years, came down from Newfoundland as a lad of sixteen and began his shipbuilding career, first as an apprentice under Isaac Webb, where he learned his trade, and then as a mechanic in the yard of Smith and Dimon, where he was employed at the same time as John W. Griffiths, another former apprentice of Isaac Webb. In 1839 McKay served briefly as a foreman in the Brooklyn Navy Yard. Then in 1840, the year William Webb took over his father's yard, McKay moved to Newburyport, Massachusetts, where he and a partner opened their own small shipyard. Four years later, however, Enoch Train, a wealthy Boston merchant, impressed with McKay's innovative designs, made it financially feasible for McKay to establish his own shipyard in East Boston. It was here in the years 1850 to 1856 that some of America's most renowned clippers were launched.

While both lore and tradition tend to associate the clipper ship primarily with Donald McKay, and particularly with his justly acclaimed *Flying Cloud*, a larger number of the most successful clippers actually came from the yard of William H. Webb in New York. Though Webb was the better businessman, more adept at finding and filling orders that would lead to a successful launch and a sure profit, he was never quite McKay's equal at publicizing his own name and romanticizing his products. Nevertheless, authorities are nearly unanimous in asserting that Webb built better ships. Even Richard McKay, nephew and biographer of the famous Boston builder, has written, "... William H. Webb stood foremost among those master-mechanics who developed the clipper ship from its inception until it reached the acme of perfection, say in 1853, when he designed and built the Young America." McKay goes on to note that "... the run, bow, sheer, masting, and general appearance of a Webb clipper were always the subject of favorable comment, no handsomer appearing vessels being seen anywhere" (1).

We have already noted some of the major characteristics by which the true clipper, or "extreme clipper," as it is sometimes called, are defined: a sharp entrance, with a concave bow and wide flare to divert waves from the deck forward; long, narrow hulls with a length-to-beam ratio of about six to one; minimal sheer and straight sides for much of the midsection; a long run with a rounded counter stern rather than the traditional square transom. Clippers varied in the shapes of their hulls. Some were built with steep deadrise, producing a V-shaped hull which certain shipbuilders and masters believed gave higher speed. Others believed a flatter hull, with little deadrise, allowed both greater speed and more cargo space. Experience seemed to show that the V-shaped hull sailed better in light airs, such as might be encountered in the China trade, whereas in heavy weather, as is usually found off Cape Horn, the ships with flat deadrise proved faster and did not fall off to leeward as readily. Naval architecture being an art in which the final word is compromise, the need to stow the maximum amount of cargo in a hull having the greatest possible speed on a given trade route—while providing sufficient strength to keep hull, spars and rigging intact in the worst of conditions—was what usually dictated a ship's design. Webb, as we have seen, leaned toward the flatter hull. The one characteristic which all clippers had in common, of course, and which contributed to their majestic appearance at sea, was far taller masts than were previously thought possible, and spread upon them an enormous yardage of sail.

We have also seen that many of the design characteristics of the clippers, in one form or another, had appeared on various vessels long before 1850. The sharp entrance and lean hull lines were used on the small "Baltimore clippers" of the early part of the century, but these ships were not designed to carry heavy cargos over long distances. The concepts of a concave bow, minimal sheer, and a rounded stern, though considered risky at the time by some, were developed as early as the 1820s and 1830s by Isaac Webb and later by William Webb as well, in such ships as *Helena* (which established records close to those of the later clippers), or *Montauk*, all of which proved unexpectedly fast in service. The later *Houqua*, built by Brown and Bell, and *Rainbow*, designed by John W. Griffiths and built by Smith and Dimon, while still closer in some respects to the clipper concept, would nevertheless, like Webb's *Helena* or *Montauk*, which they resembled, be classed as "proto-clippers."

The Black Ball Line's packet ship *Isaac Webb*, named for William Webb's father, is shown in this painting as she must have appeared on many of her Atlantic crossings to Liverpool—under her foresail, topsails, spanker and a pair of jibs. *Isaac Webb* served the line from 1850 to late 1880, when she was abandoned at sea.

Although many of the principles which together produced the clipper ship were first attempted in practice by either Isaac or William Webb, the first to publish these principles, along with an impressive collection of statistics to prove their practicality, was John W. Griffiths, both in his book Treatise on the Theory and Practice of Shipbuilding (New York, 1851), and in the marine periodical, The U.S. Nautical Magazine and Marine Journal. But it was primarily the success of his *Rainbow* that led Howland and Aspinwall to ask him to design a second ship for them to be built again by Smith and Dimon. The result was the famous *Sea Witch* of 1846, which, as we have noted, was the first ship whose collective characteristics gave her claim to be called a "clipper" (2). Commanded by Robert H. Waterman, she made the return leg of her first round trip to China, Hong Kong to New York, in seventy-seven days, a record never exceeded by a sailing ship (3).

By 1850, with the Gold Rush at its height and the China trade burgeoning, shipyards were besieged with orders for tonnage of every sort. The yard of William H. Webb as well as the other yards along the East River, already busy building steamships for the coastal as well as the California trade, were now also receiving requests for as many clippers in the style of *Sea Witch* as they could provide. Through the next few years, as Morrison describes it, "The shipyards were run under high pressure, some yards working overtime . . ." (4). And while yards such as McKay's were equipped only to build clipper ships, and others, such as Brown and Bell or Smith and Dimon, after a few early successful clippers, preferred to devote themselves almost exclusively to steamships, the one yard that was able to accept commissions for all sorts of vessels, and might at the same time have a large steamship, a traditional Atlantic packet, and a tall-masted clipper on the ways at the same time, plus, in other areas of the yard, small fishing boats or coastal schooners under construction, was that of William H. Webb.

Howard Chapelle, a former marine curator at the Smithsonian Institute and a consistent admirer of William Webb, describes the difference between Webb's yard and that of his closest competitor quite specifically:

[Webb's] record as a shipbuilder shows him to have been extremely versatile in design; turning out fine steamers and small vessels, as well as packets, clippers, and freighters in great number. Apparently his interest in ship designs was not confined to one type. (5)

Chapelle makes the same point again in a different context:

McKay also appears a less able manager in that he failed to obtain a range of building contracts each year, whereas Webb in a year might build two or three clippers, a steamer, a freighter, and a bark, and might even build a couple of fishing sloops. (6)

The yards of William Webb and Donald McKay both built their first full clippers in 1850, with Webb's *Celestial* coming off the ways on 10 June, in a double launch with the steamship *Alabama* for Mitchill's Line to Savannah; and McKay's *Stag Hound* six months later on 7 December.

From the first, Webb resisted building what would be called an "extreme clipper." It was always his stated objective to hold to a compromise among speed, capacity, and strength, and in the process he did fairly well at obtaining all three. Webb clippers were generally characterized by a full, but never bulbous, midsection, as necessary for both capacity and stability, and *Celestial*, his first clipper, is an excellent example of the Webb model (7). Ordered by the firm of Bucklin and Crane, *Celestial* was the first clipper built especially for the China trade. With a length of only 152'-6", she was not a large ship, but she was fast.

McKay's *Stag Hound* at 210' was larger and, with her more graceful sheer and sharper entrance, far more impressive. She also had more deadrise—that is, a more V-shaped hull—than *Celestial*. With these lines, *Stag Houna* too was a fast ship, but her poor carrying capacity limited her earning power.

One is tempted to ask to what extent Webb and McKay influenced each other and also to what extent either was influenced by the writings of John W. Griffiths. Since all three men were products of the yard of Isaac Webb, where many of these concepts were first attempted, and since all three would have known each other during their formative years, one must conclude that some exchange of ideas must have taken place among the three, especially considering the deep commitment of each to the art of shipbuilding.

Chapelle cuts short any speculation that great differences existed among them:

It must again be observed that, contrary to general belief, the basic design principles of Griffiths, Webb, and McKay were very much alike. All three men used the sharp entrance, full midsection, with very moderate deadrise, some hollow in the ends, straight buttocks, and the midship section placed close to the mid length of the load line in some ships. . . . In basic design they differed only in the emphasis placed on the run, and the exact placement of the midsection, afore or abaft mid-load line length. (8)

In general, Webb was the most conservative of the three, preferring, as a rule, a less sharp entrance and a flatter hull, and thus achieving greater speed without significant loss of carrying capacity.

The year 1851 was easily the most active for clipper launchings. In that year alone, William Webb completed five clipper ships: *Gazelle*, *Challenge*, *Comet*, *Invincible*, and *Swordfish*, with as many as four simultaneously on the stocks. In the same year he also completed the elegant steamships *Golden Gate* for Pacific Mail and *Florida* for Mitchill's Line, both already mentioned; as well as the packet ships *Isaac Bell* and *Great Western*.

Webb led off the year with the triple launch on 21 January of the clipper *Gazelle*, the big steamship *Golden Gate* for Pacific Mail, and the packet *Isaac Bell* for Fox & Livingston's Union Line to Le Havre. Considering the expense of a launching, with its many invited guests, champagne and punch, and food, plus the fact that neither Webb himself nor any of his men could do much other work on the day of a launch, one suspects that the idea of three launches on the same day was not so much to provide a spectacle for publicity as to save time and money by getting three of them out of the way in one shot.

*Gazelle* was the least typical of the Webb clippers in that she had a V-shaped hull with a sharp deadrise, a characteristic of which Webb did not approve but which was imposed, he claimed, by her prospective owners. Sitting at her pier shortly after her launch, however, she apparently attracted the eye of James Gordon Bennett, editor of the *New York Herald*, whose article is quoted by Richard McKay:

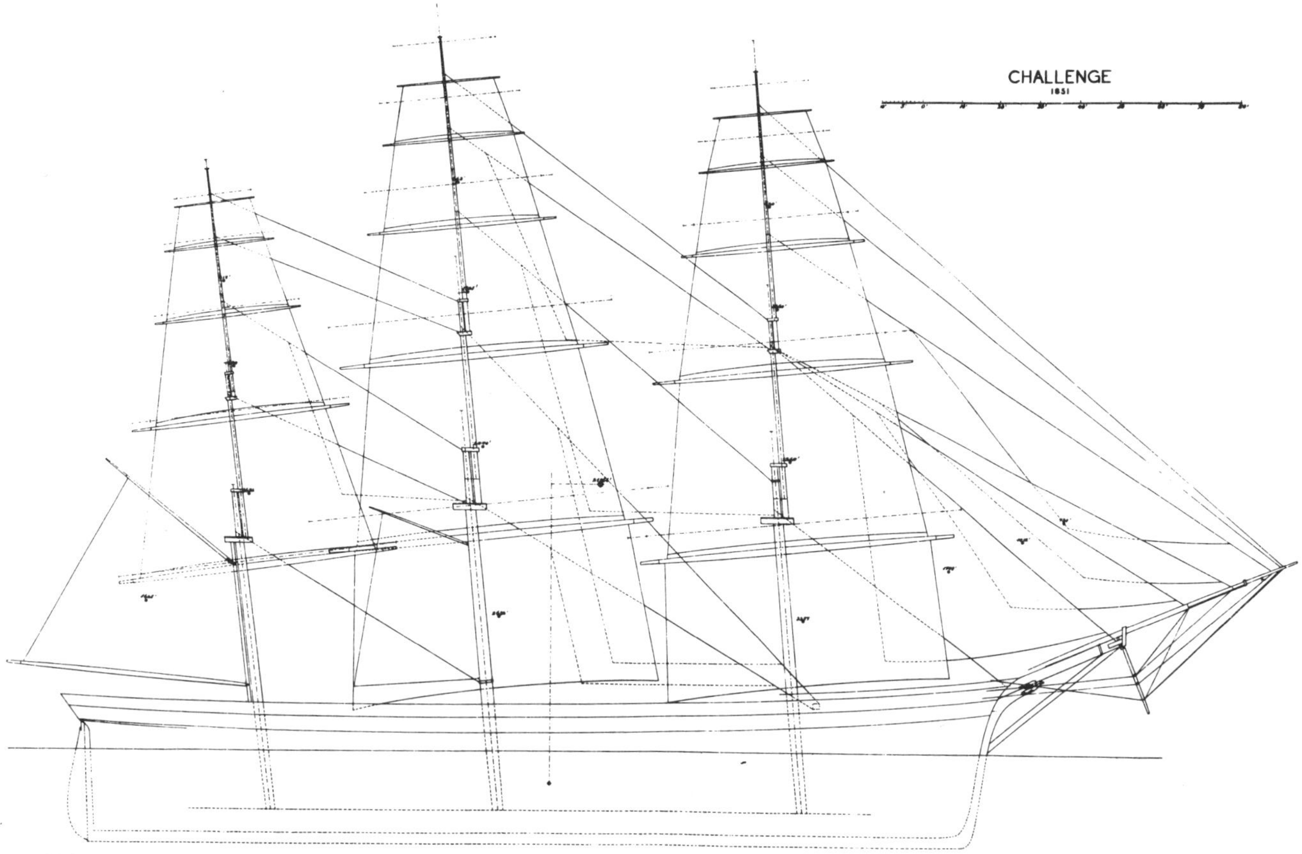


This sail plan of the extreme clipper *Challenge* shows the vessel after her rig was cut down following her first voyage around the Horn. Dashed lines show the original height of her masts and position of her yards.

We yesterday paid a visit to this beautiful craft, lying at Peck Slip, where she will take freight for California and China. We have seen many of this kind of vessel lately and all of the first class; but we have never met with one that came nearer to our idea of a skimmer of the wave than does the 'Gazelle.' She was constructed upon the same principle as the far-famed 'Celestial,' and by the same builder (Mr. Wm. H. Webb, of this city), but is much sharper—everything, excepting due and proper regard for strength, being made secondary to speed. . .

The 'Gazelle' has a dead rise of forty inches—nearly as much as any ship ever built—with very fine lines below, which gives her the power of sailing very fast in light winds; at the same time . . . her deep keel will enable her to carry a very great press of canvass in strong winds. . .

She has but few mouldings outside, and her round, swelling sides, small and beautifully turned cutwater, and light airy stern—the outline of which is perfection itself—gives her hull a most neat and graceful appearance. (9)

Next off the ways was Donald McKay's most famous ship, the fast and beautiful *Flying Cloud*, launched at East Boston on 15 April. By late May, when she had finished fitting out, she was towed down to New York to take on passengers and cargo. Then on 2 June, as crowds gathered at the Battery to watch, *Flying Cloud* set sail for San Francisco. On this first voyage, she clocked up to eighteen knots, a phenomenal speed for a sailing ship, and arrived in San Francisco in three hours short of ninety days, the best run up to that time. From San Francisco, she headed out across the Pacific to China.

As *Flying Cloud* was preparing to sail on her maiden voyage, an even larger clipper was nearing completion at the yard of William H. Webb. Late in the previous year a representative of N. L. and G. Griswold, one of New York's most prosperous trading firms, had approached William Webb and asked him to build them a clipper that would be the largest, the grandest, and, if possible, the fastest in the world. Since they knew he would build what they wanted, they made it clear that Webb could name his price. And, as a sign of their intent to other New York firms such as Howland and Aspinwall or Grinnell, Minturn and Co., whose houseflags adorned other clippers then sailing in various seas, this ship would be named *Challenge*.

As *Challenge* began to take shape in his yard, William Webb gave her construction very special attention, choosing each piece of timber with care. In this case he even designed the rigging and sail plan himself. *Challenge*, which was going to be over 230 feet long, and measure over 2000 tons, was not only considerably larger than McKay's *Flying Cloud*, she was the first three-decked clipper and at the time the largest merchant ship afloat. She also had the sharpest entrance of any of William Webb's clippers.

In order to be sure to get the most from their ship, the Griswolds secured Robert H. Waterman, erstwhile captain of both *Natchez* and *Sea Witch*, and a man famous for driving his crew as hard as he drove his ships, as her commander. Thus while Webb was doing his best to design and build as nearly perfect a ship as possible, he was haunted by the almost daily appearance of Waterman, who felt he had a right to have some say in how "his" ship was built. In the end the hull was Webb's, but Waterman increased the height of the spars and the spread of the sail over Webb's objections (10).

As his high-sparred *Challenge* was nearing completion, three other clippers, each also destined for fame as record-breakers on the California run, were rising on the stocks at the yard of William H. Webb: *Comet* and *Invincible* on either side of *Challenge*, and *Swordfish* on the uptown ways. And within sight of the Webb yard, *Hornet* was building at Westervelt and Mackay and *Trade Wind* at Jacob Bell's (11).

By the time *Challenge* was launched on 5 May, she had cost $120,000, but the Griswolds had gotten their money's worth. Chapelle describes her as a "very handsome ship" (12), and available drawings bear him out. She had only slight sheer, and with her deck clear, except for a short poop, she had exceptionally clean lines. Since she was larger than other ships along the waterfront, contemporary accounts marvelled that her bowsprit reached over the roofs of the shops along South Street (13). After Waterman had redesigned her rigging, her mainmast stood 230 feet high, the same as the length of the ship itself; and she carried an astounding 12,780 square yards of canvas. Waterman was out for a record, as well he might be, for the Griswolds had offered him a bonus of $10,000 if he could beat *Flying Cloud*'s record by bringing *Challenge* into San Francisco in under ninety days (14).

Finally on sailing day, 13 July, a towboat pulled *Challenge* down the East River and out into the open harbor. Here, even before she set sail, Waterman fired his first mate—only a taste of what was to come—and hired on the spot another mate (rowed over from an incoming ship), whose reputation for brutality made him fear to go ashore. Only when this exchange had been negotiated did *Challenge* finally unfurl her great spread of sail and head south into the Atlantic (15).

July was not the best month for a sailing, since it meant rounding stormy Cape Horn in what was a stormy season in those waters. In spite of the magnificent vision *Challenge* presented as she unfurled her voluptuous sails after clearing New York harbor, the rest of the voyage was apparently more like a nightmare. As the details of this trip have been terrifyingly described in A. B. C. Whipple's book, The Challenge (New York, 1987), we need not repeat them here. Needless to say, one of the first problems to arise was that the lofty masts ordained by Waterman snapped near the top in the first gale, so that *Challenge* had to be rerigged while underway minus the extra sail.

As expected, *Challenge* encountered raging seas and icy winds once she started rounding Cape Horn. Sailors, some shanghaied and others merely seeking passage and therefore totally untrained, were sent aloft to change sails at every whim of the captain, even when the ship was being tossed by the sea into a heel of up to forty-five degrees and the rigging was encrusted with ice. As a result, with speed given priority even over human lives, at least three men fell from the rigging to their deaths, though in the case of one man, his companions insisted he had been pushed by the first mate. Two more died of dysentery, and one, who spoke no English, according to the testimony of his fellow sailors, from physical abuse, also from the first mate (16).

With the severe storms encountered in rounding the Cape, *Challenge* lost so much time that, even pushing his crew to the extent described, Waterman failed to get his bonus. *Challenge* took 108 days to reach San Francisco, which, coming so soon after *Flying Cloud*'s fantastic run only a few weeks earlier, came as a disappointment to her owners and to William Webb as well as to Waterman. On arrival in San Francisco, Waterman not only failed to get his bonus, he came very close to getting lynched by an angry crowd gathered on the waterfront, once they had heard the stories of his disregard for the lives of his crew. Only by departing from the company offices rather ignominiously by way of the roof rather than through the front door did he escape the wrath of his accusers. Subsequently brought to trial at his own request, however, Waterman was exonerated of wrongdoing for lack of evidence (17).

Waterman wisely took this opportunity to retire from the sea, and *Challenge* sailed for China commanded by John Land. On the return, Land brought *Challenge* from

*Young America* is here shown at a New York pier at an undetermined date. Her first voyage, which commenced on 10 June 1853, was the precursor of a career that was to last until February 1886. On the 19th of that month, under the name *Miroslav* and flying the Austrian flag, she sailed from the Delaware capes for Trieste and was never heard of again. In her latter days, she crossed and recrossed the Atlantic carrying petroleum to Europe and empty barrels back. (Courtesy of The Mariners' Museum, Newport News, Virginia)

Shanghai to San Francisco in a record thirty-four days. Several other record runs across the Pacific attest to her remarkable speed (18).

Before *Challenge* had even been launched, however, William Webb had turned his attention to the more im-

pressive clipper, *Comet*, ordered by the firm of Bucklin and Crane, which had been so satisfied with *Celestial*, they had returned to Webb for an even larger and faster model. Neither *Comet* nor any of Webb's later clippers adhered so strictly to the sharp entrance and long run—

that is, the "extreme clipper" lines—of *Challenge*. From this time Webb seems to have concluded that the cargo space lost by so sharp an entrance and long run was not compensated for by any significant increase in speed. And he was apparently right, for *Comet* proved faster than *Challenge* and indeed faster than most other ships of the era. Richard McKay describes *Comet* as "in all quarters conceded to be one of the most beautiful, fast, and generally successful sailing ships ever launched from any shipyard" (19). But the highest praise comes from Morrison:

History has handed down to us that the 'Flying Cloud' was far in advance of all clipper ships of her day in making record time. She was undoubtedly a vessel having fine lines for speed, and was most ably handled in all her voyages as shown by the record; but there was one other American clipper ship of the same period that the record has shown to be her equal, and this vessel was the 'Comet,' a New York built vessel whose record for high speed over a long period of time is equal to any vessel of her class. (20)

*Comet* was launched on 10 July, three days before the sailing of *Challenge*. On 1 October she sailed on her own maiden voyage to California. Although *Comet*'s trip was made in warmer weather, she also encountered storms, so that her sailing time of 108 days was exactly the same as that of *Challenge*. On a later return trip in 1854, however, *Comet* sailed from San Francisco to New York in seventy-six days and seven hours, which remained the record for that run.

Later in the year 1851 *Invincible* and *Swordfish*, two ships which William Webb preferred to label "moderate clippers," were also launched from the yard. Since Webb had by now concluded that the finer lines of the "extreme clippers" were not essential for speed, these ships had proportionately longer midsections, and somewhat shorter ends. *Invincible*, built for the firm of James W. Phillips, measured 2100 tons, making her roughly the same size as *Challenge* or *Comet*. *Swordfish*, designed for the China trade, was, at 1150 tons, considerably smaller, but clipper or not she was fast. At one time *Swordfish* held the record for the Shanghai to San Francisco crossing (21).

On her maiden voyage to the Pacific *Swordfish* found herself a participant in a well-publicized race. By 1851 it was understood that, although clippers were being built in many yards in America, the two builders best known for producing fast sailing ships were Donald McKay and William H. Webb. William Webb, whose primary concern was in designing and building fine ships, was not much

interested in gambling or in engaging in competition of any sort. But many of the men who gathered in the bars or clubs of New York or Boston took special pride in the respective marine specimens of their own cities. And it was among these people that a lively rivalry, often including wagers of fairly large sums, developed between the supporters of Donald McKay in Boston and William H. Webb in New York.

*Swordfish* sailed out of New York on 11 November 1851, and, as it happened, McKay's recently-completed *Flying Fish* had sailed from Boston only a few days earlier. Needless to say, the ardent backers of Webb or McKay took advantage of this opportunity to place some heavy bets on their favorites. The two youthful captains, David Babcock of *Swordfish* and Edward Nickels of *Flying Fish* were told of the wagers and each offered a fine bonus if he won, so that the high spirit of competition followed the two flyers as they made their way down the Atlantic.

At the Equator, McKay's *Flying Fish* was four days ahead and seemed almost certain to win. But Babcock apparently knew the better passages through the Doldrums, for by the time they were rounding Cape Horn the two competing clippers were often racing side by side. When she reached the Pacific, however, Webb's *Swordfish* was already a full day's sail ahead, a lead she maintained all the way to San Francisco. Webb's *Swordfish* swung through the Golden Gate on 10 February 1852, completing the run from New York in just over ninety days, not even one full day longer than *Flying Cloud*'s record of the previous year. *Flying Fish* did not appear until the seventeenth. When this news reached the East the New Yorkers were elated, but many Boston bettors were ruined (22).

The year 1851 is notable, not only as the year of *Flying Cloud*, *Challenge* or *Comet*, but also as the year the sailing yacht *America*, designed by James and George Steers, was born at the yard of Smith and Dimon. Although she had been built expressly to race in the annual regatta of the London Royal Yacht Club, British yachtsmen, learning in advance of her speed, disqualified her ostensibly on the grounds of having too many owners. *America* raced anyway and easily sped far ahead of the British contenders, impressing all with her spectacular performance. At the finish, Queen Victoria presented the "Hundred Guineas Cup" to *America*, and it was carried off to New York, where it reposed for over a century in the hallowed halls of the New York Yacht Club. Yachtsmen around the world still compete for this legendary trophy, henceforth called the "America's Cup."

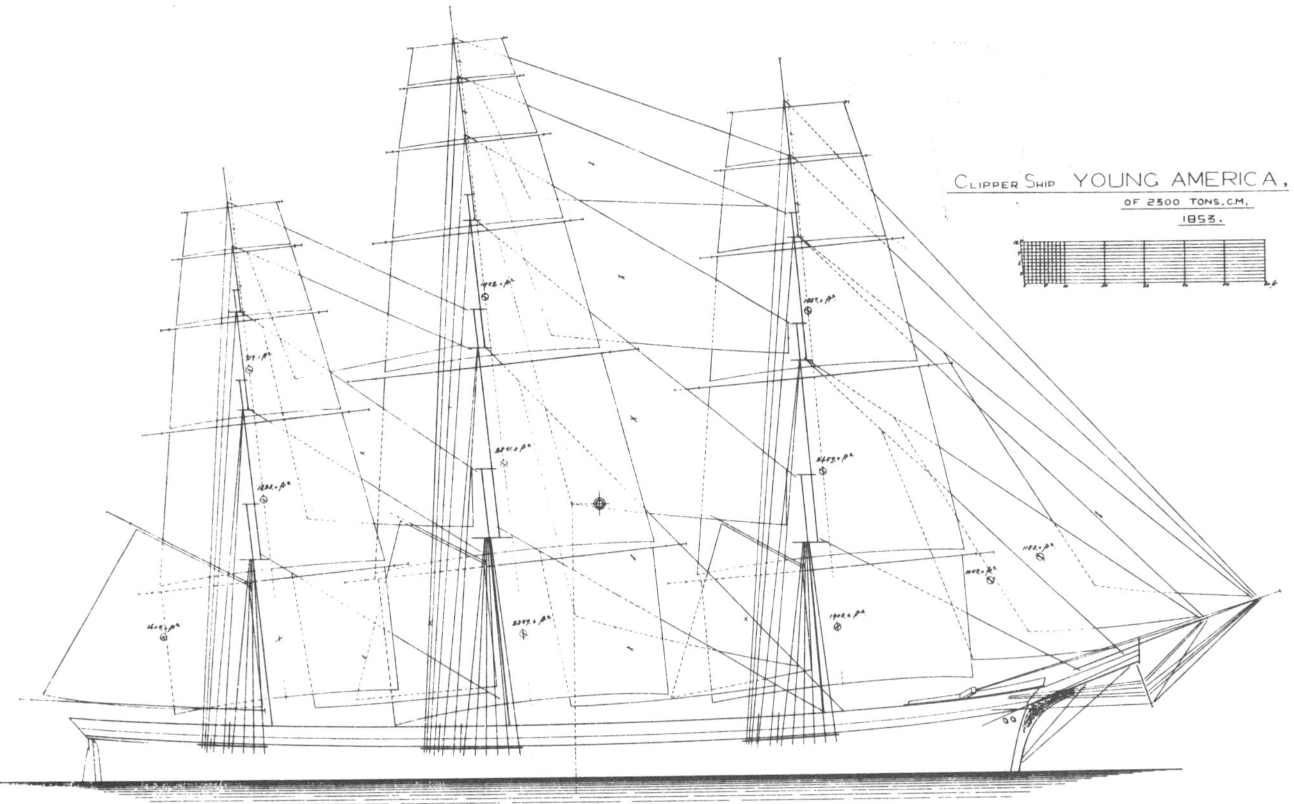



The sail plan of William H. Webb's masterpiece, the extreme clipper *Young America*, built in 1853 for George B. Daniels, appeared in the Book of Plates. Note that skysails are shown on her lofty fore and mizzen masts, whereas her main mast is fitted with a rarely seen (or used) moonsail. Compare this rig to that shown in the preceding photograph of the ship, where she has apparently been rerigged with upper and lower topsails.

In 1852, the year that Franklin Pierce of New Hampshire was elected President, a saddened America mourned the deaths of three of her most loyal statesmen, Henry Clay, Daniel Webster, and John C. Calhoun, all within a few months of each other. It was also the year in which the burning of the steamboat *Henry Clay* in the Hudson River off Yonkers, with a tragic loss of life, awakened the federal government to the need for imposing severe safety regulations on the building and operation of steamships and steamboats.

For some reason 1852 was a relatively quiet year at the Webb shipyard, with *Flying Dutchman* the only clipper completed. *Australia*, built for the trans-Atlantic service of Williams and Guion, is sometimes listed as a clipper, but, although she had some of the fine lines associated with clipper ships, William Webb himself called her a packet, a designation which seems borne out in his drawings. Two large coastal steamships, each about 220′ in length, were also produced by the Webb yard that year, *Augusta*, for Mitchill's Line to Savannah, and *James Adger* for Spofford and Tileston's line to Charleston.

Taking advantage of the lull in ship orders, William Webb spent the winter of 1851-1852 in Russia, selling the Tsar on a plan for a warship which will be discussed in a later chapter. By 1853 when Webb returned, the yard was again busy. In this year Webb completed two large steamships for Pacific Mail, also to be discussed in a later chapter. He also built the clipper *Flyaway*, the bark *Snap Dragon*, the brigs *Josephine* and *Fanny*, and the packet ship *John Bright*. But the great pride of the yard in 1853 was the clipper *Young America*, generally considered William Webb's masterpiece. With her long lean lines and tall spars, and with a somewhat more graceful sheer than was typical of earlier Webb ships, *Young America*, racing through the sea with a full head of sail and bone in her teeth, represents the epitome of the magnificent American clipper ship.

Built for the trade to California and China, *Young America* cost her owners $140,000. But with her speed she earned as much back for them in a few voyages, and then continued her fast passages across the Pacific for another thirty years before being sold to Austrian operators.

The completion of the graceful *Young America* only a few months after *Sovereign of the Seas*, Donald McKay's second most successful clipper, again whetted the betting impulses of the New York and Boston backers of Webb and McKay. On 7 May 1853 McKay's *Sovereign of the Seas* arrived in New York after an amazingly fast run of only eighty-four days from Honolulu. The following day Webb's *Comet* sailed into New York Harbor having just made a record passage of eighty-three days, eighteen hours, from San Francisco. And the day after that, on 8 May, *Flying Dutchman*, another Webb clipper, came in from San Francisco after a run of just eighty-five days.

The arrival of two Webb ships on successive days after record runs started the New York backers boasting of the prowess of their native-built ships. When the Boston people had heard all they could take, they let it be known that they would make $50,000 available to the winner of a race between any Webb-built clipper and any McKay-built clipper from New York to San Francisco.

William Webb was not the type of man to risk money in wagers. But George Daniels, one of the owners of the new *Young America* was game, and he persuaded the reluctant Webb to go along. Webb and Daniels then not only accepted the Bostonians' challenge but specified that the race should be between *Young America* and *Sovereign of the Seas*, and that each of them would put up an additional $10,000 on *Young America*. Alas, William Webb never got the chance to prove whether *Young America* could beat McKay's clipper, for on 18 June, before the race could take place, McKay sold *Sovereign of the Seas* to a British company which sailed her off for a route between London and Australia (23).

Later in 1853, *Great Republic*, by far the largest of all the clippers, was completed by Donald McKay. At 4555 tons, she measured twice the tonnage of even the largest of Webb's clippers; and at 335 feet in length, she was half again as long. She also carried four masts, one more than was usual for a clipper. This behemoth, which cost $300,000 to build, was not constructed in response to an order from any shipping firm. It was built on speculation—something William Webb always refused to do and never needed to do—by Donald McKay himself, who borrowed to the limit of his capacity to finance her, so certain was he of her ultimate success.

Throughout the summer of 1853 Boston papers reported the daily progress of the great ship, destined to be the pride of their city, rising at McKay's yard. Her launch date on 4 October was declared a state holiday in Massachusetts, and whole schools of children, plus families from many miles around, a total of 50,000 spectators, came to watch. After the launching, the ship was towed down to New York for her final fitting out. Then, still under the ownership of Donald McKay, *Great Republic* was moved to one of the commercial piers to take on a cargo of wheat for Liverpool.

At midnight on 26 December, the day after Christmas, a fire started in a bakery shop on Front Street. With gale winds blowing, flying sparks set off fires on several ships docked across from South Street. Before long wind-blown debris had lit up the furled sails high on the four lofty masts of *Great Republic*. When the firemen arrived, the water from their hoses could not possibly reach this high, so that fiercely burning rigging and sails falling from above soon had the whole ship in flames, with little hope of saving her.

After she had been smoldering at her pier for several days, *Great Republic* was examined to see what chance there was of salvage. But the wheat in her hold, thoroughly soaked by the hoses, had expanded and had sprung her timbers in several places, so that the carefully crafted hull could never be the same again. In time A. A. Low and Brothers purchased the remains, thus saving McKay from total ruin, and, after removing the charred third deck and equipping her with a lower rig, this firm found her still to be a fairly fast as well as spacious cargo ship. McKay, however, never fully recovered from this disappointment (24).

William Webb and Donald McKay were not the only builders of clippers in the 1850s. Many clippers came from other yards, both in New York and in New England, but these are outside the purview of this study. Though there were some ships built along clipper lines at other yards after 1853, *Young America* was the last extreme clipper built by William Webb (though he built a few ships of a modified clipper design over the four succeeding years). By this time the Gold Rush had so completely abated there was no longer the pressing demand for fast

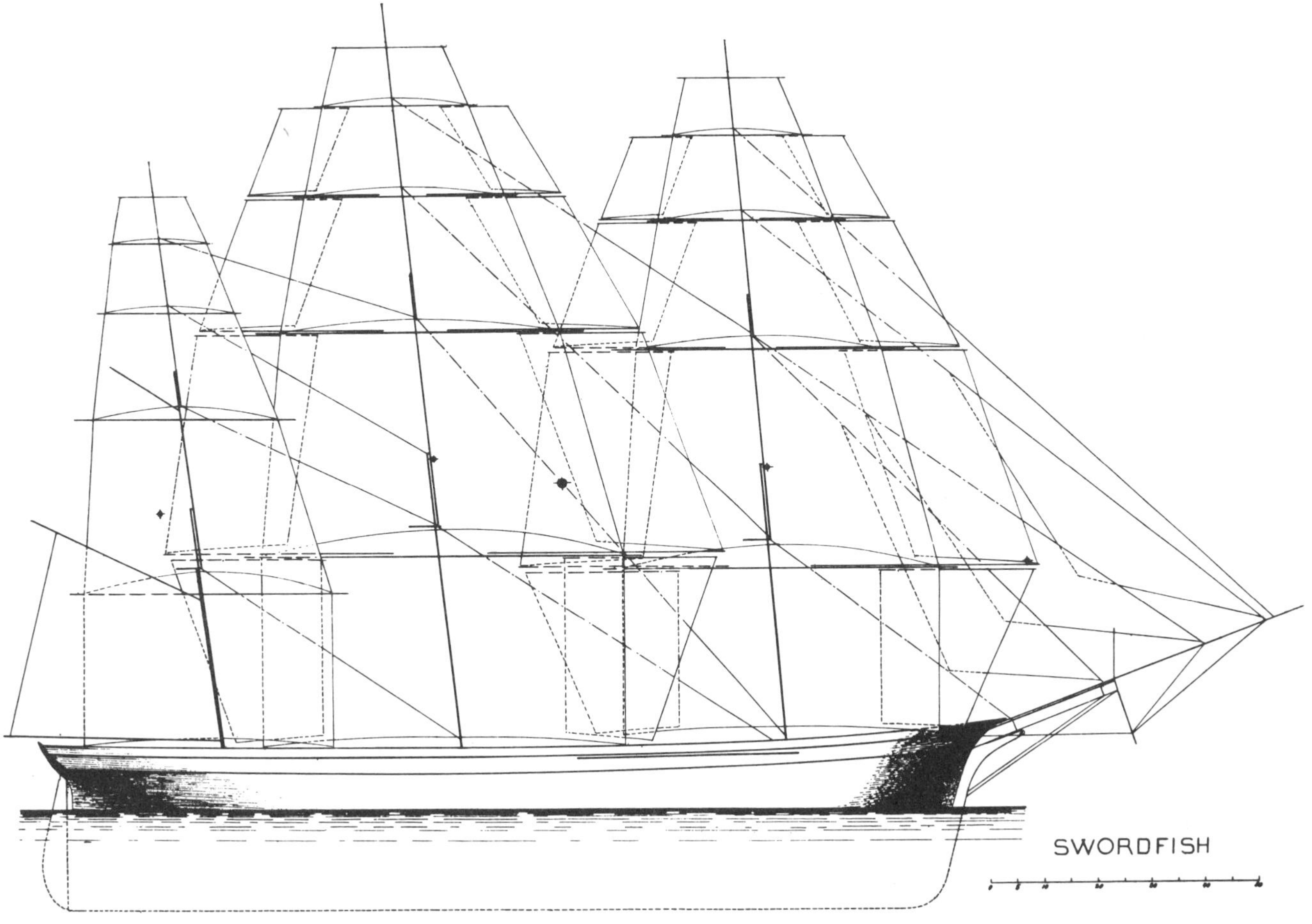



The sail plan of the medium clipper *Swordfish* shows nearly every piece of canvas that her first master, David S. Babcock, had at his disposal—everything but her staysails. Babcock and *Swordfish* may have used every sail they had on her first passage to San Francisco in 1851, on which her time of 91 days bested that of Donald McKay's *Flying Fish*.

passages to California. And by now even the merchants trading with China were finding that the increased speed of the sharp-hulled clippers over that of more conventionally designed vessels was not sufficiently profitable to warrant the loss of cargo space.

William Webb was one of the first to sense that the time had ended for the great clippers which had won him such repute. One day, after *Young America* had left his yard and was loading for her maiden voyage to the West, Webb wandered down to have a last look at her. Catching sight of the mate, Webb called up to him, "Take good care of her, Mister, because after she's gone there will be no more like her" (25).

# VIII

# New York Shipyards and the Economic Decline of 1856-1857

BY 1856 the great boom in clipper ship building was over, and, in fact, American shipping in general had started a decline, at first barely perceptible, but falling rapidly and finally bottoming out in the economic depression of 1857. One cause was the overenthusiastic response to the westward movement. By 1856, while many American shipbuilders, geared to the demands of 1851, were still producing clipper ships, the Gold Rush had essentially ended. Traffic had now evened out to a fairly normal flow of businessmen or serious settlers, most of whom could be accommodated aboard the big sidewheel steamships sailing via Panama.

Another factor that seems to have taken American shippers unawares was that the British, who had easy access to iron and coal, and whose scattered empire permitted the establishment of coaling stations in various remote areas of the globe, had begun producing iron-hulled steamships for their long-haul traffic. These early steamers were actually no faster than the clippers, and they certainly had little of a clipper's beauty or glamour. But they were far more dependable. The steamer lines could reasonably publish schedules in advance for both sailing and arrival dates. One did not have to take bets on arrival times which could vary by as much as a month.

A third problem was that men of the sort who crewed these lofty clippers were no longer so easily available. Reading any of the many memoires of those who sailed aboard them, one quickly senses that the glamour of a clipper was largely in the eyes of a somewhat distant observer. With speed always the first priority, in fact the reason a clipper existed at all, life aboard was often as close to hell as this earthly life dared approach. Only those who could not make a living elsewhere, recent immigrants not yet absorbed into the system, or men desperate to get to California by any means available, would sign aboard a clipper, and even these were often scooped out of waterfront bars or bawdy houses in a drunken state at sailing time.

By 1856 America was busy filling in the vast continent it had but recently acquired and at the same time developing its own native commerce and industry. With so many better opportunities beckoning, clipper captains were finding far fewer desperate souls willing to climb the rigging in a freezing gale at the threat of a cudgel or belaying pin. For those men on the other hand sincerely attracted to the life of the sea, the lumbering steam packets sailing from London or Liverpool provided a far more orderly regime and attracted more professional crews.

In short, by 1856 the day of the clipper was over. Many shipbuilders along the coast, caught with more clipper hulls on their stocks than could be sold, simply went out of business. A few, such as Donald McKay, had earned enough of a reputation to continue building modified sailing vessels for such shippers as had survived in the declining market. But even as early as 1853, as we have seen, when William H. Webb launched his beautiful *Young America*, he knew that this was his last extreme clipper.

Not only in 1856, when the clipper boom was ending, but even throughout the six-year period when his yard was producing more clipper ships than any other New York shipbuilder, William H. Webb was active in building other kinds of vessels: traditional packets for the Black Ball Line, coastal steamships for Mitchill's Line to Savannah, and big wooden sidewheelers for the Pacific Mail Steamship Company, of which Webb was now an increasingly active member of the board.

As early as 1850 William H. Webb, noting the sorry state of the American Navy during the 1846-1848 war with Mexico, decided to try his hand at building naval vessels. During 1850 and 1851, at the very time he was also designing and building five of the largest, fastest, and most successful clipper ships ever to sail, Webb was at his drawing board trying to produce a design for an ideal warship. In so doing, Webb noted that a large vessel dependent for propulsion on sidewheels precariously suspended on a transverse iron shaft amidships and hanging outside the hull, i.e., away from the only part of the ship providing protection, presented the enemy an unnecessarily vulnerable target. So, although propeller design was then still in its infancy and still rarely in actual use, one of Webb's primary considerations was that his warship be propeller-driven. It was also his contention that cannon-ports cut into a hull after it was built weakened the integrity of the hull. On ships strengthened with crossed-iron bracing, as all of Webb's larger ships were by this time, cutting cannon-ports would be almost impossible. Thus Webb's warship was designed fully-armed from the outset.

In 1851 Webb took his warship design to Washington with an offer to build such a vessel for his country's navy. But, probably in part because the President was then Zachary Taylor, the general who had just won the victory over Mexico with land forces, our national government showed no interest in Webb's offer.

Undaunted, as usual, William Webb then decided to try to sell his ship abroad and promptly sent an agent to tour the capitals of Europe with his design. Sometime in 1851 the agent returned with the suggestion that the Tsar of Russia might be interested. The Tsar at this time was the ultra-conservative Nicholas I, a much younger brother of the tsar who had fought Napoleon, and the last surviving grandson of the great Catherine. Since Nicholas was one of the few monarchs still around who remembered the upheavals caused by the French Revolution and Napoleon, he considered it his obligation to keep Russia strong and thus to serve as the bastion of a conservative system which he identified with order. The Tsar's elder son, Alexander, was then in training to become his father's successor. A second son, Constantine, was given the job of supervising the imperial navy with the rather grand title of General Admiral ("General" here being a rather poor translation of the Russian word for "Chief" and not, as might be assumed, a military title) and a uniform to match.

Encouraged by these reports, Webb sent his agent back to St. Petersburg to negotiate a contract. However, the agent's first audience, it appears, had been with Constantine, who was genuinely enthusiastic about acquiring Webb's ship. Neither had cleared the matter with Tsar Nicholas, who, on the second visit, informed Webb's agent that he was not interested in having a Russian naval vessel built in a foreign shipyard.

Most shipbuilders, one would assume, on receiving so stiff a rebuff from the Tsar of Russia, would have accepted defeat and sought satisfaction elsewhere. Not William H. Webb. In the late Fall of 1852, with over a thousand workmen hammering, sawing, and caulking hulls in his teeming shipyard, William Webb took ship for St. Petersburg to confront the Tsar of Russia himself. Once in the imperial capital, Webb was warmly received by the Grand Duke Constantine, who was genuinely pleased to have the matter of the warship revived. Webb himself never reached Tsar Nicholas, but the plans for his vessel did, and by the time Webb was en route back to America in the Spring of 1853, he was carrying a contract to build a warship for Russia.

William Webb, contract in hand, returned to his yard in the Spring of 1853 in time to supervise the completion of several major ships then on the stocks in his yard including the handsome *Knoxville* for Mitchill's Savannah Line, one of the few steamships designed by Webb with a pronounced clipper bow, and the beautiful *Young America*, which was launched that June.

Also in the yard when Webb returned was the big steamship *George Law*, the first ship built by Webb for the United States Mail Steamship Company, and named for its president. While not nearly so well-managed or so prosperous as Pacific Mail, its western counterpart, United States Mail was also doing a good business on its Atlantic leg of the route. And, though William H. Webb, who was closely associated with Pacific Mail, had not built any of the earlier steamers for this company, during a brief period in 1852 when the two lines were on reasonably cordial terms, United States Mail had asked Webb to build a steamer for them that would be the equivalent of the successful *Golden Gate*, recently built by Webb for Pacific Mail, in size, speed, and accommodations. When completed, *George Law*, at 278'-3" and 2141 tons, was, in fact, even somewhat larger than *Golden Gate*. Though launched on 28 October 1852, shortly before William Webb left for Russia, it took so long for the Morgan Iron Works to install the two oscillating engines and then for Webb's yard to complete the interior hull structure and the upper deck joiner work, that *George Law*, now the flagship of United States Mail's growing fleet, did not make her maiden voyage to Panama until 20 October 1853, nearly a full year after her launch.

This fine new steamship had barely begun her service to Panama, however, when George Law himself, apparently yielding to pressure from other stockholders, sold out and left the company. From this time it appears that, to an even greater extent than before, the people in control of United States Mail were more interested in milking this prosperous business for whatever profits it could be made to yield while traffic was still at its peak than in supplying the paying passengers with the kind of dependable and respectable steamship service supplied on the West Coast by Pacific Mail. As a result, the operation of United States Mail soon deteriorated to the point of becoming an embarrassment to Pacific Mail, whose business was almost wholly dependent on the traffic fed them at Panama by its Atlantic counterpart.

At about the time that *George Law* sailed on her maiden trip, Pacific Mail's magnificent *San Francisco*, launched at Webb's yard with considerable ceremony in June 1853, came back to the yard from the Morgan Iron Works. Through the Fall the Webb yard worked on finishing the elegant joiner work of her cabins. Then, early in December, the completed vessel was brought around to the Pacific Mail pier on the North River to take on passengers and cargo for her long trip around the Horn to San Francisco before starting her regular service on the West Coast between Panama and California.

At 281′-5″ in overall length, *San Francisco* was considerably larger than *Golden Gate*, the most recent steamer built for the company, and but a few inches short of the great Collins Line ships, then the largest steamships in American registry. With cabin space for 350 passengers and steerage quarters for a thousand more, she had a passenger capacity more than four times that of *California* or *Panama* which Webb had completed for the company only four years before, and which had been considered at the time more than adequate for the service. When installing her machinery the Morgan people had determined that, with two powerful oscillating engines together producing 2000 horsepower, she did not need to carry the big thirty-six-foot sidewheels then usual on ocean-going steamships. With wheels only twenty-five feet in diameter, which therefore did not project much over the superstructure, plus the usual straight and clean lines typical of ships designed by William H. Webb, *San Francisco* had an even neater and more graceful profile than her predecessors in the fleet.

*San Francisco*, with her two tall stacks placed quite far. apart, was a majestic sight as she steamed proudly out of New York harbor on 22 December. There was some comment that she was overloaded, for in addition to her 750 regular passengers, Pacific Mail had agreed to transport seven companies of the Third U. S. Artillery to California, thus more than doubling the number of people aboard.

Two days out of New York, on Christmas Eve, *San Francisco* was battling the usual winter seas off the Carolinas when it became apparent that the rapidly increasing gale winds she was encountering were turning into a full-scale hurricane. That night, as *San Francisco*'s officers were doing their best to keep her on course in the mountainous waves, a fractured piston in an air pump suddenly left the engines powerless. With the pilots no longer able to keep her headed into the wind, *San Francisco* was left wallowing broadside to the seas, and taking water rapidly as giant waves crashed over her superstructure. The bilge pumps kept working, and so also did all able-bodied passengers, so rapidly was water boarding the ship. While all human and mechanical power available was being exerted to keep the vessel afloat, suddenly, with a thundering crash, the ship was struck by a mammoth wave that swept the upper deck clean, taking with it cabins, lifeboats, and 195 passengers and crew members.

The now terrified survivors saw some small hope the following morning, Christmas Day, when, in a somewhat calmer sea, the crew managed to get the engines turning again. But on the 26th, with the seas again growing steadily more threatening, the engines once more ceased to function, this time permanently. And again the ship and its survivors were at the mercy of the sea. One ship passed close by that afternoon, but in these crashing seas a transfer was impossible, so the ship sailed on. By this time, as though the more than a thousand passengers and crew still surviving aboard the beleaguered liner had not already faced trials enough, several passengers were discovered to have come down with cholera, which, as we have already seen, was a highly contagious disease that could easily spread. *

On 28 December, the bark *Kilby* came in sight, but again the seas were too rough to transfer passengers. *Kilby* stood by, however, and on 29 December her boats managed to take about 100 passengers from *San Francisco* aboard. But again the seas grew too high to risk further transfers, so *Kilby* too sailed on her way. The next day, 31 December, the ship *Three Bells* came alongside and this time had to stand by five days before the weather permitted her to lower her boats. By now, 4 January, nearly two weeks had passed since *San Francisco* had left

* One cannot help feeling a certain sympathy, even in retrospect, for passengers afflicted with a disease that causes violent vomiting while they are caught aboard a powerless vessel being tossed about by a hurricane!

This rare lithograph from the Webb Institute collection shows *Astoria*, the only merchant screw steamship built by William H. Webb. Her propeller could be disconnected and raised clear of the water when she was under sail. Built in 1855, she was coppered on the stocks before launching—a practice not often followed when building wooden vessels.

port. Finally *Three Bells* boarded 230 of *San Francisco*'s passengers and took them back to New York. That same day the ship *Antarctic* also came alongside and took all of the remaining passengers as well as the captain and the crew with her to Liverpool.

When a rescue party sent out from Washington discovered the abandoned vessel, the Webb-built hull was still intact and still afloat having survived a hurricane and violent seas. Unfortunately, however, it had to be scuttled to prevent its becoming a menace to navigation. Thus one of William Webb's largest and most beautiful ships, one which had taken over a year to build, did not even finish the first leg of her maiden voyage. But here the fault was clearly with her machinery (one source has suggested that the small size of the sidewheels placed too great a demand on the engines) and with the owners for not subjecting these engines to more complete trials before

sending the overloaded ship into the ocean, not with any weakness or fault in the construction of the hull (1).

About the time of the tragic sinking of *San Francisco* early in 1854, United States Mail and Pacific Mail faced further competition when Cornelius Vanderbilt started a new line of steamships, operating on both oceans, between New York and California via Panama. In May 1853 Jeremiah Simonson had completed the large steamship *North Star* for a new trans-Atlantic line Vanderbilt was planning. Instead of starting the new line immediately, however, Vanderbilt, his family, and a select group of guests, had boarded the *North Star* for a private cruise of European and Mediterranean ports, leaving the operation of his Nicaraguan Steamship Company to his partners, Charles Morgan and Cornelius Garrison. When Vanderbilt returned in September, he was informed that in his absence Morgan and Garrison had managed, by buying a certain amount of stock from private investors, to wrest control of the company away from him. Furious but undaunted, Vanderbilt responded by starting yet another line of steamships to California, this time, like the United States Mail and Pacific Mail, via Panama. The fast and luxurious *North Star*, obviously, was available for the Atlantic leg of the run, on which she made her first sailing out of New York on 20 February 1854, on the very same day that *George Law* of United States Mail also sailed for Panama and *Star of the West* sailed for Nicaragua. For the Pacific side of his new line, Vanderbilt simply bought up another independent line that had been started there by Edward Mills, with steamers much inferior to those of Pacific Mail (2).

With three lines now running between New York and California via Central America, and with one of them operated by one of the most professional cut-throats in the steamship business, rates were slashed regularly until it was almost cheaper to travel by sea to California than to stay at home. The rate wars were ruinous to all parties, and once again the matter was settled when Cornelius Vanderbilt cornered his rivals into paying him handsomely to abandon the field. In September 1854, Pacific Mail, United States Mail, and Nicaragua Steamship combined their resources to buy off Vanderbilt, paying him $800,000 for his ships and another $100,000 for a promise to stay out of the California trade. The greater part of the payoff was assumed by Nicaragua Steamship, which in the process became heir to Vanderbilt's inferior Pacific fleet (3).

What was in the long run to prove a severe blow to American shipbuilding took place on the Atlantic that same month. The Collins Line steamship *Arctic* sailed from Liverpool on 21 September 1854 with a full load of passengers, including the wife and two of the children of Edward K. Collins. Six days later, by which time *Arctic* had made most of her way across and was steaming slowly through a fog about fifty miles southeast of Newfoundland, the small French steamship *Vesta* suddenly materialized through the mist ahead too late to avoid a collision. After the crash, *Vesta*, with her bow stove in, seemed in danger of sinking, and *Arctic* offered to take her passengers. But *Vesta* refused assistance and quickly disappeared into the fog headed toward St. Johns, while *Arctic* started her big wheels turning and proceeded on her way. Nearly an hour later word was sent up from the engine room that *Arctic* had started taking water rapidly. The captain immediately turned his ship toward Newfoundland which was only a few hours distant, but in less than an hour

the bow was so far down that it was no longer possible to make headway. The fact had to be faced that *Arctic* was sinking. By the time abandon-ship operations finally began, the situation had become acute, for *Arctic* was going under very fast. Several boats loaded with women and children were simply smashed against the side of the ship by the heavy seas. In a desperate last-minute effort, some of the men constructed a make-shift raft by tying ship's spars together and laying parts of the cabin structure on top of them. The raft was barely afloat with seventy-seven people upon it when *Arctic* slid into the sea. Forty-eight passengers and crew were later picked up by passing ships. The raft, however, was not discovered for several days, by which time only one of the people aboard it was alive. In all 322 died, including Edward Collins' wife and children (4).

The highly-publicized sinkings of *San Francisco* in January and of *Arctic* in September of the same year, each with a tragic cost in lives, and both involving steamships of lines supported by lucrative government mail contracts, helped bring about a change of heart on the part of the American people, or at least of their representatives in Congress, concerning the wisdom of granting subsidies to American steamship lines.

With orders for ships clearly on the decline by 1854, William Webb was again actively seeking contracts to keep his yard operating. Webb had expected in 1854 to be busy building the big warship for the Russian Navy, but events had intervened which made it necessary to postpone this project indefinitely. In the Fall of 1853, a minor conflict between the Russian and Ottoman empires had escalated into a war. Then in February 1854, the British and French governments, fearing the possibility of Russian influence extending to the Mediterranean, had joined the war—known to history as the Crimean War—on the side of the Ottoman Turks. Thus when Webb made plans to start laying a keel for the Russian warship, the American government requested that, in the interests of maintaining a scrupulous neutrality, he not at that time consider building a naval vessel for one of the belligerent powers.

Frustrated in what was to have been his major project for the year, Webb's largest construction job in 1854, as it turned out, was not a ship and was not built in his yard. The New York Balance Dry Dock Company had been formed in 1848 and had at the time constructed a dry dock 210 feet long. Within a few years, however, this dock was proving too small to service many of the ships then being built. So the company solicited bids for the construction of a larger dock, and the contract was awarded to William H. Webb. Since his own shipyard was not large enough for this job, Webb secured property in Williamsburgh across the river and took a group of his workmen over there to build the dock. When it was completed in October 1854, the dry dock was 325 feet long and large enough to accommodate any ship then afloat. Its tanks were emptied or filled by twelve pumps all activated by a single steam engine.

On each side of the dock, about six feet within the outer timbers, and extending from the bottom to the top of the dock, a very heavy and strong longitudinal truss or hog frame, formed of large uprights, top and bottom chords and large iron bars crossing each other diagonally, the whole being strongly secured to the bottom of the dock, cross trusses, diagonal braces, and top deck frame. The hog frame is planked on the inside, thus forming water-tight tanks the whole length of the dock on each side and bottom. (5)

Although William Webb had never experienced any complications launching a ship in all of his many years as a shipbuilder, he did have trouble with the Balance Dry Dock. When the day for the launching came, and all the guests and dignitaries had assembled, and the somber speeches had been delivered, the workmen ceremoniously removed the blocks holding the great creation on land, and all waited for it to glide gracefully toward the sea. But no. The Dry Dock simply sat there. One must remember that here Webb was dealing with a construction far heavier and wider than a ship and that he was not in his own yard with his own equipment. The problem, quite simply, was that the dock was too heavy and the ways not steep enough for its weight to carry it down. The following day, with battering rams brought into play, the dock finally had the grace to leave its perch and move, slowly at first, down into the river.

Once in place and in service on the Williamsburgh side of the river, the Balance Dry Dock built by William Webb served New York shipyards for many years and continued to do so after it was sold in 1890 to the Erie Basin Dry Dock Company (6).

The new year, 1855, brought more problems. By January 1855, the Panama Railroad, in which William Webb had invested and which was to become a very profitable property, was finally completed. Its construction had proved both more difficult and more expensive than anticipated. As we have seen, the company was close to bankruptcy at one point. Before the completion of the railroad, the difficulty of crossing Panama by canoe and burro was driving an alarming number of passengers to the Nicaraguan route. But by January 1855, the mangrove swamp that had once been the island of Manzanillo had been converted into a functioning seaport and transfer terminal named Aspinwall (today, Colon); miles of jungles had been cleared; alligator-infested swamps and rivers had been bridged; rails had been laid and the trip across Panama reduced from several days to just four hours. But the cost had been appalling. The $8,000,000 had been a risk that was eventually rewarded, but the 6000 lives lost paid no dividends (7).

On 5 February the directors of the railroad company and their families and guests (but not including William H. Webb) sailed from New York aboard *George Law* for Aspinwall, arriving there on 15 February. On their arrival, they boarded the first train to make an official run and crossed to Panama. Once in Panama, these people, who must have been getting somewhat worn by now, were treated to the usual round of speeches and toasts which were requisites of openings in those days, all followed by a banquet and grand ball at the Panama Hotel in the evening (8).

One effect of the completion of the Panama Railroad was to hasten the end of the Nicaraguan route which by then was already in trouble. Its ships were never as well maintained or its service as respectable as Pacific Mail's. Its main draw was the shorter trip, an advantage considerably reduced with the opening of the railroad across Panama. The sinking of three of their ships in two years with the loss of a total of 155 lives gave them some unpleasant publicity, although the total deaths in these three accidents still did not equal the number who were washed from the decks of *San Francisco* when she foundered in 1854.

The line also had internal problems. Cornelius Vanderbilt had been picking up stock where he could and eventually controlled enough votes to unseat both Morgan and Garrison. These two men, however, had an ally in William Walker, an unsavory American adventurer who had taken a private army to Nicaragua and was by 1854 in control of that country. By simply annulling the charter of the Accessory Transit Company to carry passengers across Nicaragua, Walker effectively put Vanderbilt out of business. Meanwhile, Morgan and Garrison founded a new company, bought some new ships, got the charter from Walker, and had their own line running again. Vanderbilt, now thoroughly angered, decided he too could get into Central American politics, and in this venture found a willing ally in, of all people, George Law, his erstwhile partner in Long Island Sound steamboating.

Their object was to get Walker out of Nicaragua and to this end Vanderbilt discovered that there was a planned invasion of Nicaragua by Costa Rica merely waiting for him to finance it. Late in 1856, Costa Rican troops moved into Nicaragua, and Walker was forced to leave. (He was later shot as a traitor at the orders of a Nicaraguan tribunal.) Morgan and Garrison were again out of business, but the political situation he had helped create was now too volatile for Vanderbilt to get the charter either. Consequently, for the time at least, the Nicaraguan Steamship Company ceased to exist and, for a few peaceful years, the only lines to California in operation were United States Mail on the Atlantic and Pacific Mail on the Pacific (9).

Though sailing was smooth temporarily for Pacific Mail, there were other problems in the country as a whole by this time that were to have their effect on the shipbuilding business. By 1856 the tensions between the Northern and Southern states were becoming acute. The Compromise of 1850, allowing the admission of California as a state, had only plastered over the deeper divisions. Since then, the near state of war between pro-slavery and anti-slavery settlers in Kansas and Nebraska had reawakened the tensions. In the election of 1856, the creation of the new anti-slavery Republican Party virtually obliterated the old Whig Party, with the result that both lost to the Pennsylvania Democrat, James Buchanan.

One result of the national tensions was that the Southerners in Congress began to voice their resentment against policies of the federal government favoring Northern business interests. In particular, they cited the mail subsidies to Northern transportation companies such as the Collins Line, United States Mail, or Pacific Mail.

Another problem was the growing economic depression which culminated in the stock market panic of 1857. New York shipyards, already suffering from the fact that too many ships had been built during the Gold Rush, were hard hit as well by unexpected increases in the costs of production. These increases too often appeared only after a builder had already contracted for a ship at a given price. In the early fifties, for instance, the average cost of timber went up by 25 per cent. Also, after a series of strikes in the yards, the minimum wage for shipyard workers rose from $2.00 per day to $3.00 per day. Shipbuilders were often caught at the mercy of the strikers, since their contracts for ships usually included a stiff financial penalty if the ship were not delivered by a specified date.

Largely as a result of these problems, many of the older shipyards ceased to exist during the fifties. When David Brown died in 1848, Jacob Bell continued the yard of Brown and Bell alone for several years—we have already noted that he constructed two of the big Collins Line steamships during this period—but Bell too decided to quit in 1852 and sold the yard to the new firm of Roosevelt and Joyce.

The only yards along the river other than Brown and Bell which might be considered to have been on a par with that of William H. Webb were those of William H. Brown and of Smith and Dimon, both of which went out of business at this time: Brown in 1853; Smith and Dimon the following year. Another casualty of the depression of the mid-fifties was the yard of Westervelt and Mackay, successors to the yard of Christian Bergh, one of the oldest on the river. Westervelt and Mackay, announcing that it had become "financially embarrassed," went out of business in 1856, though S. G. Bogart took over the yard temporarily, just to complete existing contracts (10). Jacob Westervelt, after giving up his shipyard, went on to serve one term as Mayor of New York. According to one source, the Democrats also asked William Webb, who by now had a well-earned reputation as both an honest and successful New York businessman, to run for mayor. But, as should be obvious, Webb's work in his shipyard kept him far too busy to consider adding public office to his duties.

One rather sad indicator of the straitened condition of the shipbuilding world at this time is a letter from John W. Griffiths, now senior partner in the designing firm of Griffiths and Bates, to William Webb requesting a loan of one thousand dollars and suggesting a repayment plan that would extend over an eighteen-month period. Thus it appears that at a time when William Webb had become a millionaire, John W. Griffiths, a man whose name was well-known at the time as a major designer of clipper ships, was reduced to asking for a year and a half to repay a loan of a mere thousand dollars. Although Griffiths had been an apprentice of Webb's father and very likely a frequent guest in their home when William was still but a boy, it is rather pathetic to see this letter addressed "Dear Sir," and signed, although clearly in John Griffiths' own handwriting, "Griffiths and Bates" (11).

Another serious loss to the shipbuilding world in 1856 cannot be attributed to the depression. On 25 September, George Steers, one of the two sons of Henry Steers, was driving a pair of horses from the city to Glen Cove, Long

The Great Balance Dock, built in 1854 and so called to differentiate it from the older, smaller Little Balance Dock, was built by William H. Webb, but it is neither conventionally included in lists of his ships nor was it built at the yard at the foot of Sixth Street, where there was insufficient room to build this massive structure. In this painting, the Collins Line's mammoth steamship *Adriatic*—the largest built in the United States—is shown on the dock sometime during her long fitting out period. A typical packet ship lies to the left. The dock's working location was near its building place in Williamsburgh.

Island, to bring home his wife who had been visiting there. On the way, his horses got out of control and began to run wild. Steers jumped from the cart, but in so doing struck his head and severed his spinal cord at the neck. Completely paralyzed, he died soon afterward. The Steers yard in Williamsburgh, one of the few to survive the depression, continued under the management of James Steers, George Steers' brother.

During 1856, the year that saw many other yards go out of business, William Webb's yard was busy building the largest sailing ship it had constructed in its long history: the mammoth packet cargo ship *Ocean Monarch*. *Ocean Monarch* could in no way be confused with a clipper; she was a full-bodied ship designed for the coastal shipping firm of Stanton and Frost to carry as much cargo as practicable, and most sources assert that she had the largest cargo-carrying capacity of any packet ship ever built. Considering that on one occasion she loaded 7000 bales of cotton at Charleston and that on another she carried 120,000 bushels of wheat to Liverpool, the claim is not hard to believe.

Among several other smaller vessels completed at the Webb yard that year was the big steam-driven towboat *William H. Webb* designed, like his earlier *Goliah* (which had by then joined the Gold Rush ships in California), for towing sailing ships to and from the open sea and also for salvage work along the nearby Atlantic Coast. *William H. Webb*, whose engines were said to be almost as powerful as those of the Collins Line steamers, was described at the time as the largest and most powerful vessel ever designed purely for towing.

It was also in the year 1856 that news of yet another maritime tragedy struck panic into the nation. On 29 June 1856, the Collins Line steamship *Pacific* sailed from Liverpool on a regularly scheduled run. But the ship never arrived in New York. No trace of *Pacific* was ever discovered; no survivors ever reached shore to tell the story. She simply sailed and disappeared. Since other ships crossing at the same time reported sailing through a field of icebergs, the best assumption is that *Pacific* struck an iceberg and sank, taking everyone aboard down with her.

With half of its government-subsidized fleet lost at sea in accidents costing hundreds of lives, the reputation of the once highly-touted Collins Line, the line that had been intended to wrest dominance of the New York-Liverpool trade from the rival British Cunard Line, was now lost beyond reviving. And gone with it, of course, was any inclination the American government once had for giving financial support to an American merchant marine.

At the time of the loss of *Pacific*, the Steers yard in Greenpoint had recently launched a new steamship for the Collins Line, to replace the lost *Arctic*. This was *Adriatic*, ** which was even larger and more powerful than the original quartet, and now by far the largest steamship built in America. Interestingly, *Adriatic* was built with timbers collected by William Webb when he was starting to build the warship for the Russians, but which he had sold to Steers when the American government had obliged him to give up the project. Though Steers completed the big liner in 1856, so many adjustments to her engine were necessary after the trial trips that she was not ready for a maiden voyage for over a year. By the time *Adriatic* entered service, she had time for only one round trip to Liverpool before the mail contract of the Collins Line expired in 1859. Needless to say, the contract was not renewed, and the three surviving Collins Line ships, including *Adriatic*, were simply sent to layup.

The following year still another tragic accident, this time to the steamship built by William Webb for United States Mail, still further discouraged government support

** A painting of *Adriatic* fitting out in William Webb's Balance Dry Dock now hangs in the Reading Room of the Webb Institute of Naval Architecture.

of American shipping. On 3 September 1857, the steamship *Central America* (formerly *George Law*; the name had been changed after Law ceased to be associated with the company) sailed from Panama with 593 passengers and $1,590,000 in gold specie. Five days later, on 8 September, she cleared Havana and headed northward. The following day, *Central America* was encountering unusually high winds and heavy seas. By 11 September, she was bucking a full-scale hurricane. At noon that day, it was reported to the bridge that water was leaking into the boiler room, and by late afternoon the water had risen far enough to douse the fires, leaving *Central America* without power and at the mercy of a raging hurricane.

Throughout that night, as had happened on *San Francisco* in a similar position a few years before, passengers and crew spent the night bailing as well as they could with buckets. The following day the seas were calmer, so that when the bark *Marine* hove to, *Central America* was able to get all of the women and children into boats and transferred to the bark. There was an ugly incident, however, when, as the last boat was loading, *Central America*'s chief engineer, who was supposed to be a responsible officer, but who, one must admit, was in a better position than most to know how serious their plight actually was, jumped from the ship into the lifeboat and, suddenly baring a large knife, refused to allow any more passengers into the boat. One explanation not offered in the contemporary account could have been that the lifeboat was already overloaded and the chief engineer had to take this extreme measure to keep it from being swamped and thus losing even more lives.

That evening, after *Marine* had long passed out of sight, the seas turned rough again, and during the night of 12-13 September, *Central America*, still taking water, finally filled and sank, taking most of the remaining passengers and all of the specie with her into the ocean. The following morning the Norwegian bark *Ellen* passed by and picked up the few people still clinging to pieces of wreckage. But in all 420 lives had been lost (12).

The sinking of *Central America* was for all practical purposes the final blow to the reputation of United States Mail. By this time the newspapers had already been running stories, mostly true, of overcrowding, poorly maintained ships, untrained crews, poor food, and filthy accommodations. With the mail contracts coming up for renewal in only two years, the sinking of *Central America*, following so soon after the losses of *Arctic*, *Pacific*, and *San Francisco*, made it fairly evident that its mail contract, like that of the Collins Line, would not be renewed.

With the end of the mass migration to California, the rising competition from British iron-hulled steamships, and the retreat of our federal government from a policy of supporting the American merchant marine with mail subsidies, American shipping, by the time of the nation-wide economic depression of 1857, had suffered a blow from which in some ways it was never to recover. The once-busy shipyards along the East River, which had been operating at full capacity in the early years of the decade were gradually beginning to disappear. But the yard of William H. Webb at Sixth Street, while never again as busy as it had been between 1850 and 1853, was never-theless, as a result of the unflagging energy, determina-tion, salesmanship, and, let us not forget, shipbuilding skill, of its proprietor, still operating and still in excellent financial condition.

# IX

# Webb Rides Out a Depression, 1857-1861

**B**ETWEEN the depression year of 1857 and the start of the American Civil War in 1861, a period in which lack of business was forcing many New York shipyards to close down, William H. Webb managed, by adapting to new kinds of work, to keep his yard operating at nearly full capacity. It was also about this time, when Webb was just past forty, that he was becoming recognized as America's leading naval architect and shipbuilder as well as a highly respected member of the New York business community. Whenever his name appeared in a newspaper, as it frequently did in these days, it was preceded with almost Homeric consistency by, "the eminent New York shipbuilder. . .". An article in the *New York Herald* that year was particularly laudatory:

Although still a young man, Mr. Webb has constructed about one hundred and twenty vessels of every description, among which are some of the largest ships and sea steamers of the day; and it is worthy of remark, and due him to state, that upon the recent classification of vessels by the underwriters, those of his construction were found entitled to the mark of extra confidence. (1)

One of the Webb yard's first orders in 1857 was to build another ship for United States Mail to replace the lost *Central America* (ex-*George Law*), this one named *Moses Taylor*, also for one of the executives of the company. Since United States Mail was by then probably well aware that it would not be in business much longer, the company kept costs down on Moses Taylor by installing a second-hand engine. And, as the size of the ship was limited by the power of this engine, Webb had to design *Moses Taylor* as a somewhat smaller vessel (244 feet overall length) than *George Law* or some of the other more recent steamships of this line.

Early in the year William Webb also became interested in a project for a new revenue cutter for the port of New York. In 1856 a group of New York merchants and shippers had petitioned Congress to authorize the construction of a powerful steam-driven vessel for the United States Revenue Marine Service (predecessor of the United States Coast Guard). The Revenue Marine had the responsibility for assisting ships in peril in waters along the American coast, but the sailing vessels they were then using for the purpose were too small and too slow to be of much help in emergencies. The shippers who signed the petition estimated that in the years 1854 and 1855 alone, at least a thousand lives and $10,000,000 worth of cargo had been lost for want of satisfactory rescue vessels (2).

The petition, which had the full support of Captain Alexander Fraser, commander of the Revenue Marine, specified that the ship should be steam-driven, should be powerful enough to tow disabled vessels into port, and should be equipped to care for any injured passengers or crew members it might take aboard. Another stipulation was that the ship should be fast enough to catch suspected slavers who might be trying to slip out of the port of New York undetected.

In March 1857 Congress responded by passing an act authorizing the building of the revenue ship requested and appropriating $150,000 for the project. The act also called for shipbuilders to submit bids for constructing the vessel (making this the first time the government had ever awarded a contract by competition). The board appointed to examine the various plans submitted was in this case not made up entirely of political and military functionaries but included several prominent naval architects. Notable among them was Samuel Hart Pook, next to Donald McKay the most highly-regarded ship designer of New England, and especially noted for his successful clipper ships.

In all, eighteen bids were submitted, each with deck plans, written specifications and half-models. John W. Griffiths submitted three separate bids. William H. Webb, who in better times might not have considered entering such a competition, submitted only one. Apparently some of the shipbuilders submitting bids, given the hard times most must then have been facing, beleaguered the committee with high-level influence peddling. But ignoring these pressures, for which they were later lauded by the press, on 13 July 1857 the committee announced that the plan submitted by William Webb had been chosen by unanimous vote (3).

An article in the *New York Herald* approving the decision of the committee also gives some sense of William Webb's standing in the New York business community:

... in addition to the opinion of the committee, [Webb's] selection has the endorsement of all the underwriters of this city in addition to that of the principal merchants. (4)

Howell Cobb, Secretary of the Treasury and a frequent guest at the White House, suggested naming the new cutter *Harriet Lane* for President Buchanan's niece. Harriet Lane, who had been orphaned at eleven and subsequently raised by her uncle, was at the time, since Buchanan was a bachelor, serving, most graciously according to contemporary reports, as hostess at the White House.

Even before Webb had laid the keel for *Harriet Lane*, he received word from the imperial government of Russia that they not only wanted him to move ahead with the warship ordered four years earlier but also wanted him to build a smaller corvette as well. The Crimean War had ended with the Peace of Paris during the summer of 1856, thus releasing Webb from the strictures against building a ship for a belligerent. Also, Tsar Nicholas had died in 1855 during the conflict and had been succeeded by his son, Alexander II, whose natural conservatism was considerably tempered by practicality. (It was Alexander II who finally liberated the Russian serfs in 1861, anticipating the Americans by four years.) The new Tsar fully supported his brother Constantine's request to have these two naval vessels built in America by William H. Webb.

Early in May 1857 the keel for the Russian corvette, named *Japanese*, was put down in the Webb yard next to the nearly-completed *Moses Taylor*. Then in July work was started on *Harriet Lane*. Samuel Hart Pook was assigned to Webb's yard to supervise her construction as an agent for the government. He was assisted by Captain John Faunce, of the U.S. Revenue Marine Service, who was to be *Harriet Lane*'s first commander.

Finally in August, soon after *Moses Taylor* was launched, Webb was ready to lay the keel of the large Russian warship, to be named *General Admiral* in honor of the Grand Duke Constantine. The Russian ambassador to the United States, Baron de Stoeckel, was eager to host a major ceremony on 21 September, the Grand Duke's birthday. As Webb did not wish to wait so long to lay the keel, the event was planned instead on the occasion of the raising of the sternpost. At the ceremony Baron de Stoeckel solemnly placed, into a mortice in the keel, a silver plate inscribed (in Russian):

The seventy-gun ship General Admiral was begun in the presence of Baron de Stoeckel, Russian minister at Washington, September 21, 1857, in New York, after the plans of William H. Webb, American shipbuilder. (5)

Once the obligatory liturgies over the ship had been pronounced, a large party including, in addition to Baron de Stoeckel, who was hosting the affair, and William H. Webb; Fernando Wood, the incumbent mayor of New York; the shipping magnates Charles H. Marshall and Henry Grinnell, both good friends of Webb's; Samuel Hart Pook; and a great many Russians, repaired to the Clarendon Hotel for an all-afternoon banquet, accompanied, needless to say, by many speeches, some disguised as toasts.

Although Baron de Stoeckel probably did not know it, the silver plaque he had so solemnly placed in *General Admiral*'s keel already contained a misstatement. Without bothering to check with the Tsar or his imperial brother, William Webb had decided that the ship he was building would be more effective with forty larger cannon than with the seventy smaller ones in his original design, and he was building her accordingly, in the process changing the official designation of the vessel from a ship-of-the-line to a very large frigate.

As *General Admiral* began to rise on the ways at William Webb's yard, she also began collecting a fair number of superlatives. At 327 feet overall and measuring an unheard of 6000 tons, she was not only the largest warship in the world, she was also, in tonnage, the largest wooden-hulled ship built in America (although *Adriatic* at 354 feet and the Commodore's new *Vanderbilt* at 340 feet, were both longer). Nor was the interior of *General Admiral* to be limited to the Spartan quarters usually associated with warships. As flagship of the Russian fleet, *General Admiral* would often carry either the Grand Duke Constantine or the Tsar himself aboard. Thus she was to be equipped with an imperial suite of palatial grandness in addition to the usual crew's quarters.

*Japanese*, launched on 16 November, was the first of the three ships off the ways. A moderate-sized corvette only 214 feet in length, her wooden hull, like those of *Harriet Lane* and *General Admiral*, was braced with diagonal iron straps connecting her frames. And for protection against ramming, she carried thick exterior iron

"Rolling Mose"—the steamship *Moses Taylor*—was built for Marshall O. Roberts' United States Mail Steamship Company in 1857. She was propelled by two vertical beam engines formerly in the unsuccessful steamship *El Dorado*, built by Thomas Collyer in 1850. The two walking beams may be seen in this lithograph. (Courtesy of The Mariners' Museum, Newport News, Virginia)

plating about six feet wide around the upper level of her hull. After her launching, accompanied by the usual ceremonies, she was towed up to the Novelty Iron Works where two oscillating engines were installed to power her single propeller. By Spring *Japanese* was ready for her trial trips, and after their successful completion she sailed off to Russia under her own steam.

*Harriet Lane* came down the ways just three days after *Japanese*, on 19 November, and was towed up to the Novelty Works for the installation of the two inclined engines which would supply the extraordinary power re-

quired by the Revenue Marine. By early February *Harriet Lane* was back at Webb's yard for her final fitting out, which was completed in time for her to be inspected, approved, and officially commissioned on 28 February.

In spite of the serious business for which she was designed, *Harriet Lane* was an exceptionally beautiful vessel. She was only 180 feet in length, but with her graceful sheer, more pronounced than was usual on Webb-built ships, and her jauntily raked single stack amidships, this sidewheel cutter was as pretty and perky underway as a small clipper. As usual with Webb's ships, her hull was strengthened with crossed-iron straps between the frames. On her deck she carried three cabins for the captain and the other officers, while below was a large bunk room for sixty to eighty crew members as well as accommodations for the passengers or crews that might be rescued from vessels in distress. And in the hold she carried a large assortment of engine parts to have available should she be called upon to assist a steamship with a mechanical breakdown.

When *Harriet Lane* was nearing completion, William Webb heard that Mohammed Pasha, a Turkish admiral, was in town with a group of his officers to inspect American naval vessels. With his eye ever alert for a possible contract, Webb quickly arranged with the Revenue Marine for *Harriet Lane*, although she had not yet made her official trials, to make a special trip on 11 March for the benefit of the Turkish officers. At the last minute he heard that some Russian officers were also in New York, so he invited them along. Whether Webb was merely naive or whether, which seems likely, he had such complete confidence in his own diplomatic talents that he dared risk inviting both Russian and Turkish officers together just two years after the unusually bitter Crimean War, we can never know. In any event, *Harriet Lane* took Webb and his naval guests for a run of several hours out into the Atlantic and back. No new contracts, however, resulted from this impromptu outing.

On 24 March *Harriet Lane* finally went out for her official trials and easily fulfilled every expectation of the Revenue Marine. A New York Times reporter who sailed with her that day was also impressed, commenting,

Her hull is exceedingly beautiful, and she is so strongly built that there is not the least perceptible vibration on deck when her machinery is in motion. (6)

With her trials successfully completed, in April 1858 *Harriet Lane* was officially assigned to duty out of New York harbor and based at the Brooklyn Navy Yard.

The construction of the great hull of *General Admiral*, took much longer to complete. Work on the Russian warship continued through the summer of 1858 during which newspapers were regularly reporting the series of debates between Abraham Lincoln and Stephen A. Douglas over the constitutionality of slavery. The launch of *General Admiral*, which was scheduled for 21 September 1858, again on the Grand Duke's birthday, just one year after the official raising of the sternpost, turned out to be one of the grandest affairs ever staged along the New York waterfront. Through the morning of the twenty-

first, as the mammoth hull stood ready to be released from the ways, 3000 invited guests climbed aboard to join her for the occasion and 10,000 more crowded into the yard to watch from the shore. At exactly noon, when William Webb gave the signal, Baroness de Stoeckel swung a bottle of champagne over her bow and spoke something to the great ship in Russian; then, miraculously, the largest wooden hull yet built in America began sliding slowly into the river, while Dodsworth's band attempted to blare out some appropriate tune over the deafening cheers of 13,000 voices.

It was indeed an occasion to celebrate. The *New York Times* reported that *General Admiral* entered the water so gracefully that those on board were not even aware she had done so. Once the ship was seen safely to the river, the approximately 500 men who had built her were invited to a banquet arranged for them in the molding loft by William Webb. Seated at two long tables, the men were treated to a feast of ham, roast beef, potatoes and assorted vegetables. After the meal each man received one glass of champagne, and, for those who required a chaser, there were several vats of beer available. After the meal, the grateful workers raised a rousing cheer for William H. Webb. One would like to assume that this pleasant affair continued for most of the afternoon, after which these deserving yard workers returned to their homes for a needed rest. But no, not at the William H. Webb yard in 1858! When the mechanic's bell tolled at 1:00 sharp, the men dispersed and returned to their work on the hull of the steamship *Yorktown*, then under construction in the yard for the New York and Virginia Steamship Company's Norfolk run (7).

Through the winter of 1858-1859 *General Admiral* remained at the Novelty Iron Works, which installed two oscillating engines that together produced up to 2000 horsepower to drive the single 19′6″ propeller. By April 1859 *General Admiral*, with her engines in place, was back at the Webb yard for the completion of the superstructure. By May the ship was ready for inspection. The total cost had been $1,125,000, an enormous price for a single ship in 1859, though apparently the Russian government had gotten what it wanted and was willing to pay for it. A board of American naval officers sent to examine the ship found her workmanship and materials "fully equal to those of any vessels constructed by our government,"

. . . and we may mention that in regard to the location of beams relative to the ports, she is superior, from the fact of the armament having been determined before building the vessel. We therefore recommend in future that the armament be always determined in our service before building the vessel. (8)

This observation, one might have assumed, could be taken for granted, but apparently in 1859 it was not. The report concluded by noting that "the construction of the *General Admiral* reflects great credit on Mr. Webb." There was considerable truth in this modest observation, as probably no other ship contributed so much to William Webb's international reputation as *General Admiral.* An article in the *New York Herald* at about this time noted that:

. . . in the construction of these vessels [William H. Webb] has, by the preference given to him, been virtually endorsed by the Russian government as well as our own as the very first naval architect in this country. (9)

On 4 May 1859, with William H. Webb and the Russian naval supervisor as the only passengers aboard, *General Admiral* was towed down the East River as far as Governors Island and then unshackled and let loose. Cruising down the New Jersey coast, she more than lived up to every expectation, handling "like a pilot boat" in spite of her size, and reaching a speed of twelve knots without even opening the engines all the way.

When *General Admiral* was ready to sail for Russia in June, William Webb elected to sail with her so that he could deliver the ship in person. With him on the trip were his wife, Henrietta Amelia Webb, and their two sons, William E. Webb, then aged fourteen, and Marshall Webb, then aged twelve. On 16 June, sailing day, *General Admiral* was stationed in the harbor off Governors Island. Webb, his family, several other passengers, and a large number of friends boarded the towboat *James A. Stevens* at the Battery at 2:00 P.M. to be taken out to the big ship.

Aboard the vessel the party was served a farewell banquet. About four o'clock the guests were returned to shore, and then promptly at 5:00 P.M. *General Admiral*, with William Webb and his family aboard, moved slowly down the bay, gradually picking up speed as she pointed her prow out into the Atlantic.

*General Admiral* made an exceptionally fast crossing, part of it under canvas alone with her single screw lifted entirely out of the water to reduce drag. With her engines running at full speed, she made up to fifteen knots, and even under sail she often moved along at twelve knots or better. On 27 June, she turned into the harbor at Cherbourg after a crossing of only eleven days, ten hours. After a week in the French port refueling and taking on provisions, she headed back into the Channel and on toward the Baltic Sea and Russia.

On 15 July *General Admiral* finally pulled alongside her pier at Kronstadt, the Russian naval base on the Neva River near St. Petersburg. Waiting eagerly on the dock to meet her was the Grand Duke Constantine, who greeted William Webb warmly and beamed with satisfaction at the sight of the ship. The Grand Duke's genuine enthusiasm can be sensed in a later report by an American official then stationed in St. Petersburg:

The Grand Duke Constantine came up to me last night at a great ball and asked if I had been on his ship, the "General Admiral." I said yes, and then he said that she was everything he could desire. . . and [that] she was the finest ship afloat. (10)

To register his own satisfaction with the new flagship of the Russian naval fleet, Tsar Alexander II presented William H. Webb with a solid gold snuff box encrusted with diamonds (11).

While the Webbs remained a few weeks in the Russian capital, attending a series of receptions in their honor, most offered by members of the small American community living there, the Grand Duke took his new ship on a trial trip in the form of a state visit to England. His real object, of course, was to show off this new pride of the Russian fleet to the British, who were not only regarded as Europe's leading maritime nation but whose Navy had also recently defeated the Russians in war. The British press, however, does not reflect very much enthusiasm about the appearance of this huge Russian naval vessel, probably because at that time they were more concerned with Brunel's gigantic *Great Eastern*, then almost ready for sea trials. This ship, of course, although she was built in Britain of iron and therefore did not contest *General Admiral*'s claim as the largest wooden vessel built in America, was, with an overall length of over 600 feet, to eclipse almost any other maritime record in size for several decades to come.

Once *General Admiral* had been safely delivered, William Webb, still accompanied by his family, decided to make some visits in other European capitals in search of new orders. It would seem that in 1859, for the first time since he started his business in 1840, Webb had no contracts waiting on his desk at home. Although none of the still-divided German states were in a position to order a ship of such a size, and England and France, of course, preferred to build their own (though the British later built two rather unsuccessful vessels, both specifically modelled from *General Admiral*), Webb did find a very interested

[From a Lithograph by Endicott and Company.]

THE " GENERAL ADMIRAL."

The Russian frigate *General Admiral*, shown in this lithograph, was ordered in 1853, laid down in 1857, launched in 1858 and delivered the following year. The two Italian warships completed a few years later resembled *General Admiral* in profile. William H. Webb pursued foreign naval contracts for many years, but never completed a domestic warship. *Dunderberg*, ordered by the United States Navy, was never delivered to them.

customer in the person of Count Camillo Cavour, the Prime Minister of Italy in Turino.

Count Cavour had earlier been Prime Minister to Victor Emmanuel, then king of Piedmont, a small state tucked under the Alps in the northwest corner of the Italian peninsula. Since 1848 Cavour had been the strongest advocate of a plan for all of the neighboring small Italian states to join Piedmont in creating a new kingdom of Italy, with Victor Emmanuel as king. With Italian nationalism then creating a strong sentiment favoring a united Italy, this policy had the enthusiastic support of most of the populations of these states. There was a problem, however, in that two of the largest Italian states, namely Lombardy and Venezia, were then considered part of the Austrian Empire, which had no thought of giving them up.

Napoleon III, then Emperor of the French, decided it would be in the interests of France to support Cavour in his plan to unite Italy. So in the Spring of 1859 France had joined Piedmont in a short war to wrest Lombardy and Venezia from the Austrians. The war ended with Lombardy at least being ceded to the Italians at the peace of Villafranca, which was signed while the Webbs were in St. Petersburg. A few weeks later the new kingdom of Italy was at last a reality, with Victor Emmanuel now recognized as King of Italy, and Count Cavour as his first Prime Minister.

With this enlarged country to defend and Venice still

This remarkable painting of the clipper *Challenge*, flying N. L. & G. Griswold's distinctive house flag, shows her with topgallants and royals furled. Her skysail yards have been lowered as she roars along in what must have been a stiff breeze.

to be won, Cavour was more than ready to consider creating an Italian Navy just at the time that William H. Webb arrived in Turino with a plan to help him achieve one. Cavour, it turned out, had already seen the plans for *General Admiral*, and, with relatively little discussion, he told Webb that he wanted two ships of the same type for Italy. In view of the great pride Count Cavour must have felt now that his policies had succeeded in creating the new Kingdom of Italy, one can understand his wanting one of these warships to be named *Re d'Italia* ("King of Italy"), but why he asked that the other be named *Re Don Luigi di Portogallo* ("King Louis of Portugal") is not now clear.

With this request, but no signed contract, for two more large warships, William Webb and his family took ship back to America, arriving in New York in late August. Webb thus returned to the United States just a few weeks before the ten-year mail subsidies granted to United States Mail and Pacific Mail in 1849 were about to expire. With the accidents, the mismanagement, and the poor maintenance of their ships by United States Mail, it was generally understood that their part of the contract would not be renewed, and also that, without the subsidy, the company would not be able to remain in business. Under these circumstances, it should not come as a surprise that during the summer of 1859 Cornelius Vanderbilt, like a vulture awaiting his feast, decided that this was the time for him to have another steamship line to California by way of Central America, ready and operating, in case the government should wish to award him the mail contract. Somehow Vanderbilt managed to make a peace of sorts with his former partners, Morgan and Garrison, and to repurchase the ships which had been his in the first place, but which, since the loss of the charter to cross Nicaragua, had been idle (12).

Pacific Mail now realized that if United States Mail did go out of business, which by the fall of 1859 had become a virtual certainty, Vanderbilt's would be the only line left operating on the Atlantic side, which meant that he was very likely to get the contract. The problem was compounded by the fact that the Commodore also had a line on the Pacific side and might well have enough friends in Washington to help him grab both contracts. Pacific Mail, therefore, decided that its best move was again to start a line of its own on the Atlantic route to get themselves on at least an equal footing with Vanderbilt before the contracts were awarded in September. For this purpose a new company, the North Atlantic Steamship Company, was formed by a combination of Pacific Mail and Panama Railroad interests (13).

William Webb, already a major stockholder in both Pacific Mail and the Panama Railroad, now found himself holding a considerable interest in the new steamship company as well. Returning from Europe about the time the company was being formed, and noting that his shipyard no longer had enough business to require his full energies, Webb now began to devote part of his time to the North Atlantic Steamship Company. As early as 1850 when he was starting to build the clipper *Celestial*, Webb had taken on Christian Metzger as his "Assistant Shipbuilder." By the time Webb sailed for Russia aboard *General Admiral* in June 1859, he had already begun to reduce his own role to that of designing ships and soliciting contracts, leaving most of the supervision of the yard to Metzger. It was at this time that Metzger was also granted power of attorney, so that he could manage the business of the yard in Webb's absence. With this trusted assistant in the office at Lewis and Seventh Street, William Webb now increasingly devoted his time to his other affairs, most particularly to the management of the series of steamship companies, beginning with the North Atlantic Steamship Company in 1859, with which he became involved.

The immediate need of the North Atlantic Steamship Company in the early Fall of 1859 was to find steamships of a quality that could successfully compete with Vanderbilt's. Also, with the contracts due for renewal in September, there was no time to build new vessels or even to renovate older ones. The company needed ships that were then in operating condition. For this situation a ready solution was provided by Brown Brothers, a New York investment firm. Brown Brothers, it turned out, had invested heavily in the Collins Line and, therefore, held a lien on its recently idled steamships, *Atlantic*, *Baltic*, and the newer and larger *Adriatic*, all then languishing in layup at the line's New York pier. Thus, just at the time the North Atlantic Steamship Company was seeking ready-made steamships of high quality, Brown Brothers (who had by then purchased these vessels at auction to save what they could of their investment) happened to be seeking profitable employment for the fastest and most elegant steamships yet built in America.

As the negotiations worked out, the three idle Collins Line steamers became the fleet of the North Atlantic Steamship Company, with Brown Brothers in the process acquiring considerable stock in the Pacific Mail Steamship Company. Of the Collins Line ships, only *Atlantic* and *Baltic* were placed in regular service to Panama. *Adriatic*, although virtually new, was too large and too expensive to operate profitably, and thus made only a few trips for the line when one of the others was temporarily out of service for overhaul or repair. Even *Atlantic* and *Baltic*, neither of which had been built with economy in mind, were coal eaters, though, by this very fact, they were also fast and regularly completed the run to Panama in two or three days less time than Vanderbilt's ships (14).

As of September 1859 when the 1849 contracts were due to expire, it was still not clear what the government planned to do. The confusion was only compounded when at the last minute the government announced that the mail contracts for both the Atlantic and Pacific routes had been awarded to Daniel H. Johnson, a man quite unknown in steamship circles. But the mystery was soon resolved when a few days later Daniel Johnson sold the lucrative contracts to Cornelius Vanderbilt. Now Vanderbilt had the mail contracts for both oceans. United States Mail, as expected, immediately ceased operations. When many of its officers, however, now joined Vanderbilt, the possibility of some earlier agreement, not only between Vanderbilt and the officers of United States Mail, but also with highly-placed government officials, could not be ruled out.

Even though they had failed to get the mail contracts for either ocean, the officers and stockholders of both Pacific Mail and North Atlantic Steamship, among whom William H. Webb was an influential voice, decided that, with their better ships and superior service, they could still make a reasonable profit. Most of the stockholders, in any event, also held an interest in the Panama Railroad, which still had the exclusive rights for transporting mail and government supplies across the isthmus (15).

Not surprisingly, the fall of 1859 saw a bitter rate war between Vanderbilt and North Atlantic Steamship on the one coast and Pacific Mail on the other. There was ample business on all of the lines, but with the low rates prevailing, all were losing money (16). For a few days in November, it appeared that the matter had been resolved when Vanderbilt offered to buy Pacific Mail for $2,000,000. This was a good price and would have meant a pleasant profit for the stockholders, the majority of whom were prepared to accept Vanderbilt's bid. But when Vanderbilt asked for guarantees that the directors of Pacific Mail would never again engage in the California steamship business, William Aspinwall persuaded his colleagues to reject the offer (17).

Negotiations were begun again early in 1860, this time more successfully. According to an agreement reached in February 1860, the North Atlantic Steamship Company ceased operating, the elegant *Atlantic* and *Baltic* were again placed in layup, and Vanderbilt was given a free hand on the New York-to-Aspinwall part of the route. On their part, Pacific Mail purchased Vanderbilt's Pacific fleet and were left to operate the Panama-San Francisco leg of the route without competition from the Commodore. Vanderbilt still held both mail contracts but agreed to pay Pacific Mail for carrying mails on the West Coast. While these terms were financially acceptable to all concerned, there was one part of the agreement which gave Pacific Mail stockholders serious and legitimate concern: namely, that Vanderbilt now became a major stockholder in Pacific Mail (18). Unfortunately, we do not know on which side of the issue William Webb stood. In any event, Aspinwall, who favored the agreement, managed to get it approved by his board, and, since rates went up again as soon as the agreement was signed, it can be assumed that even the opposing stockholders were eventually placated.

Even before reaching the agreement with Vanderbilt, the North Atlantic Steamship Company had been considering more profitable employment for their largest and newest steamship *Adriatic* and had decided to use her to start a new line between New York and Le Havre. A great deal of work on her engines was necessary before she could be serviceable, but by early Spring she was ready, and she made her first sailing for Le Havre on 14 April (19). The commander of *Adriatic*, still the largest, fastest, and most elegantly appointed liner in American registry, was Joseph J. Comstock. Comstock had started his career as captain of one of the early experimental steamboats running between New York and Providence, Rhode Island. After a short stint as captain of a California clipper, he had returned to Long Island Sound to become the first commander on the famed Fall River Line when it began service in 1847. After several years with this line, he left in 1854 to assume command of the Collins Line steamship *Baltic*. It was apparently when *Baltic* was subsequently in the service of the North Atlantic Steamship Company that Comstock came to know William H. Webb, for whom he was to command many ships in later years.

By the time *Adriatic* began service on the Le Havre route, the North Atlantic Steamship Company had reached its agreement with Vanderbilt and was no longer operating between New York and Panama. Since Vanderbilt, who now sat on the board of the line's parent company, Pacific Mail, also had his own steamship line between New York and Le Havre, it was necessary to schedule *Adriatic*'s sailings for dates that would not com-

Pacific Mail's steamship *Constitution* was notable for a number of reasons—built in 1861, she was the first ship built for the company since 1854, and the first from the Webb yard since the ill-starred *San Francisco* was delivered late in 1853. Her profile was unusual in that she was one of the few Pacific Mail sidewheelers to have boilers forward and aft of the engine. (Courtesy of the Mariners' Museum, Newport News, Virginia)

pete with Vanderbilt's ships. In any event, the life of this new line was not long. After only a few round trips, this majestic liner was laid up for the winter in November 1860. By the time the season of 1861 was due to open, the Civil War had intervened, *Adriatic* had been sold abroad, and the service was never resumed (20).

Late in the year 1860, Webb had also received orders for two more American coastal steamships: *Constitution* for Pacific Mail and *Mississippi* for Mitchill's Line to Savannah (all of whose ships had been built at the Webb yard since its first ship *Cherokee* in 1848).

By 1860, however, other events were taking place which were to have their effect on the shipbuilding business. In the presidential elections of 1860, sectional differences so divided the dominant Democratic Party that it put up two separate candidates, one northern and one southern, thus ensuring the election of the minority candidate of the new Republican Party, Abraham Lincoln of Illinois. Between Lincoln's election in November 1860 and his inauguration in March 1861, several Southern states, unable to accept a President so outspokenly opposed to the extension of slavery into the unsettled territories or a party so dedicated to the interests of northern business, opted for secession from the American union. Then in February 1861 these secessionist states joined to form the Confederacy.

Once sworn into office in March 1861, Lincoln determined to bring these states back into the Union, by force if necessary. As an indication of his intent, he decided to send supplies to Fort Sumter, a union fortress on an island in the harbor of Charleston, South Carolina. The ship used for this mission was *Star of the West*, one of the original vessels built by Jeremiah Simonson for Vanderbilt's line to Nicaragua. When local installations in Charleston fired on the Union fort, the American states had started their long and destructive Civil War.

# X

# The War Years, 1861-1865

VIRTUALLY alone among the established shipyards along the East River, the yard of William H. Webb was not sitting idle when war broke out in April 1861. At the time two large hulls were under construction: the great *Constitution* for Pacific Mail and the smaller *Mississippi* for Mitchill's Savannah Line. Webb was also daily expecting confirmation from Count Camillo Cavour for the order to build two large frigates for Italy.

When the war first started, no one expected that the fighting would become serious or that the war would last very long. This, of course, was not to be the case. When the state of Virginia opted for secession on 17 April, the border of the Confederacy was now right across the Potomac from Washington. The subsequent panic over the immediate need to defend the capital turned the minor military skirmishes into a full-scale bloody war between the northern and southern sections of our nation. Although this war, even as it escalated, could not be characterized as a naval war, it nevertheless created an immediate and enormous demand for tonnage, thus bringing the American shipyards, temporarily at least, out of the slump of the late 1850s.

Much of the early action of the war took place either in eastern Virginia along the Potomac or in the Tidewater areas of Virginia, North and South Carolina, or Georgia. Since the obvious way to convey Union volunteers to these areas was by sea, hundreds of passenger and cargo vessels of all types and sizes, some of them Webb-built ships, were chartered by the War Department, usually for fairly short stints, for service as transports.

Also, Union strategists early sensed that the Confederacy, which produced mostly export crops, particularly cotton and tobacco, would be dependent on imports for both arms and food, and on exports for cash. Lincoln's generals, therefore, decided from the outset of the war, to cut Confederate contact with the sea by setting up a strong naval blockade around all Southern ports. Though sound and river steamers, however inappropriate, were often conscripted during this emergency for transports, the blockade, obviously, called for ocean-going vessels with more satisfactory sea-keeping qualities. Thus even more Webb-built ships were to be found in the blockade fleets than among the transports.

Then too, since the American Navy had received scant attention from our land-oriented population since the War of 1812, and since a fair share of the Navy that existed had been located in Southern ports when the conflict began, there was now a great demand for new naval vessels as well as for the immediate conversion of merchant ships for military service.

With this sudden demand for transports, blockade ships, and naval vessels, the American shipyards devoted to the building of wooden-hulled ships, so recently on the brink of obsolescence, were now unexpectedly revived to enjoy a short era of prosperity for the duration of the war. Not only were many of the older established shipyards resuscitated for this purpose but also many new yards appeared, especially in Philadelphia or New England, to profit from this demand for tonnage.

Although William Webb made many offers to produce ships for the Union cause, the fact that his yard was already fully occupied with contracts Webb had arranged during the Depression meant that most of the war contracts were taken up by yards in greater need of them. Also, with the building of *General Admiral*, and the subsequent eager interest in his designs by several European courts, Webb had established himself as a designer of warships with an international reputation. And, therefore, while he informed the Navy Department that he was willing to build military vessels of his own design for the Union cause, building ships to specifications presented by others was something he had never done and would not do.

A case in point arose when, early in the war, the Navy announced that it wanted Northern shipyards to submit bids for the building of several 3000-ton steam sloops. William Webb responded and was presented with the plans and specifications drawn up by the Navy Department. Webb's frustration and his unwillingness to produce a ship to a design he knew to be inferior can be seen in his response to the Secretary of the Navy:

... a vessel of 3000 T. has not sufficient capacity to accommodate the machinery and stow the coal required by the proposal to the best advantage; nor is a vessel of this tonnage with the above requirements susceptible of dimensions likely to secure the least draft of water or the greatest speed; but ... one increased in nominal tonnage—about 3500 tons, could be made to attain, with the above requirements, the best results without much additional cost to the department. (1)

From the tone of this letter one can more easily understand a statement printed many years after the war in the *New York Herald* that the Union Navy could have had the advantage of many more ships built by William H. Webb had the Navy Department not insisted on so many restrictions and specifications requiring supervision by department personnel (2).

Two bills passed by the Congress early in the war, which, with military events dominating the news, did not command much attention at the time, nevertheless were to have a profound effect on William Webb's shipbuilding operations in the long run. One of these bills was the Morrill Tariff, a high tariff on imports, passed in 1862. In the years before the Civil War, several Northern congressmen, wishing to provide protection for the new American industries then developing in the northern states, mostly in southern New England, had proposed raising the tariff to discourage competition from foreign

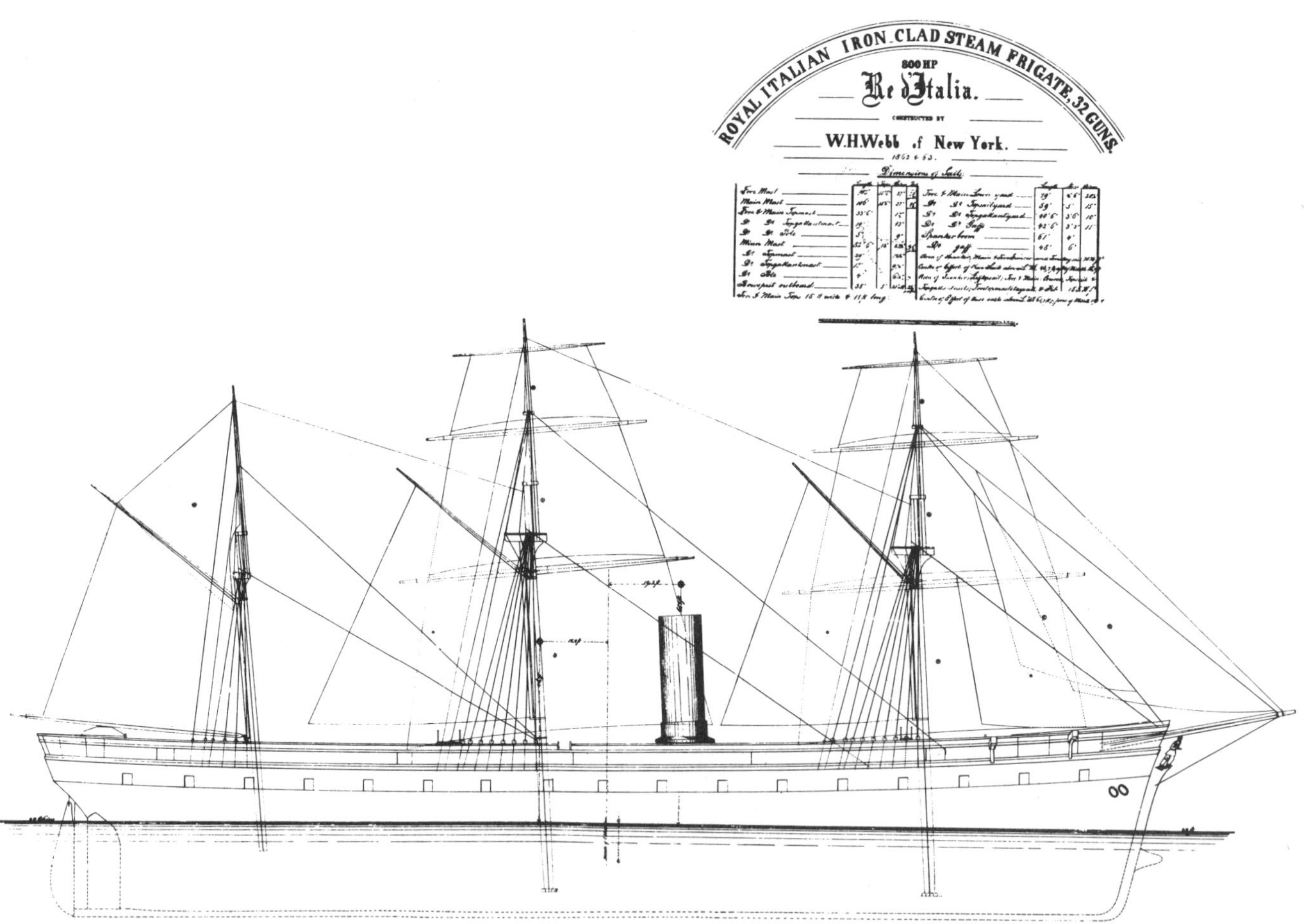

Profile of the steam frigate *Re d'Italia* from Mr. Webb's Book of Plates. This vessel and her sister, *Re Don Luigi di Portogallo*, took part in the Battle of Lissa in 1866, but the capability of the vessels under fire was not proven convincingly because of tactical errors on the part of ill-trained crews. *Re d'Italia* was rammed and sunk during the battle.

imports. But these proposals had been consistently blocked by Southern Senators, whose primarily agricultural constituencies depended on imported manufactures and who would suffer should European nations retaliate by placing high tariffs on American agricultural exports. When the Southern states seceded, of course, many of these agriculturally-oriented lawmakers left the Congress. Now firmly in control of the Congress were members of the newly-formed Republican party, whose commitment was not only to the prevention of the extension of slavery but also to the protection of the interests of northern business. With the Southerners out of the Congress, the northern Republicans easily passed the Morrill Tariff in 1862 raising import duties from 5 percent to 10 percent. Later revisions eventually raised the tariff to an astounding 47 percent by 1869.

While most Northern manufacturers approved these measures designed to give them a virtual monopoly of the American market, there was a negative side as far as the builders of both wooden and iron ships were concerned. This enormous 47-percent tariff added to the inflationary spiral of prices and wages that accompanied the conflict and continued after its end, which led to United States shipbuilders being noncompetitive when compared to their counterparts in other shipbuilding countries. The cost of building ships in America (and of operating American-flag ships) has remained prohibitively high in relation to the world market since the 1860s. This was, in William Webb's opinion, the primary cause for the demise of shipbuilding in New York in the years following the Civil War.

Another act passed by the Republican Congress, also in 1862, was the Pacific Railway Act. This act granted vast public lands in the still largely-unsettled areas of the Louisiana Purchase and the Mexican conquests to provide both for a right of way and for the financing of a transcontinental railroad. A second act in 1864 doubled the land grant to the railroad companies involved. Although no one questioned the advantages to our country of a transcontinental railroad, this railroad, once completed, would obviously render the Pacific Mail line—a company in which William Webb was a major stockholder and for which he built a great many large steamships—virtually obsolete.

The effects of these two acts of 1862, however, were not to be felt until many years after the war. For the time being, Webb's shipyard continued to be as busy as ever, especially building ships for Pacific Mail, which had not as yet begun to anticipate the extent of the impact the Pacific Railway bill would have on its operation.

William Webb was also actively devoting his services to the cause of the Union. Although in the first crucial years of the war the Union did not deign to order any vessels from his yard, Webb nevertheless gave freely of his time and talent throughout the war, travelling often over to the Brooklyn Navy Yard or to Washington to advise the government on the design of naval vessels, the reconstruction of merchant ships for military uses, or the building of wharves or maritime repair facilities. According to the *New York Herald*, "During the Civil War [Webb's] knowledge was of incalculable aid to the Navy Department" (3). An article in the *New York Times* was even more emphatic:

During the rebellion, no shipbuilder in the North was more active in assisting the Union cause. He gave the advantage of all his experience to the Secretary of the Navy and designed several of the first men-of-war built after Fort Sumter was fired upon. (4)

Among the first casualties of the war were the steamer lines connecting New York with various Southern ports, such as Mitchill's Line to Savannah, all of whose ships had been built by William H. Webb, or Spofford and Tileston's line to Charleston. Other lines, however, such as those running to Europe or South America, as well as Vanderbilt's Line to Panama, and therefore also Pacific Mail's leg of that route from Panama to San Francisco, continued sailings throughout the war. In fact, traffic on Pacific Mail flourished to such an extent that the line continued its program of renewing its fleet right through the war years. Thus, beginning with *Constitution*, which was under construction at his yard when the conflict began, William Webb's yard was kept busy throughout the war years building new steamships for Pacific Mail.

When *Constitution* was launched on 25 May 1861, she was the second largest commercial vessel ever built in America, at 342 feet 6 inches only a few feet short of the great *Adriatic*. As soon as Novelty had installed her walking-beam engine and she had completed her trials, *Constitution* was chartered to the War Department as a transport. With this charter bringing the company as much as $2000 per day, the stockholders of Pacific Mail cannot have suffered too greatly in being deprived of her services. The department returned the ship after one year, however, and in June 1862 *Constitution* at last sailed to the Pacific to begin her regular service between Panama and San Francisco.

The Italian frigate *Re Don Luigi di Portogallo* is shown just prior to her launching on 29 August 1863. This photograph is of great importance because it is the only known view of the Webb shipyard, which extended along the East River from Fifth to Seventh Streets. The frigate was built in the extreme northern part, next to the wooden fence that separated the Webb premises from those next door. The Pacific Mail steamer *Colorado* is in frame to the right, near the foot of Sixth Street.

By the time the hull of *Mississippi* was completed in July 1861, Mitchill's Line was no longer operating. This ship, therefore, was purchased on the stocks by the Navy for $200,000 and completed as a warship. Since *Mississippi* was not exactly an appropriate name for a Union naval vessel at this time, she was launched as USS *Connecticut*.

Meanwhile, William Webb waited for a confirmation of the orders for the two ships of war from the recently-constituted Kingdom of Italy. Unfortunately, however, the first news from Italy that summer was that Count Cavour had died unexpectedly on 5 June, before signing the contract with Webb for the two frigates. Now the matter had to go before the Italian parliament, where it faced a formidable challenge. During the fighting for a united Italy, British warships had provided the nationalist troops with protection and cover, if not actual support. Now that Italian unification had been achieved, British shipbuilders, backed by the considerable influence of their government, were at pains to persuade the inexperienced Italian parliament that any new Italian warships should be built in British yards. Not one to let an opportunity slip away so easily, in July 1861 William Webb again took ship for Europe to confront the Italian parliament in person. Unfortunately, no record exists to show just how he managed it, but Webb returned to New York in August with a signed contract to build the two Italian frigates.

Now that the Italian government had been won over, Webb next had to travel to Washington to get permission from the federal government to build warships for a foreign country at a time when our own nation was at war. This permission was duly granted, and on 21 November the keel of *Re d'Italia* was laid at Webb's yard, with the keel of *Re Don Luigi di Portogallo* following a few weeks later. One might note that these were the only new keels laid at the Webb yard through all of 1861, a year in which tonnage was in extreme demand.

On 9 March 1862, with the frames of the two Italian frigates just beginning to take shape, an event took place which led William H. Webb to re-evaluate his whole concept of warship design: namely, the battle in the Hampton Roads between the Confederate ship *Virginia* (formerly USS *Merrimack*) and the Union ship *Monitor*.* In this, the first battle in history between two ironclads, the result was a draw: neither ship could win; on the other hand, neither could lose. From this time William Webb realized that any new warship must be ironclad, and he so advised in the case of the Italian frigates then under construction. Once Webb had redesigned the Italian ships to include thick iron plating around their hulls, he found difficulty finding high-grade iron, shaped to his specifications, anywhere in America; so he had to have the iron plating he needed shipped from France.

Shortly after the *Monitor-Virginia* battle, William H. Webb received a communication from the Navy Department asking him to submit plans for a large ironclad warship. Actually, Webb had already begun to consider how he might design the "indestructible" ironclad war-

ship. Nevertheless, his yard already had as much work as it could handle, so Webb responded that he was not able to comply with the Navy's request. Another reason for Webb's refusal was that he had by this time already submitted several plans to the Navy, all of which had been rejected for one reason or another. Now busy with private contracts as well as with the Italian frigates, Webb was not willing to spend the time required to design a large ironclad, if the Navy would not accept his plans when he submitted them.

This time, however, the Navy persisted. The Department needed to develop a superior ironclad warship, and they knew that William H. Webb was the only American shipbuilder capable of designing or building the ship they needed. The second request, therefore, came in the form of a personal letter from Gideon Welles, the Secretary of the Navy. This time, Webb accepted.

As he had done a decade earlier when he first conceived the plans for *General Admiral*, William Webb retired to his drafting table to design what he believed would be the best possible warship, this time based on concepts he now believed necessary for a ship planned from the start as an ironclad. The design Webb produced was for a large, powerfully-armed ship, completely sheathed in iron. His plan also called for a "bomb-proof deck," consisting of heavy oak beams covered with a half-inch layer of iron plating, and over the plating a decking of pine five inches deep. At the bow of his ship Webb provided a sharp ram, fifty feet in length, of solid wood covered with thick iron plating. Webb also designed his warship with a double hull, an innovation at the time: an outer hull of live oak sheathed with iron and an inner hull, also caulked and waterproofed, so that, should the outer hull be punctured in the course of battle, the ship could still remain afloat. The space between the two hulls was arranged to serve for the storage of coal.

In May 1862 William Webb again travelled to Washington, this time to lay the plans of his proposed warship before the Navy Department with an offer to build it for the Union. He also carried with him a letter from Charles H. Marshall, who had been a leader in the organization of the Republican Party in New York, introducing Webb to the President, "His Excellency, Abraham Lincoln":

I take the liberty of introducing to your acquaintance my esteemed friend Mr. William H. Webb, a citizen of this city and one of the most eminent . . . naval architects that our country can boast of. . . .

Mr. Webb has, I believe, submitted some plans for vessels of war and harbor defenses and he now visits Washington to make

---

* On the same day as the *Monitor-Virginia* battle, though in a different arena of the war, Lieutenant Henry B. Hidden, the twenty-three year old son of Henrietta Webb's brother, was killed in action.

himself acquainted with the heads of respective departments, and to offer his services in the construction of any vessels which the government may require.

I add with great pleasure that Mr. Webb possesses great means and is a gentleman in whom the utmost reliance may be placed. . . .                                        Very respectfully
                                                    C. H. Marshall (5)

ɪᴛ ɪs not now known whether William Webb actually met Abraham Lincoln on this visit. But he did spend a great deal of time with Gideon Welles going over plans for the proposed warship. While Welles was himself enthusiastic about Webb's plans, there were many in the halls of power who were not convinced that a ship so laden with iron could float, let alone prove maneuverable in battle. Others balked at the enormous cost, and in the end the Navy Department rejected Webb's proposed ship.

Though William Webb returned to New York thoroughly convinced of the futility of trying to work with the Navy Department, he soon received another personal letter from Gideon Welles, stating that the Secretary had personally overridden the decision of the Navy Department and instructing Webb to begin building the ironclad warship immediately. This great ship, the keel for which was now laid down alongside the two Italian frigates, the ship in which, of all William Webb had constructed during his long career, he took the greatest pride, he now named himself: *Dunderberg* ("Thunder Mountain").

No sooner had construction on the great ironclad begun than engineers from the Navy Department turned up at the yard with lists of specifications to be met and modifications to be made. William Webb, who knew that his original design was correct, responded by stopping work on the ship. The deadlock once again was resolved by Gideon Welles who issued an order stating that Navy Department officials should allow Webb to build the warship according to his own specifications (6).

The Pacific Mail sidewheeler *Colorado* as she appeared when completed by William H. Webb in March 1865. Built for the company's coastwise line, she was fitted with a mizzen mast and her overhanging guards removed when she inaugurated the transpacific service in 1867. The paddleboxes of many sidewheel steamships were ornately decorated. Those of the Pacific Mail were no exception—with scrollwork surrounding a central seal and the letters "P.M.S.S.Co." Compare this lithograph with the photograph of *Golden City*. (Courtesy of The Mariners' Museum, Newport News, Virginia)

About this time, with three naval vessels already under construction at his yard, William Webb received orders for two more large steamships for Pacific Mail, both to be on a scale similar to that of the recently-completed *Constitution*. One, to be named *Golden City* (ordered as a replacement for the beautiful Webb-built *Golden Gate*, which had burned at sea off the coast of Mexico on 27 July 1862), was built by Webb himself, making this the fourth large hull then under construction at the yard. The second, *Sacramento*, was subcontracted to Webb and Bell over in Greenpoint.

The magnificent *Golden City* was launched on 24 January 1863. Over 340 feet in length and measuring 3373 tons, *Golden City* was as large as *Constitution* and thus also one of the largest ships in the world. One author described her as, "One of the finest wooden-hulled vessels ever built in America. . . .a faultless model and very fast" (6). She apparently outdid even her well-appointed predecessors in interior elegance, boasting fifty-six staterooms, many with twin beds or four-posters rather than berths, as well as gilded mirrors and silk hangings in her public rooms. Although *Golden City* was similar to *Constitution* in size and accommodations, she was structurally quite different and in exterior lines more nearly resembled the larger steamships built later for Pacific Mail. Whereas *Constitution*, like some of the line's earlier vessels, had two tall thin stacks, one ahead and one aft of the side-wheels and walking beam, *Golden City* carried but one thick stack, nicely raked, quite far forward. And, while *Constitution*, like most of the more recently-built vessels of the line, had her cabin deck (called Main Deck) covered only toward the stern, with a railed sun deck above it, *Golden City* was the first to have her Main Deck covered all the way to the bow, a practice followed on most subsequent steamers built for Pacific Mail.

By 18 April 1863 *Re d'Italia* was also ready for launching. At 285 feet she was somewhat shorter than *General Admiral*, though about as heavily armed, and redesigned now to carry iron plating from six feet below the waterline to a point nine feet above, i.e., just short of the gun-ports. The lower strakes of this sheathing were to be put in place when the ship went to the Novelty Works for her engine installation; the upper strakes, however, were to be carried as ballast and fastened into place after she arrived in Italy. As the usual crowd of over 10,000 began wandering into the Webb shipyard on the day of the launch, early arrivals were invited to inspect the hull of the great 378-foot *Dunderberg* taking shape alongside *Re d'Italia*.

At 9:50 A.M. yardmen were detailed to start knocking out the chocks holding *Re d'Italia* in place. Then exactly at 10:00 William Webb gave the signal, and the wife of the Italian consul in New York, who was aboard the ship rather than on a platform, leaned over the bow and christened the frigate with a foaming spray of Asti Spumante as the vessel slid slowly down into the river.

Even before *Golden City* was completed, Pacific Mail had ordered another similar vessel to be built by William H. Webb. This was *Colorado*, the keel for which was laid at the yard on 6 June 1863. No sooner had Webb started work on *Colorado*, however, than Pacific Mail, pleased with the performance of *Golden City*, was back with an order for three more ships of the same type as *Golden City* and *Colorado*, though, with a length of about 320 feet, somewhat smaller. By this time, with *Re Don Luigi di Portogallo*, *Colorado*, and *Dunderberg*, all then under construction in his yard, William Webb felt he could take on only one of these contracts. One, therefore, *Arizona*, was awarded to Henry Steers. A second, *Montana*, was subcontracted to Webb and Bell. The third, *Henry Chauncey*, was built by William H. Webb, though her keel could not be laid until October, after *Re Don Luigi di Portogallo* had cleared the stocks.

Preparations for the launch of *Re Don Luigi di Portogallo* occupied the Webb yard in the early days of July 1863, at the time when news was reaching New York of the costly Union victories at Vicksburg and Gettysburg. But when the draft riots exploded in New York on 13 July, literally turning the lower parts of the city into another Civil War battlefield for several days, Webb wisely postponed the launch lest the great crowds assembled at his yard erupt into violence.

One of the results of the New York draft riots of July 1863 was the founding of the Union League Club of New York, an organization in which William Webb was to become an active member. During the course of the Civil War there were many influential people in the North who did not see either the abolition of slavery or the preservation of the Union as causes for which such great sacrifices needed to be made. To oppose these anti-Union sympathies, which some saw as seditious, other Northern businessmen who were staunch supporters of Lincoln and of his policies, formed organizations called Union League clubs in many American cities. Although William Webb is not listed among the charter members, several of his friends, such as George Griswold, Robert Minturn, and Charles H. Marshall, all merchant shippers close to Webb socially as well as in business, were among the original

[From a Lithograph by Endicott and Company.]

THE " DUNDERBERG."

This line engraving from Harper's New Monthly Magazine of July 1882 was taken from an Endicott lithograph. When *Dunderberg* was delivered to the French Navy in 1867, Mr. and Mrs. Webb and their two sons were aboard for the transatlantic trip to Cherbourg.

organizers of the League in New York in 1863. Webb himself seems to have joined later, probably in 1865, the year his friend Charles H. Marshall served as the League's third president.

As a member of the New York Union League, William Webb would have come to know other members such as George Bancroft, then America's leading historian; William Cullen Bryant, the noted poet and newspaper editor; James Roosevelt, father of Franklin D. Roosevelt; or, after 1873, the American financier, J. Pierpont Morgan.

By the late summer of 1863 the series of Union victories had somewhat restored general confidence in the Union cause and helped quiet the violence of the July draft riots. Thus Webb was able to schedule the launch of the second Italian frigate, *Re Don Luigi di Portogallo*, which turned out to be a quiet affair by the usual standards, for 29 August.

On 12 November *Re d'Italia* finished receiving her engines and the lower strakes of iron plating at the Novelty Iron Works and was ready for her first trials. With William Webb aboard and Joseph Comstock in command of an otherwise Italian crew, the frigate made a fast run down the bay and several miles into the ocean, returning that evening to the Bremen Docks in Hoboken, where she remained several weeks having her cabin work completed. It was here at the Bremen Docks that William H. Webb hosted a gala reception aboard the vessel on the evening of 2 December. Guests received embossed invitations announcing that their company was requested aboard the "Royal Italian Frigate Re d'Italia" from 3:00 to 9:00 P.M., with a supper to be served at seven and Dodsworth's Band again dutifully supplying the music.

A shakedown cruise for *Re d'Italia* was scheduled for the day after Christmas. Actually Comstock sailed her out of the Hoboken pier late Christmas night and anchored off Ellis Island so the ship would be positioned for an early start the following morning, a Saturday. About midnight William Webb and a group of American naval officers invited to observe the trials went out to the ship aboard a harbor tug. On board tea and snacks were served, so that the party did not get to bed before 2:00 A.M. Nevertheless, the entire complement was awakened

A model of the ironclad ram *Dunderberg* forms the centerpiece of a collection of Webb artifacts and memorabilia owned by Webb Institute. Many consider *Dunderberg* to be the famous shipbuilder's most important construction. She never served the United States Navy, to which she was originally to have been delivered, and, as far as is know, she never fired a gun in her unfortunately short French naval career as *Rochambeau*.

to a drumroll at 6:00 A.M. (well before dawn at this chilly time of year), and at the first sign of sunlight, *Re d'Italia* lifted anchor and headed down the bay toward the ocean.

Although Webb-built ships had a habit of performing well on trials, *Re d'Italia* had other ideas, and it was not long before William Webb probably wished that he had not invited so many observers or even that he himself had stayed home to enjoy a longer Christmas with his family. First, as soon as she reached the open ocean, *Re d'Italia* developed an uneasy roll, which was only accentuated when Comstock had to negotiate several rapid turns to avoid large ice floes in her path. After about an hour of these maneuvers, the Italian pilots announced that they would go no farther, so Comstock brought the ship about and sailed her back into the Lower Bay where she anchored for the night.

On Sunday morning the 27th, *Re d'Italia* again sailed out into the Atlantic. This time, although a cold drizzle kept spirits dampened most of the day, the sea was calmer and Comstock was able to get *Re d'Italia* cruising at almost twelve knots, her full anticipated speed. But the ship had been underway less than an hour when suddenly a deafening crash emanated from below, after which the engines stopped altogether. An hour later, by which time the after engine had been partly dismantled, it turned out that, as with *Panama* fourteen years before, someone at the engine works had left a small piece of wood in the steam line. Once this wood had worked its way into the cylinder of the after engine, it had sheared off parts of the valves, which in turn had severed two bolts connecting the piston to the piston rod. In short, the after engine was a mess.

So the great *Re d'Italia*, with William Webb aboard, returned to New York from her trials under sail. Or at least she headed toward New York. By 2:00 A.M. on Monday morning the crew managed to get the forward engine turning and the ship was able to move again under power though only at around six knots. Then about 9:00 in the morning, with the ship still well down the New Jersey coast, a heavy northeaster blew up, so that *Re d'Italia* could make no headway at all. The best she could do was to head into the wind and keep the one good engine turning, just to stay in place for the next twelve hours.

By evening the wind had shifted to the south and, with sails once more unfurled, *Re d'Italia* again attempted to reach New York. But by midnight she was enveloped in a thick bank of fog. There was now apparently some disagreement between Captain Comstock and the Italian officers about which direction the ship should be taking in the fog, which Comstock, it appears, resolved by turning the ship over to the Italians so he could get some overdue rest. Moments later the matter of direction ceased to be an issue when *Re d'Italia* crunched hard aground on a sand bar, which by the welcome dawn of Tuesday morning turned out to be about two hundred yards off Long Branch, New Jersey, or about twenty miles south of the entrance to the Lower Bay. And here they all sat for two days before tugs came down from New York to tow them back into port.

By 11 February *Re d'Italia*'s engine had been repaired and a trial trip as far as Norfolk, Virginia (now back in Union hands), was scheduled. She sailed out of the harbor on 17 February, once again with William Webb aboard and Joseph Comstock in command. And once again *Re d'Italia* rolled and once again there were ice floes in her path, but this time Comstock kept her moving southward, now easily exceeding twelve knots. The ship had sailed only about one hundred miles, however, and was somewhere off Cape May when the Italian crew announced that the ship did not need any further trial and that they were ready to go back. One suspects that by this time neither Webb nor Comstock was prepared to argue, so early in the evening of 17 February, *Re d'Italia* returned to New York harbor, this time, at least, under her own steam (7).

Finally on 9 March 1864, with Comstock still in command, but Webb now safely ashore, *Re d'Italia* set sail for what was to be her homeland, the first ironclad ship of any nation to cross the Atlantic. Her roll persisted, according to Comstock's later report, but it was not se-

rious. She reached Naples on 2 April, after lengthy stops at Madeira and Gibraltar. On arrival, the already well-tried crew was sent back to America aboard another ship to fetch the second frigate.

With considerably less ado, probably because of the choice of a kinder season, *Re Don Luigi di Portogallo* completed her trials with a successful three-day cruise late in August and then early in September followed her sistership across the ocean to Italy. An article in the *New York Herald* noted that, after all the work, William Webb had probably lost money on the Italian frigates. His estimate of $150,000 each to Cavour was made in 1861. But during the American Civil War, with the rampant inflation, increased costs of materials, and rising wages, the two ships probably cost him at least $200,000 apiece to build.

By the time the two Italian frigates had been completed and safely delivered, Webb's latest ship for Pacific Mail, *Colorado*, one of the finest ships ever built at the yard, was nearing completion. At 340 feet, *Colorado* was about the same size as *Constitution* or *Golden City*, the two largest Pacific Mail ships. In profile, she closely resembled the more recently completed *Golden City*, for, like this steamship, she had but one raked stack placed forward. Additionally, like *Golden City*, her Main Deck was covered to the bow, though on *Colorado* the covered area toward the bow was also enclosed, whereas on *Golden City* it was open.

*Colorado* was launched on 21 May 1864, at the time when General Grant was wasting the lives of thousands of Union soldiers in the Wilderness Campaign. But installing the engines and finishing the cabin work took nearly a year, so that *Colorado* was not ready for her long voyage to the West Coast until April 1865, by which time her total cost, given the wartime inflation, had escalated to $1,000,000.

The launch of *Henry Chauncey* followed in October 1864, and now the only hull left in the yard of William H. Webb was *Dunderberg*, the ship which had always been Webb's first priority and which now received his full attention. By the Spring of 1865, Webb's great ship, the largest, fastest, and most powerful ship of war in the world, was nearing completion, and the time was approaching to set the date for her launch. Then on 9 April 1865 General Robert E. Lee surrendered to Ulysses S. Grant at Appomattox Court House. America's deadliest war was mercifully over. But also, the victorious Union had no further use for Webb's *Dunderberg*.

The steamship *Montana* was launched in February 1865 by Eckford Webb as a sub-contractor to his brother William. She is shown in the wooden drydock at Hunter's Point, San Francisco. The deck arrangement of this enormous sidewheeler, similar to many others in the Pacific Mail fleet, can clearly be seen in this photograph. (Courtesy of The National Maritime Museum, San Francisco)

# XI

## *Bristol*, *Providence*, and *China*, 1865-1867

THE end of the Civil War in America was also the end of the reprieve the war had brought to American shipyards. With so much of the former American merchant marine appropriated for military purposes, a large share of American overseas trade passed into the hands of enterprising British companies during the war. And now that America had abandoned its policy of subsidizing its merchant marine, and that our government, and American capital as well, were dedicated to internal continental development and to building railroads, American shippers were no longer in a strong position to compete with British or other foreign shipping companies which were supported by subsidies from their governments.

At the same time, with the wildly spiraling rises in wages and costs of materials that accompanied the war, plus the near doubling of the prices of necessary imports as a result of the tariff, shipbuilding in America was becoming increasingly unprofitable. Also, since American yards had been producing to capacity for four years to meet the demands of a nation at war, the end of hostilities left a glut of idle tonnage in American harbors and unfinished hulls in American shipyards. Then during the spring and summer of 1865, the Navy put 219 surplus steamships up for auction and the War Department, another 140. Since most of these ships were sold at scandalously low prices, there was obviously little call for new shipbuilding for some time after the war.

When the war ended, William H. Webb had only one hull still on the stocks in his yard, the great ironclad *Dunderberg*. With the government now no longer interested in acquiring powerful—and expensive—new warships, the Navy informed Webb that it intended neither to take possession of *Dunderberg* nor to make the payments that were to be due at her completion. Nor could the American government countenance Webb's disposing of the largest and most powerful warship in the world to any foreign government. In short, the ship William Webb considered the pride of his career had become simply surplus. However, since work on the great ironclad was already winding up when the war ended, Webb completed

*Dunderberg* at his own expense and set the time of her launch for Saturday, 22 July 1865, at 9:00 A.M.

The launch of this large and strange-looking craft drew a crowd of about 20,000 spectators. People began arriving about 7:30 in the morning, and by 8:30 there were crowds not only in the yard but also on neighboring wharves and steamers, on nearby housetops, and even on small boats out in the river. Webb had issued 2000 tickets for special guests to be aboard *Dunderberg* for the launch, and, according to the *New York Evening Post*, on the morning of the launch these tickets were selling at scalper's prices (1). As usual Dodsworth's band was on hand to strike up the proper martial music for the occasion.

Not a few of the spectators had come because they wanted to watch this great iron ship sink to the bottom as soon as it reached the river. Even experienced yard hands were certain her enormous bulk would draw far more water than anticipated and create something like a small tidal wave as she lowered herself into the river.

At exactly 9:00 William Webb gave the signal for the chocks to be removed and, slowly at first, so slowly in fact that those on board did not realize for some time that the ship was moving, the great 378-foot *Dunderberg* slid toward the river. Again, the *Post* describes the event:

The Dunderberg entered the water in a manner surprising everyone conversant with launches. She cut the water like a knife, and did not raise a swell of a foot in height. This fact proves how beautiful are her lines and faultless her form . . . the subject of common remark among experienced ship-builders and ship-owners who know the importance of this demonstration.

The effect was beautiful in the extreme and attracted much attention. No ship ever entered the water with more ease and grace, and amid such thundering plaudits, and no man could receive more commendation than did Mr. Webb on this occasion. (2)

Once *Dunderberg* had been launched on 22 July, there were no ships at all building in Webb's yard. Finally, after his yard had remained idle all summer, on 5 October 1865, William Webb signed a contract with the Merchants' Steamship Company to build two large overnight steamboats for service on Long Island Sound between

99

New York and Bristol, Rhode Island. The sound or river steamboat, designed to run from New York up the Hudson or through Long Island Sound to some port in southern New England, or from Boston up to the coast of Maine, was in the Nineteenth Century one of the most popular forms of transportation in America. Literally hundreds of these vessels, some of them small excursion steamers but many as large and as commodious as any of the ocean-going steamships of the Collins Line or Pacific Mail, although of a very different type, were built in the shipyards of New York in the years before the Civil War. Some yards, such as Bishop and Simonson or Smith and Dimon, did a very good business specializing in this type of vessel. But William H. Webb, although known for the variety of ships he was able to produce, had not, as of 1865, ever built a sound or river steamboat. When he did, however, the sisterships *Bristol* and *Providence*, which he designed and built for the Merchants' Steamship Company in the winter of 1865-1866, were to be the largest, fastest, most elegant, and certainly the handsomest steamships on Long Island Sound. Aside from *St. John*, a 400-foot steamer recently completed for the Peoples' Line (which, as a New York to Albany nightboat, operating exclusively on the placid Hudson River, did not need the strength of hull required for a Long Island Sound steamer), *Bristol* and *Providence* were also the largest steamers of the type in America.

There were at that time four major overnight steamer lines operating out of New York through Long Island Sound to ports in southern New England: to Providence and Newport in Rhode Island, and to Stonington and New London in Connecticut. Each was a part of a combination water and rail route between New York and Boston. Although there was a direct freight line between New York and Boston, the long ocean voyage around Cape Cod made a direct passenger service between these two cities impractical. Instead, passengers boarded one of the large steamboats which left New York in the early evening and sailed overnight through Long Island Sound to one of these New England ports where, early in the morning, they transferred to a train for the rest of the relatively short distance to Boston. By far the most popular and prestigious line, and the one with the largest and grandest steamers, was the famed Fall River Line, which was still generally known by that name although it had in 1863 moved its eastern terminus from Fall River to Newport.

In May 1865 the Neptune Line to Providence, which had managed to amass a large amount of capital during the war years, both selling and chartering steamers to the Union, and the older and more established Stonington Line, decided to join in a merger for the express purpose of driving the Fall River Line out of business. With this object in mind, the new combined organization, known as the Merchants' Steamship Company, inaugurated yet another steamer and rail route through the Sound between New York and Boston, with Bristol, Rhode Island, as the port where passengers transferred from the steamer to the railroad. Since Bristol was not much farther from New York by water than Newport, but far closer to Boston by rail, passengers travelling on the Bristol Line could arrive in Boston about one hour sooner than those going by way of Newport on the Fall River Line.

With the considerable capital available to this company, its backers were prepared to spare no expense to provide their new Bristol Line with two overnight steamboats which would be not only the largest and fastest but also the most elegantly appointed steamers on the Sound, far grander even than the vaunted new steamers of the Fall River Line. To produce these vessels the Merchants' Steamship Company turned to William H. Webb. In the agreement Webb became the primary contractor with responsibility for producing the twin hulls and parts of the superstructures, as well as for delivering the completed steamers by a given date. The engines, which were to be the largest simple walking-beam engines built up to that time, were subcontracted to John Roach. The total cost of each of these identical steamers was $1,200,000, an exorbitant figure, even taking the Civil War inflation into account.

The design for *Bristol* and *Providence* produced by Webb was truly remarkable, especially considering that he had never before designed a ship of this type. With a length of 360 feet on the waterline and 374 feet overall, and a breadth of 48 feet (dimensions almost the same as those of *Dunderberg* and greater in length than any of the ships built by Webb to date for Pacific Mail), they were to be the largest wooden hulls ever built for Long Island Sound service, and, aside from *St. John*, as already noted, the largest wooden hulls produced for any overnight coastal or inland service. As was the case with most of the larger Sound or river steamboats, these long hulls were reinforced with huge arched trusses known as hogging frames, or, more commonly, hog frames.

The hulls and Main Decks of *Bristol* and *Providence* followed a pattern typical of sound and river steamers. In the hull the crew's quarters were forward, the engine and sidewheel shafts amidships, and a galley, dining sa-

*Bristol* is here depicted on what must have been an excursion charter, with flags flying from every staff and a large crowd of passengers covering her open decks. Three daring men are atop her port paddlebox. The heavy trusswork that is visible above her boat deck is the hog frame, which provided longitudinal stiffness to the shallow hulls of wooden steamboats. (Collection of Edwin L. Dunbaugh)

loon, and a room with free berths (for passengers not wishing to pay extra for a stateroom, which on overnight steamers was not included in the fare) aft of the engine room. The Main Deck, as on all steamers of this type, was cantilevered out from the hull to accommodate the sidewheels and was thus several feet broader than the hull proper. On this deck, a large area forward was devoted to cargo space, while a shorter section aft, known as the Quarter Deck, contained an entrance hall and a few staterooms. And, again, as on most Sound steamers, the deck above, known as the Saloon Deck, was devoted entirely to passengers, with a double row of staterooms on either side of a long central saloon, broken amidships by the housing for the walking-beam. The inner row of staterooms on each side opened onto the saloon, the outer row onto the open outside deck.

From here on Webb's design for *Bristol* and *Providence* showed the boldness of concept for which Webb was famous and provided the model on which virtually all later steamers on Long Island Sound, or even steamers of a similar type on Chesapeake Bay or elsewhere, were to be based. Although three of the larger Hudson River nightboats, plus one then building, carried a second full passenger deck (called the Gallery Deck) above the Saloon Deck, it had never been considered safe to do so with a steamer operating on the rougher waters of Long Island Sound, where the weight of an extra deck could prove too great a strain on a wooden hull as well as dangerously topheavy in a stormy sea. But William Webb, who had built many staunch hulls designed to hold up in any sea and in any part of the world, knew that his hulls were strong enough to carry an extra deck of staterooms and still maintain stability. Furthermore, with the big hog frames and stays strung from a series of upright stanchions, Webb considered the superstructures strong enough to omit transverse deck beams under much of the after section of the Gallery Deck, permitting an open balustraded well of about one hundred feet in length between the two stateroom decks. Thus *Bristol* and *Providence* each had a Grand Saloon two decks high, furnished as elegantly as any fine hotel, with patterned carpets, plush-covered chairs and settees, and, suspended from the dome above, three great gaslight chandeliers. A similar but shorter open well was located on the Gallery Deck forward of the engine casing.

At this time all of the larger Sound and river steamboats carried their boilers, with their towering smokestacks above them, on their cantilevered main decks outboard of the hull, a vestige of an earlier day when boilers tended to explode. But the heavy iron boilers being produced by 1865 were far more reliable. So, following the practice used on his ocean-going steamships, Webb placed the boilers of the Bristol Line steamers in the hold, with their two smokestacks athwartship rising through the rows of staterooms forward of the walking-beam. This arrangement not only helped make *Bristol* and *Providence* by far the handsomest steamships on the Sound, it also helped to lower their center of gravity and thus offset the increased weight of the second tier of staterooms they carried.

With no other work in his yard at the time, William Webb completed the hulls (with hog frames in place) and Main Decks of *Bristol* and *Providence* in record time: the keels were laid in October 1865, and the two hulls were launched within a few days of each other in April 1866. They were then towed up to John Roach's yard to have their engines installed. But before Roach had finished his part of the work or William Webb had collected his final payment, a series of accidents had driven the Merchants' Steamship Company into bankruptcy, leaving Webb with two more large vessels uncollected by the people who had ordered them. On 29 December 1865, while *Bristol* and *Providence* were still on the stocks at Webb's yard, the Stonington Line's finest steamer, *Commonwealth*, caught fire at the company's wharf in Groton, Connecticut, and burned beyond repair. As the wharf was also consumed by the fire and neither the steamer nor the wharf was covered by insurance, the combined loss to the Merchants' Steamship Company amounted to more than a million dollars.

After the loss of *Commonwealth*, *Commodore*, an inferior steamer maintained as a spare, was placed on the run with the line's other vessel, *Plymouth Rock*. Then on 17 January 1866, just three weeks after *Commonwealth* burned, *Plymouth Rock*, working her way toward New York in a blinding snowstorm, went aground near Greenwich, Connecticut. She was eventually repaired and returned to service, but this second disaster, so soon after the loss of *Commonwealth*, proved costly to the company both in cash and in public confidence. On 27 December 1866, just short of a year after the *Commonwealth* fire, when *Commodore* was blown ashore in a gale at Horton's Point, Long Island, and her hull damaged beyond salvaging, the Merchants' Steamship Company, financially exhausted, declared bankruptcy and went out of business.

*Bristol* and *Providence*, now orphaned, were then at John Roach's yard having their engines installed. Since the contract with the Merchants' Steamship Company was with William Webb, with Roach as a subcontractor, Webb was obliged to pay Roach for his share of the work whether or not he was compensated by the bankrupt steamship company. It appears, however, that Webb had already received most of his payments, though not all, from the company before it went under, so that, while he did not make his full profit on the two Sound steamers, he also did not stand to lose very much. In this he was fortunate, for, after the obligations had been met, probably including some payments to Webb, the recently-optimistic stockholders of the company received only three cents for each dollar invested.

Fortunately, just at this time an angel appeared in the unlikely guise of the portly and flamboyant James Fisk, Jr., a former New England peddler, whose profitable Civil War contracts and subsequent stock market manipulations had recently rendered him a millionaire. When Fisk heard that *Bristol* and *Providence*, the two largest and finest steamboats ever built for Long Island Sound service, were available at a bargain price, he hastened to Boston where he found the financial backing to form the Narragansett Steamship Company. Fisk himself became president, with the equally notorious Jay Gould, of Erie Railroad fame, second in command. The Narragansett Steamship Company thereupon purchased *Bristol* and *Providence* from the receivers for a reported $350,000 (3).

Part of the final payment to William Webb before the steamers were released to Fisk was in the form of Narragansett Steamship Company stock, which in this case turned out to be fortunate, for the company was successful from the start, though it strikes an odd note to see William H. Webb, a man known as a model of integrity, involved in business with two such infamous financial finaglers as Jim Fisk and Jay Gould.

As soon as the steamers were completed in June 1867, Fisk began running them between New York and Bristol, Rhode Island, as the original company had planned to do. And, even though the Stonington Line, now defunct,

The lavish interiors of the Long Island Sound steamboats *Bristol* and *Providence*, built in 1867, can only be described as pure Victoriana. It is said that the notorious and flamboyant Jim Fisk, under whose watchful eye the vessels were completed, roamed their saloons and decks wearing the full—if somewhat ersatz—regalia of an Admiral. (Courtesy of The Mariners' Museum, Newport News, Virginia)

was no longer part of the plot, the new Bristol Line was soon successful in bringing the great Fall River Line to its knees. The ultimate denouement, however, was a compromise. In 1869 the Fall River backers of the older company agreed to sell their line to Fisk if he would agree to return the eastern terminus to Fall River, rather than run to either Bristol or Newport. So, while the original corporate entity of the famous old Fall River Line had ceased to exist, the spirit as well as the name of the line continued when the Narragansett Steamship company, in which William Webb was now a stockholder, began running *Bristol* and *Providence* to Fall River, with stops at Newport in either direction. With the addition of *Bristol* and *Providence* to its fleet, the Fall River Line again easily resumed its place as the primary steamer and rail route between New York and Boston, a position it maintained well into the Twentieth Century. And every subsequent steamer of this line, all of which were designed either by

George Peirce or by J. Howland Gardner, were but further developments of the basic design provided by William H. Webb with *Bristol* and *Providence*.

By the Fall of 1865, when the hulls of *Bristol* and *Providence* were beginning to take shape in Webb's yard, there was so little demand for new tonnage that most of the other yards along the East River had become completely idle. William Webb himself had no assurance of any further business once the two Sound steamers had been launched. But Webb was to get one more major contract before the close of the year, when Pacific Mail asked him to build an ocean-going steamship larger than any he had built before for a new route the company was planning to inaugurate across the Pacific between San Francisco and the Orient.

When the Civil War ended and work began on the transcontinental railroad financed by federal land grants, it became apparent to Pacific Mail, as well as to other companies operating shipping lines between New York and California by way of Central America, that this railroad would significantly curtail their profitable operations. Thus Pacific Mail had responded enthusiastically when the government—aware that the new transcontinental railroad would provide a faster and more efficient route, not only to California but also to the Far East as well—had advertised an annual subsidy of $500,000 to a company that would provide regular monthly steamship service carrying mails from San Francisco to Japan and China. Since Pacific Mail, the only company with adequate facilities for steamship operations on the West Coast, submitted the most reasonable bid, this line, as expected, was awarded this lucrative mail contract in August 1865. The agreement signed subsequently stipulated that Pacific Mail was obliged to begin monthly sailings from San Francisco to Yokohama and Hong Kong no later than 1 January 1867 (4).

With this date less than a year away at the time the agreement was signed, Pacific Mail immediately solicited contracts for the construction of two large steamships to serve on the new route. These ships, to be named *Celestial Empire* and *Great Republic* in honor of the two nations they were to serve on either side of the Pacific, were to be similar in general plan to *Golden City* or *Colorado*, recently completed by William H. Webb, but even larger, considering the longer voyage and the enormous amounts of coal they would need to carry. To assure faster delivery, the contracts were assigned to two separate shipyards: *Celestial Empire* was awarded to William H. Webb and *Great Republic* to Henry Steers over at Greenpoint, one of the few surviving yards other than Webb's equipped to build such a large steamship.

It would appear that Webb and Steers must have consulted with each other to some extent during the construction of these two ships, for, while *Celestial Empire* and *Great Republic* were not in any sense identical, the completed ships were certainly similar in size, appearance,

This classic lithograph shows Pacific Mail's massive sidewheeler *China*, with foresail being unfurled and nearly every sail set, flying flags and pennants galore. Lithographs such as this were produced as advertising media, and the artist's impression of the vessel depicted did not necessarily represent real life. (Courtesy of The Peabody Museum of Salem)

and layout. At 376 feet overall, these two ships and the two similar steamships that followed (both built by Henry Steers), were to be the largest wooden-hulled steamships (other than *Dunderberg*) ever built in America. With these ships, the capacity of a wooden hull, carrying heavy machinery, to sail safely on a transoceanic run had been stretched to its greatest limit. The hull of Steers' *Great Republic*, in fact, showed such strain after only three years on the trans-Pacific run that she was reassigned to a safer coastal route (5), though Webb's ship remained on the China run without incident until replaced by more modern vessels twelve years later.

Soon after William Webb and Henry Steers began construction of these two large steamships in the early spring of 1866, Pacific Mail decided to order two more vessels of similar design. When the names *America* and *Japan* were chosen for the second pair, Webb's *Celestial Empire*, to conform, was renamed *China*. *Great Republic*, however, retained its original name.

*China* was not only the largest, but also in every way the finest—and, as it turned out, the last—passenger steamship ever built at William Webb's yard. She differed from the earlier Pacific Mail steamships built there in that, like *Dunderberg*, she was built with a double hull (up to the wale) with the frames of both hulls braced by diagonally-laid iron straps. Her single, vertical walking-beam engine of 1500 H.P., provided by Novelty Iron Works, was designed more for dependability than for speed. A more powerful engine might have reduced the long three-week passage by a few days at best, but it would also have consumed far more coal, thus using valuable revenue-producing cargo space.

The first- and second-class cabin quarters, together accommodating about 500 passengers, were designed to provide the comfort and elegance passengers of means had come to expect on a long ocean voyage. Many of the staterooms were fitted out as private sitting rooms, while the deeply-carpeted public rooms reflected all of the ornate elegance of this opulent era. The steerage dormitories below decks, however—designed to carry another 800 to 1000 passengers, usually Chinese workers travelling to America, or, to a lesser extent, returning home—presented a considerably more austere atmosphere.

Soon after the keel for *China* had been laid, construction was halted for several weeks when a group of ship's carpenters and caulkers went on strike in late April. Their demands included higher pay per hour and an eight-hour working day. Most of the New York shipbuilders were prepared to seek some sort of compromise, not because they agreed with the yard workers' demands, but because their contracts, which at best were hard to come by these days, usually provided for financial penalties if their ships were not delivered by the date specified. Some also expressed concern for the families of workers who were not involved, but whose pay had been stopped because the strikers had kept them from going to work (6).

William Webb was willing to consider some compromise on the matter of increased pay, though he did not agree with the demand, stating that if workers received higher pay, they would only be tempted to take extra time off. But he was adamant in refusing to consider a reduction of the work day from ten to eight hours, on the grounds that it would then be impossible for shipbuilders to complete vessels under construction within the contracted times. Webb, therefore, was active in organizing other shipbuilders and persuading them not to deal with the strikers. He went so far as to offer financial help in the form of short-term loans to shipbuilders who might otherwise not be in a position to hold out against the strikers.

By late May, the strike had been settled by offering the caulkers and carpenters a pay increase of twenty-five cents an hour. But William Webb won his point, and the work day remained at ten hours.

When *China* was finally launched on 8 December 1866, to the muffled applause of a thoroughly rain-soaked audience, William H. Webb probably did not suspect that hers would be the last launch of a major passenger vessel from his yard. But when the large and graceful hull of *China* was towed away in the rain to the shops of the Novelty Iron Works, there were no other ships standing in the stocks in William Webb's yard, and no other contracts waiting to be filled.

Rare indeed are photographs of early steamships on floating drydocks. Here, Pacific Mail's magnificent *China* has just received her copper sheathing in the sectional dock in New York's East River. The date is May 1867. The assemblage of ladies and gentlemen on the dock floor suggests that the occasion was important. Could the man in top hat standing at the vessel's forefoot be Mr. Webb? (Courtesy of The National Maritime Museum, San Francisco)

# XII

# The North American Steamship Company, 1867-1868

DURING the summer of 1866, while *China*, probably Pacific Mail's most successful steamship, was being built at his shipyard, William H. Webb was among a group of the company's officials and stockholders, including William H. Aspinwall, its founder and former president, planning to start a separate competing steamship line of their own operating on both coasts. More specifically, this group purchased a steamship company already in existence, the Central American Transit Company on the Atlantic side, and its subsidiary, the People's Opposition Line on the West Coast, which had been operating inferior steamers between New York and San Francisco, transferring passengers and cargo via Vanderbilt's former route through Nicaragua rather than through Panama. Once purchased, the entire operation was renamed the North American Steamship Company (a name very likely chosen because the owners already had on hand a large supply of blankets, towels, dishes, and tableware marked with the initials of the pre-war North Atlantic Steamship Company). The group's plan was to upgrade the steamships and the service of North American, so that it could compete favorably with Pacific Mail.

The moving spirit behind the formation of the North American Steamship Company was William H. Webb, who seems also to have supplied most of the financing to get the new line started. And, as there seemed at the time little to keep this chronically active businessman occupied at his shipyard, Webb now delegated most of the management of the yard to his foreman, Christian Metzger, while he himself assumed the presidency of the North American Steamship Company with offices at 54 Exchange Place. Charles A. Dana served as Vice President of the company, and I. W. Raymond as the line's San Francisco agent. William H. Aspinwall, Gustavus Fox, Moses Taylor and other former officials of Pacific Mail sat on the Board of Directors.

The series of events that led these men to start their own intercoastal steamship line began soon after the arrangement of 1860 whereby Cornelius Vanderbilt, left in sole control of the Atlantic route, had agreed not to compete with Pacific Mail on the West Coast. Raising the capital for the settlement with Vanderbilt, as noted earlier, had brought the powerful investment firm of Brown Brothers into a position of influence in Pacific Mail, not only as a major stockholder itself but also as the voting agent for many individual investors. Since at that time Pacific Mail was by far the most profitable American steamship line, the representatives of Brown Brothers now on its board were determined to control the company themselves and to run it in a manner that would guarantee the most dependable interest payments to their investors, with little regard for the policies of the men who had been on the board longer and who had been responsible for the company's solid reputation (1).

The differences between the older members and the representatives of Brown Brothers on the board became even more acute after the decision in 1865 to purchase Vanderbilt's Atlantic operation. From the start Pacific Mail had found the arrangement with Vanderbilt unsatisfactory for many of the same reasons they had earlier found the similar arrangement with United States Mail unacceptable: namely that Vanderbilt, although he ran several fine steamships on his route, offered such poor service, such poor food, and such poor maintenance on his steamers (sometimes referred to in the press as "floating pig sties") that the combined operation again worked to the detriment of Pacific Mail (2).

Vanderbilt, who understood that the imminent completion of the transcontinental railroad would bring an end to his profits on this route, was prepared to sell, at least for the right price. Pacific Mail, which was using its many smaller steamers built before the Civil War on a variety of very profitable shorter runs connecting remote ports along the West Coast anywhere from Seattle to Panama, was not dependent on the Panama route alone, as was Vanderbilt, for its profits. Also, the Commodore's own interests had lately transferred from steamships to railroads, in particular to the collection of short-haul lines of New York state which he was then planning to amalgamate into the New York Central System, and for which the sale of his steamship line could provide handy capital.

In order to put the company in a position to purchase Vanderbilt's Atlantic service, and also to engage in further diversification to offset the losses the company would incur when the transcontinental railroad was completed, the directors of Pacific Mail, led by the Brown Brothers group, managed to get permission to increase the capitalization of the company from $4,000,000 to $10,000,000 (3). The news that Pacific Mail would now control the entire New York-to-San Francisco route immediately raised public confidence in the company, so that the new stock was soon fully subscribed. The deal with Vanderbilt was concluded in September 1865, bringing the New York-to-Panama route and with it all of the Commodore's steamships, under the control of the Pacific Mail Steamship Company. Since much of the newly-issued stock, however, had been purchased by investors whose proxies were committed to Brown Brothers, the former management now found that control of the company had passed completely into the hands of the investment firm.

With its capitalization more than doubled, one of the first priorities of the new management of Pacific Mail was diversification, and it was therefore at this time that Pacific Mail sought and won the government contract for $500,000 a year to carry mails to Japan and China, as described in Chapter XI; and that the company ordered the two mammoth sidewheelers, *Celestial Empire* (later *China*) from William H. Webb, and *Great Republic* from Henry Steers (4). It was only a few months later, in July 1866, when the construction of *China* was just getting underway in his yard, that William Webb and his associates decided to form an independent line of their own between New York and San Francisco.

The Central American Transit Company, the line purchased by the group, had been started in 1862 by Marshall O. Roberts, a former official of United States Mail. Although none of the sources mentions it, certain clues— such as the fact that the line operated via Nicaragua rather than Panama, using facilities originally developed by Vanderbilt; and the use of the name "Peoples' Opposition Line," clearly a Vanderbilt trademark, for the West Coast part of the service—seem to suggest that the wily Commodore, who had recently been well-paid by Pacific Mail to stay out of the Pacific, may originally have been a secret participant in this operation. If Vanderbilt was involved, one could better understand why Roberts was willing to sell in 1866, at a time when the Commodore was himself actively liquidating his steamship holdings.

When Roberts first began his operation in 1862, he had only three steamships at his disposal, all of them markedly inferior to the fine steamers, such as *Golden City* or *Colorado*, being produced in that era by William H. Webb for the rival Pacific Mail line. On the Pacific side, Roberts was operating the steamers *Moses Taylor* and *America* between San Francisco and San Juan del Sur, Nicaragua. *Moses Taylor*, as we have seen, was a relatively small ship, only 246 feet in length, built by William H. Webb for United States Mail at a time when that line already knew that it was about to go out of business. The other steamer, *America*, an interesting hybrid, was, at 285 feet, somewhat larger. She had begun her service under that name in 1854 as a Canadian-owned lake steamboat operating on Lake Ontario. When the company that operated her failed in the Panic of 1857, she was brought to New York, where her superstructure was completely dismantled and her hull rebuilt in order to make her an ocean-going steamship. Renamed *Coatzacoalcos*, she was placed on a route between New Orleans and Tehuantepec, Mexico. Roberts bought her in 1861 and (thank heavens!) changed her name to *America*. After chartering her for a time to the War Department, Roberts took her to the Pacific for his San Francisco-Nicaragua run, on which she remained after the line was bought by William Webb's North American Steamship Company (5).

On the Atlantic side Central American Transit had for a time been operating only one steamer, the venerable 269-foot *Illinois*, built in 1851 for United States Mail. In 1864, however, Roberts had replaced *Illinois* with the much newer and larger *Golden Rule*, built in 1863 for Vanderbilt, but sold on the stocks to Roberts (another suggestion of Vanderbilt's possible involvement, contrary to his agreement with Pacific Mail). But on 29 May 1865, *Golden Rule* ran onto Roncador Reef in the Gulf of Mexico. All aboard were saved, but the ship broke up before she could be salvaged. Since the Civil War had ended only a few weeks earlier, Roberts was able to purchase two smaller ships at auction for his New York-Nicaragua route. One of them, *Santiago de Cuba* (229 feet in length), had been built by Jeremiah Simonson for a line between New York and Cuba, but after only a few trips had been purchased by the Navy at the start of the Civil War. The other, *Keystone State*, was older (built 1853) and at 219 feet, even smaller. Roberts changed her name to *San Francisco* (not to be confused with the unfortunate Webb-built *San Francisco* of 1853) and had her considerably rebuilt. In fact, she was still at the yards when the line was sold, so that when Webb and his associates took over the company, only *Santiago de Cuba* was operating on the Atlantic side.

Meanwhile, Marshall Roberts had ordered two new steamships from the yard of Henry Steers, both of which were still on the stocks when he sold the company to William Webb. In layout and general appearance, these steamers somewhat resembled *Colorado*, then being built by Webb for Pacific Mail, though at about 275 feet each in length, they were considerably smaller. The first, started as *Leona*, then launched as *Managua*, was renamed *Nebraska* by Webb before she entered service. The second, launched as *Nicaragua* but renamed *Dakota* by Webb, was built with an unconventional valve gear and cutoff, designed by the controversial Edward N. Dickerson, with the result that she had a long history of engine problems and served on Webb's line only intermittently between trips to the repair shops (6). When Webb and his associates purchased the company, both *Dakota* and *Nebraska* were towed in their nearly-finished condition across to Webb's yard in Manhattan for the completion of their cabin work and for modifications in their design mandated by Webb for the new service.

In addition to the five steamships which were part of the purchase from Roberts, Webb also bought one more ship to add to the North American Steamship's operation. Named *Paou Shun*, she was also a new ship, built in 1865 by Jeremiah Simonson. With her two tall stacks, one each over boilers located one forward and one aft of the walking-beam and paddlewheel shaft, she did not much resemble *Dakota* or *Nebraska*, and at 286 feet in length, she was somewhat larger. Given the name, one could assume that she was originally intended for service in China. In any event, she began her active career on a charter to the government toward the end of the Civil War. In October 1866, she was purchased by William Webb, renamed *Nevada*, and sent, along with *Nebraska* and *Dakota*, to his shipyard for rebuilding and refurbishing.

Although negotiations for the purchase of the Central American Transit Company began in July 1866 (the same month that Webb's *Re d'Italia* was sunk by an Austrian warship in a battle on the Adriatic Sea), the first advertisements for the North American Steamship Company's sailings for San Francisco via Nicaragua did not appear in the New York papers before October. But even then sailings were infrequent and irregular, with only *Santiago de Cuba* and *San Francisco* running on the East Coast and *Moses Taylor* and *America* serving in the Pacific.

Meanwhile, in the same columns, Pacific Mail was not only advertising regular bi-monthly sailings on its much larger and better appointed steamships (many of them built by William H. Webb) but also prominently displaying announcements of its new service from San Francisco to Japan and China scheduled to begin on the first of January 1867. As the date approached for the first sailing, however, it was apparent that neither of the large steamships, *Great Republic* or *China*, being built for the run would be ready in time. Pacific Mail decided, therefore, to inaugurate its transpacific service with the largest and most elegantly furnished of its coastal steamships, namely *Colorado*, built by William Webb in 1864, and to keep her on the run until the larger ships arrived from their builders. Thus in the annals of the city of San Francisco, both *California* (1850), the first steamship to enter the Golden Gate, and now *Colorado* (1864) the first steamship to sail out of the Golden Gate on a scheduled voyage across the Pacific, were products of the yard of William H. Webb, as also, of course, were many of the packets and clippers which had helped create the state of California.

William Webb may have been in San Francisco to witness the first sailing of *Colorado*, for in January 1867, the North American Steamship Company purchased the steamship *Oregonian* in San Francisco to add to its Pacific fleet, and Webb's later notes show such thorough familiarity with the details of this ship that it seems probable he travelled to California in person to inspect it. Like *Nevada*, *Oregonian* was essentially a new ship, having been built in 1866 for a line between San Francisco and Portland, Oregon. The line had not done well, however, and *Oregonian* (a name retained by Webb) was sold to North American after only a few trips. She was not placed in service immediately, but, like Webb's other new steamers on the East Coast, was sent first to the shops for some major modifications.

The process of upgrading the North American Steamship Company's service took much longer than expected. Through the early months of 1867, the same four small steamers were still making sailings on an irregular basis, while repairs and modifications to *Dakota*, *Nevada*, and *Nebraska* dragged on at Webb's yard. The work on *Nevada* was progressing reasonably well, but *Dakota*'s engines were still giving so much trouble that Webb was not certain whether she could ever be made to run. It is also sad to note that these three ships, here only for repair, were the only vessels in the yard through most of the year 1867. After the launch of *China* on 8 December 1866, there were no new vessels being built at William

**PACIFIC MAIL STEAMSHIP COM-<br>PANY'S**<br>
THROUGH LINE

# TO CALIFORNIA,

CARRYING UNITED STATES MAIL,<br>
VIA PANAMA RAILROAD.

Steamers leave Pier No. 42 North River, foot of Canal-st., at 12 o'clock. noon. as follows:

Oct. 1.—OCEAN QUEEN, Capt. E. Howes, to connect with MONTANA, Capt. G. H. Bradbury.

Oct. 11.—ARIZONA, Capt. Jeff. Maury, to connect with SACRAMENTO, Capt. Caverly.

Oct. 28.—HENRY CHAUNCEY. Capt. A. G. Gray, to connect with GOLDEN AGE, Capt. Lapidge.

All departures touch at Acapulco; those of 1st and 21st, connect at Panama with steamers for South Pacific ports; 1st and 11th for Central American Ports, and those of 1st touch at Manzanillo.

Departure of 11th each month connects with the new steam line from Panama to Australia and New-Zealand.

Steamer of Dec. 11, 1866, will connect with the first steamer of the Compa y's China Line, leaving San Francisco Jan. 1, 1867, for Hong Kong.

100 pounds baggage allowed each adult. Medicines and attendance free.

For passage tickets and all further information apply at the office on the wharf, foot of Canal-st., North River, New-York.     **S. K. HOLMAN. Agent.**

---

### OPPOSITION LINE

TO CALIFORNIA, VIA NICARAGUA.<br>
THE NORTH AMERICAN STEAMSHIP COMPANY<br>
will dispatch the fast and elegant steamship<br>
SANTIAGO DE CUBA,<br>
Jerry W. Smith. Commander,

On MONDAY, Oct. 29, at noon, from Pier No. 29 North River, foot of Warren-st.,<br>
AT GREATLY REDUCED RATES.

For passage and freight apply at No. 177 West, corner of Warren-st.     **D. N. CARRINGTON, Agent.**

The first advertisement for the North American Steamship Company's "Opposition Line to California" appeared in the *New York Times* on 12 October 1866, strategically placed below that of the opposition, the Pacific Mail Steamship Company.

H. Webb's shipyard for the first time in the nearly twenty-seven years since he had started his own business in 1840.

Finally, in early May, *Nevada*, ready to take her place on the line, sailed under her own steam from Webb's shipyard around to the North American pier on the North River at the foot of Warren Street, from which she made her first departure for Nicaragua on 9 May. With three steamers on the Atlantic side, it was possible for the first time to publish a sailing schedule, and Webb was able to advertise that one of the vessels of the North American Steamship Company would leave for San Juan del Norte every three weeks, on Saturdays at noon.

Neither the regular schedules nor the three-boat operation was to last very long, however. The next steamer scheduled to sail was *Santiago de Cuba*. But just hours before she was due to arrive in New York on her northbound trip, *Santiago de Cuba*, feeling her way slowly along the New Jersey coast in a dense fog, went solidly aground on a bar off Absecon Beach near Atlantic City. Captain Howes ordered the engineer to put the wheels full astern, but to no avail. She stuck fast.

The decision to remove her passengers was probably a mistake, as the steamer was apparently in no serious danger. Nevertheless, boats were loaded and arrangements made for women and children to be rowed ashore. The first two boatloads reached shore safely, but the third capsized and four women, one small child, and one seaman were drowned. Later in the day the remaining passengers were taken off by the Absecon Beach lifeboat (7).

As it worked out, *Santiago de Cuba* floated off easily on the next high tide and proceeded safely to New York under her own power. But she had to miss one round trip while her hull was examined and repaired.

On 1 July 1867, New York harbor witnessed an interesting procession of ships, all of which sailed that day for southern ports. First out was the magnificent new Pacific Mail steamship *China*. With her engine installation and her fine cabin work completed, she sailed out of New York that day, headed for San Francisco, with all flags flying. Following some time later was the 327-foot *Ocean Queen*, a former Vanderbilt vessel and one of the largest of the coastal steamers, making the scheduled sailing of the Pacific Mail line to Panama. Last to pass out of the harbor, and, one fears, a rather sorry third, was North American's *Nevada*, en route to Nicaragua.

In July 1867, with two steamships still under repair at his yard in New York and another in California, and with the process of improving the service of the North American Steamship Company so that the line could compete on fairly even terms with Pacific Mail needing his full attention, William Webb nevertheless found it necessary to leave for Europe for several weeks. The yard he left in the trusted hands of Christian Metzger and Hugh McClellan (though Webb wrote almost daily letters of instruction the entire time he was abroad), while Charles A. Dana was given the responsibility for running the steamship line.

Webb's trip to Europe had become necessary because he had at last been able to sell *Dunderberg* to the government of France and at a very favorable profit. Webb had been having trouble disposing of this warship, on which he had expended a great deal of his own money, ever since she was launched two years before in July 1865. Although the United States Navy had at the time refused to take possession of the vessel once the war had ended, not long later the governments of Peru and Chile, then together involved in a war with Spain, offered to buy *Dunderberg*. Webb suggested a price of $2,500,000 if delivered in New York or $3,500,000 if delivered in Valparaiso, and the response was to accept the latter. Webb then applied to the American government for permission to make the sale, but his request was refused. Other offers came from other countries, but in each case the American government would not permit the ship to fall into foreign hands (8).

In 1867, Webb received another offer to purchase *Dunderberg*, this time from the French government of Emperor Napoleon III. Strangely, when Webb wanted to sell the ship to the governments of Peru and Chile, South American nations with whom we were on friendly terms, and who were then involved in a war with Spain to reaffirm an independence which the United States, in the Monroe Doctrine, had guaranteed to protect, permission to sell *Dunderberg* was refused. But when Napoleon III, a usurper who had overthrown a republic to establish a monarchy, and with whom, after he had tried to set up Maximilian as emperor in Mexico, the United States was decidedly not on friendly terms, the American government relented and permitted the sale.*

Once arrangements had been completed, a French crew was dispatched to New York to sail *Dunderberg* to her new home across the Atlantic. But the French officers took one look at this iron-laden leviathan and refused to sail on her (9). Thereupon, William Webb announced that he himself would assemble a crew and deliver the ship to Cherbourg in person.

Once again Webb turned to Joseph J. Comstock, the former Fall River Line captain who had taken *General Admiral* to St. Petersburg, to command *Dunderberg* on her transatlantic voyage. And somewhere a crew was found, though not much of one as it turned out, who were not afraid to cross the Atlantic aboard an ironclad ship. For a warm-up cruise, Comstock and his crew first took a group of brave Independence Day celebrants for a jaunt out into the Atlantic on 4 July. Then, just over two weeks later, *Dunderberg* was ready for her trip to France.

As long as he had to go to Europe to deliver *Dunderberg*, William Webb decided to spend the entire summer there, both to take his family on a vacation and to conduct some business. As we have noted, there was not a single ship building in Webb's yard that summer, and the lack of business was becoming a matter of serious concern to him. About the only line still ordering large wooden-hulled ships in America at this time was Pacific Mail, but since Webb's defection from that company all of its new steamships were being built by Henry Steers. The smaller sailing vessels, many of which William Webb had constructed earlier in his career, were now being built almost exclusively in Maine or other parts of northern New England, closer to the sources of timber. Webb remembered, however, the orders his earlier trips to Europe had produced and now hoped that, if he could get no business in his own country, he could, especially with *Dunderberg* there as a grand advertisement, repeat his successes in getting orders for warships from the European governments which had been so favorably impressed when he produced *General Admiral* and the two frigates for the Italians. Sadly, William Webb seems not to have accepted that in the single decade since his delivery of *General Admiral*, British shipyards had begun producing large iron-hulled ships, so that, even in this short time, a wooden-hulled warship sheathed in iron, such as *Dunderberg*, was already essentially obsolete.

In any event, on 19 July, William and Henrietta Webb and their two sons, William E., now twenty-two, and Marshall, now twenty, boarded the great ironclad *Dunderberg* at her New York pier for her historic trip across the Atlantic. The voyage was marred somewhat by unusually rough seas for summer, by young William's developing an illness referred to only as an "intermittent fever," and by a minor mutiny among the firemen (resolved only by promising them more money) which infuriated William Webb, who called them "a basically dishonest crew" (10).

After averaging a respectable fifteen knots on the crossing, *Dunderberg* nosed into the harbor at Cherbourg on 3 August. When Webb met with French naval officers the next day to turn over the great ironclad warship, he was taken somewhat unawares by their apparent indifference. What he ultimately discovered was that these officers of the traditional French Navy were not enthusiastic supporters of Napoleon III; and that, when the

---

* Napoleon III had at first supported Maximilian of Habsburg, who, in trying to establish himself, with no legal claim and little local backing, as emperor in Mexico, had flagrantly flaunted the American position as stated in the Monroe Doctrine. When Napoleon subsequently dropped his support, in part because the United States had voiced strong objections, Maximilian had been taken prisoner by Mexican republicans and subsequently shot on 1 July 1867, less than three weeks before the Webbs sailed for France. Since the American view of Napoleon III was in these weeks at the lowest ebb possible, the fact that Webb was allowed to sell *Dunderberg* to the French and not to South Americans is even more difficult to comprehend.

Emperor had suggested the purchase of *Dunderberg*, they had advised against it, pointing out that, if the French Navy had needed a new ironclad warship, there were French shipyards that could have built one. In the end, apparently, the Emperor had overruled the Navy and had arranged the purchase of *Dunderberg* from William Webb on his own initiative, in the process angering the French officers, who now made their feelings very clear to Webb by giving him and his ship a decidedly chilly welcome.**

A minor crisis occurred when William Webb informed the officers that *Dunderberg*'s hold contained a collection of spare parts which the Emperor had ordered and which he had brought over to sell to them along with the ship. These, the officers brusquely informed him, they would not need, and if spare parts were called for, French factories could produce them. But when William Webb, who obviously had no other use for these items, offered to donate them to the French Navy, the offer was grudgingly accepted (11).

Though William Webb was eager to get to Paris as soon as possible to start doing business, he and his family were obliged by young William's continuing illness to remain several days in Cherbourg (12). Finally on 12 August, although William's condition had not improved, the family proceeded to Paris where they engaged a suite of rooms at the Hotel Meurice on the fashionable Rue de Rivoli (13).

Once in Paris, Webb, like many other American businessmen both before and after his time, became thoroughly frustrated by the fact that the Emperor, the entire government, and virtually anyone else with whom he might have been able to do business were out of town for the summer (14). Unable to conduct business in Paris, Webb began writing detailed letters to Christian Metzger, the foreman in his yard, and to Hugh McClellan, his accountant, both of whom had been given power of attorney to act as his agents during his absence.

---

** When the formidable *Dunderberg* (renamed *Rochambeau* by the French, in honor of the French general whose troops had fought along with Washington's in the decisive battle against the British General Cornwallis at Yorktown in 1781) became the pride of his Navy, Napoleon III offered to make William H. Webb a member of the very prestigious French Legion of Honor in recognition of his great service to the French nation. But in 1870, before this honor had been arranged, Napoleon III was deposed and in the following year another Republic created, dominated by politicians opposed to Napoleon III. As a consequence, the Republican government never made any effort to follow through on granting the honor to William Webb which had been promised by the former Emperor.

These letters show that William Webb had lost none of the talent for "attention to detail" which had marked his management style since he first took over his father's yard twenty-seven years before. In one letter he gives detailed instructions on the rebuilding of *Nebraska*. In another he tells Metzger and McClellan to get Mr. Roach to hurry with the repair of the engines in *Dakota* (15).

In a letter written from Paris on 25 August, Webb takes pleasure in noting that the French press had given very favorable accounts of the purchase of *Dunderberg*, even though the Navy Department was still "sulking and not very amiable" (16). In the same letter, Webb gives detailed instructions concerning the rebuilding of *Oregonian* in San Francisco, showing that he not only remembered clearly his original orders but also, even though it is unlikely that he had deck plans with him in Paris, was nevertheless familiar with every detail of the ship:

Authorize and direct [Mr. Raymond, the San Francisco agent of the North American Steamship Company] to cut gangway or stairway for second cabin and to cut it under the existing stairway and not where the hatch now is as he proposes and if not too late, forbid him by telegram to use the hatchway for gangway, but cut a new one as directed by me.

My letters authorize him to extend the forward guards, build a cattle pen, extra water closets, butcher shop, etc., caulk topside decks if necessary, but not on any account to dock [i.e., drydock] the ship now, for I cannot conceive that her bottom can be foul enough to make docking necessary or even very desirable. I think I positively interdicted his extending the hurricane deck forward till the ship earned enough money to defray expense of doing this, and I still adhere to this view of the matter, and as to his building a wooden cover above the hurricane deck instead of using a canvas awning, I positively forbid his doing so,....telegraph him to use awnings and conform to my instructions.

Although I did not give Mr. Raymond authority to extend the hurricane deck to the forecastle [in the manner in which Webb had constructed Golden City] I care but little as to the expense if already incurred as I know it will ultimately be advantageous to the ship, but I wished to spare this expense at present.

I wish you to show this letter to Vice President Dana.

I wish and expect my instructions to be literally obeyed, unless impossible. (17)

A letter written from Paris four days later shows that Webb was still worried about the extension of the Hurricane Deck on the *Oregonian* as well as about getting *Nebraska* and *Dakota* running and earning revenue:

Tell Mr Raymond not to build wooden cover over Hurricane Deck . . .

Continue to push Roach up to completing his contracts in time and say to him . . . that if his delay in getting his work

PACIFIC MAIL STEAMSHIP COM-
PANY'S
THROUGH LINE
# TO CALIFORNIA,
CARRYING UNITED STATES MAIL,
VIA PANAMA RAILROAD.
Steamers leave Pier No. 42 North River, foot of Canal-st., at 12 o'clock. noon. as follows:
DEC. 1.—OCEAN QUEEN, Capt. JEFF. MAURY, connecting GOLDEN AGE, Capt. LAPIDGE.
DEC. 11.—HENRY CHAUNCEY, Capt. A. G. GRAY, connecting with GOLDEN CITY, Capt. WATKINS.
DEC. 21.—RISING STAR, Capt. T. A. HARRIS, connecting with MONTANA, Capt. ——.
All departures touch at Acapulco; those of 1st and 21st connect at Panama with steamers for South Pacific ports; 1st and 11th for Central American Ports, and those of 1st touch at Manzanillo.
Departure of 11th each month connects with the new steam line from Panama to Australia and New-Zealand.
Steamer of Dec. 11, 1866, will connect with the first steamer of the Compa'y's China Line, leaving San Francisco Jan. 1, 1867, for Hong Kong.
100 pounds baggage allowed each adult. Medicines and attendance free.
For passage tickets and all further information apply at the office on the wharf, foot of Canal-st., North River, New-York.           S. K. HOLMAN, Agent.

---

## OPPOSITION LINE
TO CALIFORNIA VIA NICARAGUA.
EVERY TWENTY DAYS.
THE NORTH AMERICAN STEAMSHIP COMPANY
Will dispatch the fast and elegant steamship
SANTIAGO DE CUBA,
JERRY W. SMITH, Commander,
On THURSDAY, Dec. 20, at noon, from Pier No. 29 North River, foot of Warren-st.
AT GREATLY REDUCED RATES.
For passage and freight apply at No. 177 West-st., corner of Warren-st.           D. N. CARRINGTON, Agent.
Jan. 10, to follow, steamship SAN FRANCISCO, Capt. C. F. W. BEHM.

---

## FOR CALIFORNIA VIA PANAMA.
The Pacific Mail Steamship Company's steamer HENRY CHAUNCEY, will sail on TUESDAY, Dec. 11. at 12 o'clock noon.
Freight received as usual.
For rates, &c., inquire at freight office on Company's pier, No. 42 North River, foot of Canal-st.
WELLS, FARGO & CO.,
Sole Freight Agents Pacific Mail Steamship Co.

North American's second advertisement, which first appeared in the *New York Times* on 11 December 1866, was also placed below that of the Pacific Mail, but the latter company responded by inserting another below North American's. Note the number of Webb-built vessels mentioned.

done on the new steamers defeats my plans for business with the vessels, I shall never forgive him. (18)

In the same letter he instructs McClellan to make out checks for the monthly wages of the Niefnecker family, servants in the Webb's Fifth Avenue home.

On 19 September, with young William's health somewhat improved, Mrs. Webb and the two younger men left Paris for a tour of Scotland while William H. Webb boarded an overnight train for Berlin. The Prussian government of King Wilhelm II, then under the leadership of the indomitable Minister-President, Count Otto von Bismarck, had earlier shown some interest in purchasing *Dunderberg*, so Webb felt that a personal visit to Berlin might well produce some lucrative orders (19).

The American ambassador to Prussia at the time was the highly esteemed historian George Bancroft, whom Webb knew well as a fellow member of the Union League Club of New York, and who now obliged by providing him with a letter of introduction to Count Bismarck. Webb's letter from Berlin shows that he was very favorably impressed by his reception there (and indicates as well a certain talent for composing a run-on sentence):

Since my arrival here (without previous notice) I have received very marked and decided attention and civility on the part of govt. officials and particularly from Count Bismarck himself, who, on receipt of my letter of introduction, wrote me a note personally to call upon him the same evening, and gave me a long sitting interview, talking freely on both business and political matters, and before leaving inviting me to dine with him, when he said he would introduce me to some parties with whom he would like to make me acquainted and when going to dine with him I found many persons connected with the Govt. among them the chief naval officer located here—also our minister Mr. Bancroft, to whom of course was given the seat of honor right of Bismarck and I next. I found several speaking English, some who had visited and lived in our country. I enjoyed the evening very greatly—and I think turned it or the occasion to great account, think if this government had the funds in hand now I could secure some business, but they have not. (20)

Webb was especially pleased when Bismarck confided that he "regretted very much" that Prussia had not bought *Dunderberg*. During the evening Bismarck also asked Webb's advice about the feasibility of constructing heavily armored ships for the Prussian Navy. Webb responded by telling the Minister-President that smaller cruisers would be more advantageous, first because smaller and lighter vessels were more maneuverable in a battle situation, and also because the major Prussian naval bases and other seaports were all located on rather shallow rivers where a large heavily-armored ship would not be able to navigate. The types of vessels added to the Prussian navy, or, after 1871, the German Imperial Navy, suggest that Bismarck followed this advice from the American shipwright (21).†

---

† Since Bismarck's memoires make it clear that by the summer of 1867 he was already planning the war with France that took place in 1870, it is not impossible that one of the reasons for Bismarck's cordial treatment of William H. Webb could have been to gain as much inside information about *Dunderberg/Rochambeau* as polite dinner conversation could educe.

From Berlin William Webb travelled by train to Hamburg, Bremen, Wiesbaden, Bonn, and Aachen. (Since this trip was made four years before Bismarck's unification of the German states, each of these cities was a capital, empowered to do business separately.) A letter written on 7 October notes that Webb transacted some business at each of these cities, "moving rapidly night and day, sleeping in the cars half the time" (22). In another letter Webb notes with considerable pleasure that the French officials were dissatisfied because he had not given them the builder's plans for Dunderberg. "I shall let them remain dissatisfied," he wrote (23).

Back in Paris to reconnect with his family and to prepare for the homeward voyage, Webb wrote:

I am quite well satisfied that I could make it pay well if I could remain in Europe a few months longer—my professional reputation is high with European governments and I ought to avail of it, but steamship matters prevent it. (24)

However much courtesy and encouragement Webb may have received during this energetic tour of European capitals in the Summer of 1867, the fact stands that not a single new order resulted from it.

Finally on 29 October, William Webb, his wife, and his two sons, sailed from Cherbourg on the North German Lloyd *Deutschland* for a ten-day voyage back across the Atlantic to New York (25).

Webb's return to the North American Steamship Company's offices at 54 Exchange Place was overdue. It was time, after more than a year of operation, for the line to proceed far more actively with the planned upgrading of its service if it expected to survive at all, for as matters presently stood, there was little reason for passengers to choose one of the small and irregularly scheduled steamers of the North American line over one of the grand and beautifully furnished liners of Pacific Mail. North American had earlier lowered its fares to attract passengers, but Pacific Mail, with far greater resources, had immediately followed suit, so that the only result for the North American Steamship Company was a perilous drop in revenue. At one point while William Webb was in Europe, it had been necessary for his agents in New York to sell some of his personal stocks to pay the bills of the steamship company.

But the Fall of 1867, after William Webb's return from Europe, saw a series of changes in the operation of North American which, in only a few months, did in fact put the service of this line almost on a par with that of Pacific Mail. As of the first of November North American gave up the service through Nicaragua, with its long and un-

popular overland route, and moved its Central American terminus to Aspinwall in Panama, where the Pacific Mail steamships also docked. Since William Webb and several other members of North American's board still owned controlling stock in the Panama Railroad, Pacific Mail was frustrated in its efforts to prevent the rival line from using the railroad to transport its passengers and cargo between oceans.

At about the same time *Nebraska*, the largest and fastest of William Webb's purchases, came from the yard all freshly furnished and painted, to take her place on the New York-Aspinwall run with the smaller *Santiago de Cuba* and *San Francisco. Nevada* had meanwhile sailed out to the West Coast to join *America* and *Oregonian* whose rebuilding had also been completed. Also, through the Spring and Summer of 1868 the sailings of North American Steamship's three vessels on the Atlantic run were augmented by the occasional charters of the former Havre Line steamships *Arago* and, less often, *Fulton*, in both of which William Webb seems at the time to have owned a part interest (26).

Finally, sometime in the summer of 1868, *Dakota* finally emerged from Roach's yard, her engine running and ready for service—or at least so the line thought. With *Dakota* joining the Atlantic fleet, *Nebraska* now followed *Nevada* to the West Coast, allowing *Moses Taylor* to go into reserve as a spare. Although a relatively new ship, *Moses Taylor*, with her second-hand engine rated at 500 horsepower, was not known for speed and frequently had difficulty maintaining a schedule. *Nebraska*, on the other hand, which had sailed out to replace her, and which could apparently paddle along occasionally at a respectable twelve knots, was considered a fast steamer, at least according to an article in the San Francisco *News of the World*, which also lavishes praise on her interior appointments:

She will average a speed of 340 miles a day when working up to full power, and is looked upon as the fastest ship on this coast.
Running along the center of the deck are twenty staterooms . . . all double, with doors on each side and ventilators on top. . . .
Right aft is the ladies' sitting room, well fitted-up, private and well-ventilated. Forward of [the staterooms] is the smoke room, fitted-up with a degree of comfort seldom seen on board ship. On the Main Deck is the grand saloon, ninety feet long, twenty-eight feet wide, and eight feet high. On each side of the saloon is a row of staterooms, opening on the deck, and accommodating eighty passengers, with two bridal chambers in the forward part of the saloon. (27)

One wonders how the chief engineers of the Pacific Mail steamships, in particular of *Golden City* or *Colorado*,

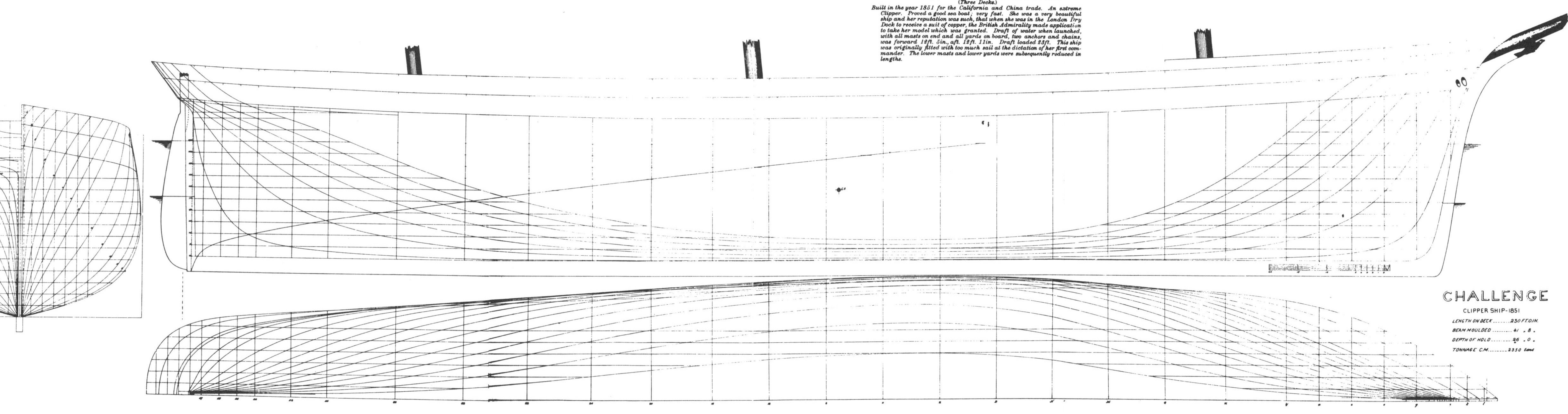


Lines of the extreme clipper *Challenge*, reproduced at a scale of about 1/16 inch to the foot.
The legendary *Challenge* was Mr. Webb's third
clipper ship, built in 1851 for the equally legendary firm of N. L. & G. Griswold.

either of which was probably capable of speeds up to about seventeen knots, greeted the dubious intelligence that the twelve-knot *Nebraska* was the fastest ship on the Pacific Coast! Mr. Raymond certainly deserves high marks as a publicity agent.

By far the finest addition to the North American fleet at this time was the graceful 300-foot *Guiding Star*, built in 1864 (for a New York-to-New Orleans run which had not succeeded) and touted as one of the speediest side-wheelers on the Atlantic coast. Webb had purchased this steamship soon after his return from Europe, but she had spent most of the winter being refurbished. Now with *Guiding Star* on the Atlantic run (still with *Santiago de Cuba* and *San Francisco*), it was no longer necessary to charter *Arago* or *Fulton*, both of which were showing their age and also proving expensive to operate.

By the Summer of 1868 the North American Steamship Company had hit its stride, with several fine and well-appointed steamers on each coast providing regular and dependable service between New York and San Francisco via Panama. At the same time, Pacific Mail, noting apparently that, with the transcontinental railroad rapidly approaching completion, its Panama service would not, in any event, remain profitable much longer, had elected to concentrate on its trans-Pacific line and its local West Coast services, and not waste its resources competing with North American on the Panama route. Consequently, Pacific Mail now reduced its Panama fleet to two steam-

The steamer *Santiago de Cuba*, built by Jeremiah Simonson at Greenpoint in 1861, is shown here as a naval cruiser during the Civil War. She had a varied career—naval and merchant, sidewheeler and propeller, steamship and schooner-barge, coastwise and transatlantic—and was owned for several years by William H. Webb's North American Steamship Company. She lasted until about 1899.

ships on each coast with sailings only once a month from each port. And now, while the sailings of North American's steamers were advertised prominently in the New York and San Francisco papers, only token notices informed the public of Pacific Mail sailings.

That summer John Roach finally managed to get *Dakota*'s engine running, so in mid-August the handsome *Dakota* proudly paddled around to the line's North River pier to make her maiden voyage to Panama.

Thus by the Fall of 1868, the North American Steamship Company, under the conscientious presidency of William H. Webb, showed all signs of having become a successful steamship service. A curt notice published in the New York newspapers on 3 October 1868, however, revealed the unhappy truth of the situation:

For the present all mails for the Pacific states and territories, British Columbia, China, Japan, and the Sandwich Islands . . . will be sent from this city by the overland mail by way of Chicago and Omaha. (28)

What this notice, published by the federal government, was announcing in essence was that the railroad running westward from Omaha, and the railroad running eastward from California, had by then reached a point close enough together that it was now more expeditious to send the mails by rail across the country than by steamer by way of Panama. Passengers and shippers too were finding the rail route preferable, even though, until the two railroads finally joined at Promontory Point in May 1869, a short connecting coach trip was necessary. Thus in spite of its fine steamers, excellent service, and scheduled sailings, the North American Steamship Company was, in fact, not doing enough business to survive, and the future, it appeared, was only to be bleaker.

By October 1868, papers on both coasts carried rumors that North American was giving up. Although the San Francisco papers published a statement by I. W. Raymond emphatically denying these rumors, at the time this statement was printed, Mr. Raymond already knew that all of the company's steamers were then making their final runs (which, for *Dakota*, was the return leg of her first and only round trip to Panama). In November 1868, the North American Steamship Company went out of business (29).

The truth was that, in its two years of operation, the North American Steamship Company, funded almost entirely from the personal accounts of William H. Webb, had been a financial fiasco. Some of the losses were recouped when Pacific Mail agreed to purchase two of the company's steamships—*Oregonian* and *America*—for use on their still successful short-haul runs on the West Coast. But most of the steamers—*San Francisco*, *Guiding Star*, *Santiago de Cuba*, and *Dakota* on the East Coast and *Nebraska*, *Nevada*, and *Moses Taylor*, now laid up in California—were retained by William H. Webb as collateral for the $1,458,000 the North American Steamship Company still owed him when it ceased operating. In fact, William Webb kept the company's corporate existence alive (until it was dissolved by court order in 1876) in the hope of collecting some part of this enormous debt—which he did not—from the other stockholders. One should not be too concerned for the personal fortunes of William H. Webb and his family, however, for, although he reported business losses of $609,407.25, his personal gross income for the year 1868 was listed at $137,026.12 (30).

Closing down the affairs of the North American Steamship Company, after expending so much energy and talent, as well as personal funds, to make it succeed, was not the hardest task William Webb had to face in the Fall of 1868. For about the same time that Webb had to accept the fact that his steamship line was not successful, he also came to understand that, although there were at that time two last ships building in his shipyard, no other orders for wooden ships were coming in, and no more were likely to come. In the Fall of 1868 the shipyard of William H. Webb was advertised for sale.

# XIII

# Closing Shop, 1868-1872

ALTHOUGH William H. Webb put his shipyard up for sale in the Fall of 1868, he did not as yet go out of business completely. There were at the time two ships under construction in the yard which were not completed until the following year. One, a small bark named *James A. Borland*, was launched in January 1869. The other, destined to be the last vessel built at the yard of William H. Webb, was, interestingly, a packet ship named *Charles H. Marshall* for the man who had been a close personal friend of both William Webb and Isaac Webb.* He was also the man who had given the young William Webb his first assignment to build a ship on his own when he was still working for his father. This last ship to be built in Webb's shipyard had been ordered by a younger Charles H. Marshall, the son of William Webb's friend and patron who had died in 1865.

Although William Webb had accepted the fact that his business was failing, it is worth noting that he had fared better than some, for by 1869 his was the only shipyard still in operation on Manhattan Island. All of the other yards had either given up or had moved, either to Green-point or to Brooklyn across the river, where riverfront property was cheaper, or to New Jersey on the other side of the Hudson.** *Charles H. Marshall*, when launched in July 1869, was both the last square-rigger and the last wooden-hulled ship built in one of the East River ship-yards of Manhattan, where, barely fifteen years before, many of the world's greatest packets, clippers, or steam-ships, had been built.

After the launch of *Charles H. Marshall*, the yard sat idle, though *Dakota* was still stored there between trips to John Roach, who continued to tinker with her eccentric valve system. In May 1870 an article in the *New York*

*Tribune* ruefully described the situation in the New York shipyards:

The Novelty Iron Works has sold off most of its machinery and tools. The Allaire Iron Works is now occupied as a stable. The Etna Iron works has ceased to make marine engines. The Fulton Iron Works are for sale. W. H. Webb's shipyard is to let. Henry Steers' yard is empty. The Continental Iron Works is almost deserted and green grass is growing in nearly all of the shipyards which, five years ago, were alive with workmen. (1)

Webb's yard sat idle for two more years, until finally in 1872 the property was sold and the yard in which so many great ships had been given life was cleared so that the area could be adapted for other uses.

Although most of the many factors contributing to the decline and ultimate collapse of the building of wooden-hulled ships in New York have already been mentioned in context, it might be helpful to restate some of them here. William Webb himself had a great deal to say on the subject. In his view the main culprit was the tariff. In a letter written to a friend in 1869, Webb stated quite categorically: "My business is destroyed by the absurd laws of high tariff, which destroys the shipping interest— the right arm....of the country" (2). In an interview printed in the *New York Times*, however, he had even more to say. Webb noted first the recent increases in the costs both of materials and of labor, adding in the process a commentary on the changing attitudes of shipyard workers:

One of the causes which produces the great dullness of trade is the unprecedented...and very extraordinary advance since the commencement of the rebellion in the prices of all materials and labor required for the construction of vessels, ranging in those composing the major part of the vessels from fifty to sixty per cent and for labor about 75 per cent to which must be added the well-known fact that less daily and individual labor is re-turned for the increased wages than formerly when less wages were paid, the evident disposition of workmen to avoid ac-countability, the unwillingness of the rising generation to ap-prentice themselves to mechanical business, and the consequent and increasing difficulty of obtaining skilled labor. (3)

---

* The contract for *Charles H. Marshall* is reproduced in its entirety in Appendix C.

** Now that Cornelius Vanderbilt's attentions had been diverted en-tirely to his growing railroad empire, even Jeremiah Simonson had been obliged to close his shipyard in 1868.

In the same vein, Webb goes on to mention that, whereas the advantages enjoyed by European shipyards of having lower wages to pay had once been easily offset by the better workmanship produced by American shipyard labor, this was, lamentably, no longer the case, because most young Americans were seeking their fortunes in the West, while the only applicants for shipyard jobs were foreign immigrants. Webb also notes that "Great Britain took advantage of our war to divert much of our foreign trade to their bottoms" (4). Only as his final observation does William Webb mention what one must see, with historical hindsight, as the major reason for the failure of the New York shipyards: namely, that by this time Great Britain was producing screw-propelled iron-hulled steamships (and even iron-hulled sailing ships) which were proving more efficient, more dependable, and certainly more durable than the wooden-hulled side-wheelers lately being built at William Webb's yard. What William Webb here only suggests but does not quite fully articulate is that by 1870 iron-hulled vessels—themselves to be to some extent superseded by steel-hulled vessels two decades later—were replacing wooden-hulled ships in any area where relatively large passenger or cargo-carrying vessels were called for.

In Europe the shift to iron had begun even before our Civil War. America's conversion came somewhat later, both because our industrial progress had been slowed by the war, and because neither our sources of iron ore nor of the fuels to treat it had as yet been developed to the extent that Britain's had been. But most marine historians are now inclined to state, if somewhat cautiously, that when William H. Webb and Henry Steers built the last several ships for Pacific Mail, most notably the four large steamships for the trans-Pacific run, large wooden side-wheelers for ocean routes were already out of date and were doomed to early obsolescence. As early as 1875, just eight years after *China* and *Great Republic* began their service on the Pacific, a requested increase in the government's subsidy to Pacific Mail was granted, but with the stipulation that the company in the future build only iron-hulled propeller-driven steamships for the route. By 1872 both of the later wooden-hulled steamers built by Henry Steers—*America* and *Japan*—had been destroyed by fire, while *Great Republic*, as noted earlier, had been removed to a safer coastal route. Thus, just five years after the line's inauguration, only Webb's *China*, of the four original wooden-hulled sidewheelers, remained on the route, and even this ship was retained mainly as a spare.

Even had he wished to convert to building iron-hulled vessels, which he did not, William Webb would not have been able to do so in his present location. Timber from the Mid-West could be brought by water across the Great Lakes to the Erie Canal and down the Hudson to New York. But the main sources of either coal or oil by this time were in western Pennsylvania. Iron ores, such as they were at that time, were therefore brought from the Lake Superior region to Erie, Pennsylvania, and from there to Pittsburgh, where the early iron furnaces were developing. The railroad across the state of Pennsylvania, completed in 1852, carried the finished iron directly to Philadelphia or Chester on the Delaware River, and it was here rather than in New York that the American shipyards building iron, and later steel, vessels were to develop. Once America had shifted to iron-hulled ships, New York was no longer a viable location for a shipyard. Thus, as we have seen, all of the Manhattan shipyards, active and prosperous only a decade before, disappeared completely in the few years between 1865 and 1869, the yard of William H. Webb last of all.

In another interview, probably from this same period, about 1870, William H. Webb was called on to comment on the future of American shipbuilding. Asked specifically what he thought could be done to revive the industry, Webb returned emphatically to the matter of the tariff:

Reduce the present high tariff on bar and sheet iron and copper, hemp and cordage, canvas, chains, and anchors, and the general equipment of vessels, all of which can be done without destroying our manufactures, simply reducing their enormous profits, and we can again successfully compete with foreign-built ships, and regain a portion of the lost trade, notwithstanding the present high prices of material and labor. (5)

In answering the question "What should be the policy of the government in order to restore American shipbuilding to its former rank among American industries?" Webb came back with a rather unexpected reply:

The government ought to provide some extra encouragement to ship owners rather than to shipbuilders. When railroad building had its boom, Eastern capitalists took their money out of ships and put it into railroads. Now that railroads are paying so poorly, it would be a good time to make the ownership of American vessels attractive to capital. Do this and American shipbuilding will take care of itself. Other nations pay bonuses for mail carrying and in other forms, and if we are to keep up with them we must do likewise. (6)

When the interviewer then asked William Webb whether he believed that American shipbuilding would "ever regain the ascendency it once enjoyed," Webb replied:

Antonio Jacobsen painted a number of Black Ball Line packets around 1915, near the end of his career and long after the last of the packets had disappeared from the sea. *Charles H. Marshall* was William Webb's last ship—the last packet—and the last square-rigged vessel built in New York. Antonio Nicolo Gasparo Jacobsen, Danish by birth and the son of a violinist, was named for three great Italian violin makers—Stradivari, Amati and da Salo. (Courtesy of The Mariners' Museum, Newport News, Virginia)

I do. Our country is placed as it were by Providence almost midway between the two most thickly settled portions of the globe, the high civilizations of Europe on the one side, and the developing civilizations of Asia and the East Indies on the other. We have some of the finest harbors in the world on both our coasts to which vessels are admitted all year around. It is therefore manifestly our natural destiny to be carriers of the earth. Then too, our people seem to have a natural-born taste or ability for building ships and navigating the seas. (7)

In 1869, the year that William H. Webb gave up building ships, he was only fifty-three years of age, still in excellent health and spirits, and by no means prepared

to be put to pasture. An article about Webb which appeared in a national publication about that time described him as a man of "gentlemanly bearing," whose "cheerful smile" and quick, lively movements were indicative of his vibrant energy. The article also called him a modest man, in spite of his known achievements, an "affable and interesting talker," and a "genial companion." In short, this was not a man who was ready to retire (8).

Thus, although William H. Webb realized that he had to give up the shipbuilding trade, which he loved, and in which he had excelled, he still had many years ahead of him and a great many other opportunities for keeping busy. In the Fall of 1868, about the time Webb decided to put his shipyard up for sale, some of his friends, notably Charles H. Marshall and Moses Taylor, both active in the organization of the Republican Party in the city of New York, put up William Webb's name as a candidate for mayor of the city. This was, in fact, the second time that Webb, by then well known in New York not only as a successful shipbuilder with a world-wide reputation but also as a man of impeccable integrity, had been suggested as a candidate for mayor. The first time, in 1856, before the Republican Party had been properly organized in the city, had been under the aegis of the then dominant Democratic Party. But, as should be obvious, in 1856 Webb's shipyard was at the height of its prosperity, and its proprietor could not then have taken the time to attend to any other responsibilities, let alone one so demanding as that of mayor of New York.

By 1868, of course, after Abraham Lincoln's two successful runs for the Presidency and the success of his administration in bringing the Civil War to a conclusion, the Republican Party was solidly established on a national level. In the elections of 1868, the year in which the New York organization asked William Webb to stand for mayor, the national party nominated General Ulysses S. Grant for President, to run against the Democratic candidate Horace Greeley, the controversial editor of the *New York Tribune*. Although the Republicans were popular nationally, and, of course, Grant won this election, a Republican contender for mayor of New York, as William H. Webb surely realized when he declined the invitation, could have been only a token candidate. The Democratic Party, itself controlled by Tammany Hall and its thoroughly corrupt "boss" William Tweed, was too well ensconced at City Hall for any Republican candidate, even William H. Webb, to have stood any chance of being elected.

By the summer of 1869, by which time his last ship, *Charles H. Marshall*, had been launched, William H. Webb began considering other outlets for his talents. His first thoughts were apparently to try to establish another steamer line, this time between New York and some port in Europe, in order to utilize the collection of steamships (now serving no purpose except to pile up dockage and maintenance costs) that had been the only return on his enormous investment in the North American Steamship Company. *Santiago de Cuba* was dispatched on two exploratory round trips, one to France and one to Denmark. During that summer William and Henrietta Webb, this time apparently not accompanied by their sons, made another trip to Europe, probably with the object of seeking a mail contract from one of the European governments. The Webbs went first to Paris, very likely sailing over on *Santiago de Cuba*'s single voyage to France. While there, however, Henrietta Webb became ill and went to stay at the Luxeuil spa in northeastern France to recuperate. While she was there, so close to the Belgian border, William Webb took a train to Brussels to talk with the Belgian government about some sort of subsidy for a New York-Antwerp service. Although the ministry involved asked Webb to return in a week, to give them "time to think about his proposal," apparently the answer was negative as the service never materialized (9).

Despite this disappointment, William Webb, it seems, had not yet given up the idea of building ships altogether, for while he was in Paris he wrote to his foreman in New York, Christian Metzger, concerning a request for an estimate to build a ship for the company of F. Alexandre and Sons, operating steamships between the eastern United States and Mexican ports. Webb was not so much interested in going back into a losing business as in having one last opportunity to get rid of leftover materials, for he writes, "I am anxious you should secure the job, if only to use up our framing timber" (10). Apparently the Alexandre Line had asked for two estimates: one for a wooden-hulled ship and another for iron. In responding to this request, William Webb shows, in two short and rather sad sentences, that he knew why he could no longer stay in the shipbuilding business: "I advise not giving an estimate for building an iron hull. We are not at present prepared for such work" (11). In any event the new Alexandre Line steamship was not built at the yard of William H. Webb.

Soon after William Webb's return from Europe, his brother, Eckford Webb, or actually the firm of Webb and Bell, located in Greenpoint next to the yard of Henry Steers, was awarded a contract to build the caisson—i.e., a large hollow wooden box—to be sunk into the East River as part of the support for the tower on the Brooklyn side of the new bridge then being built across the river to connect the cities of New York and Brooklyn. Although there is no specific evidence to this effect, it would be surprising if William Webb, who was then not otherwise occupied, had not been instrumental in securing this contract for his brother, and had not then lent a helping hand in the design of the caisson.

The caisson itself, construction of which was started on 1 November 1869, was 162 feet by 168 feet and fifteen feet high. As it had to be designed to hold up a massive stone tower and to last for something close to an eternity, its construction consisted of several layers of well-braced heavy timbers. A description of the work in the *New York Herald* adds to the suspicion that William Webb may have played some part in its design:

... it is a work of such magnitude and combining so much that is novel when contrasted with similar structures that it claims no small degree of attention from the engineering world. (12)

About the time this caisson was nearing completion, news reached William H. Webb that the beautiful Pacific Mail steamship *Golden City*, one of the most successful products of his shipyard and at the time less than seven years old, had been wrecked on the coast of Baja California, on 10 February 1870, obviously as a result of negligence on the part of her officers. Fortunately, all passengers were taken safely to shore, which was, by the nature of the event, considerably closer than the pilot had intended, but the ship and its cargo, except for some food and a considerable consignment of wine "salvaged" by anxious passengers, were a total loss.

During the summer of 1870 William H. Webb found profitable employment for at least one of his laid-up steamships. Between 1868 and 1870, J. N. Harvey and Company of New York had been operating their small steamer *Fah Kee* on a run between New York and Hamilton, Bermuda, assisted by a modest subsidy of about $1000 per round trip from the colonial government of Bermuda. But in July 1870, when their contract expired, J. N. Harvey, having received a good offer for the purchase of *Fah Kee*, gave up the service. When the Bermuda government then advertised that this subsidy would be offered to anyone who could place on the route a steamer of no less than 500 tons burthen and with accommodations for no fewer than forty first-class passengers, and who would schedule a departure from each port at least every other week, William Webb, with several surplus ships at his disposal, any of which could more than fill these requirements, decided to make a bid for the subsidy.

Although one other shipper also tendered a bid, the subsidy was awarded to William Webb, who decided to place *San Francisco* (formerly *Keystone State*) on the route. With first-class accommodations for ninety-six passengers, *San Francisco* easily met the requirements of the Bermuda government; and though her best speed was less than ten knots, she could nevertheless manage the short run to Bermuda and back in the two-week time limit.

*San Francisco* made her maiden voyage for Webb's

WILLIAM H. WEBB'S OLD OFFICE.

In July 1882, *Harper's New Monthly Magazine* published this pen drawing of William H. Webb's office in an article entitled "The Old Ship-Builders of New York." By that time, New York's once important shipbuilding industry was nearly extinct, and any surviving shipbuilder was an "old ship-builder." The industry never regained its former importance.

Bermuda Line in late July 1870, arriving in Hamilton harbor on 3 August. All went reasonably well, with *San Francisco* sailing regularly once a week from either Hamilton or New York until 5 December 1870, when, leaving Hamilton harbor, she ran onto a shoal, in the process starting several timbers on her port side and taking enough water to douse her fires. Immediately raising sails, however, *San Francisco* made her way to shallow water before settling. Here passengers were removed and taken back to Bermuda. After divers had plugged her more serious leaks, she was towed to a sandbank where her cargo was removed. She remained there until 11 January 1871, when she was finally towed to the drydock at Bermuda's Royal Navy Yard. Later that month, after more temporary repairs, *San Francisco* deadheaded back to New York for a more complete overhaul and recoppering.

By March 1871 *San Francisco* was back on the run. The line, however, was never profitable, and William Webb, who by now was involved in a new venture in the Pacific, was eager to get out of the contract. *San Francisco* continued running to Bermuda through most of 1871, but in December Webb allowed her engine and boiler certificate to expire, so he would not be obliged to operate the ship through the lean winter months. By March 1872, *San Francisco* had been inspected and was back on the run again until December. By that time the combination of chronic leaks and unfavorable comments in the Bermuda press forced Webb to replace her temporarily with a chartered steamer (since his other ships had by then all been transferred to the West Coast). Then in February 1873, the Bermuda government, by now thoroughly disenchanted with the poor service offered by *San Francisco*, finally did William Webb the favor of cancelling the contract (13).

At about the same time that he started running *San Francisco* to Bermuda, William H. Webb, frustrated in his attempts to gain mail subsidies from European governments, decided to try to run his other surplus steamships on a new line across the Pacific. The Pacific Mail Line had managed to negotiate a very sizable subsidy from the American government for carrying the mails to Japan and China. Why not, reasoned Webb, request a similar subsidy to carry mails to New Zealand and Australia, since no regular steamship service then existed between the United States and these British dominions?

Webb then made applications to the governments of Australia, New Zealand, and Great Britain, as well as to his own government, for a mail subsidy. But the response was disappointing. The British government, not surprisingly, would not consider giving a mail contract to a foreign-owned steamship line. The government of Australia simply did not respond at all. The American government considered the matter and promised to give an answer once it had investigated the circumstances. Finally, the government of New Zealand came up with an offer of $150,000 per year for delivering mail from the United States via a regularly scheduled steamship service. This grant, Webb knew, was not sufficient to support his steamship line, even when revenues from passengers and cargo were added. But it served as encouragement and a harbinger of more to come. In short, the New Zealand subsidy persuaded Webb to go ahead and start his line.

Late in 1870 William Webb went to California (no evidence exists to show whether he travelled aboard a Pacific Mail steamer or by rail!), where he officially established the United States, New Zealand, and Australia Steam Navigation Company, which was advertised locally, however, simply as the "Webb Line." Though all of the ships and most of the capital were his, Webb took on as a partner Ben Holladay, operator of a line between San Francisco and Seattle, to serve as the line's West Coast agent, since Webb, of course, would continue to live in New York once the line was safely in operation. Before returning to New York, Webb went to inspect *Nevada*, *Nebraska*, and *Moses Taylor*, the three steamships still in layup in San Francisco which he planned to use on his new line and found them "really in excellent condition" (14). Since Webb wanted to use *Dakota* as well on this run, and, since she would be taking crossings up to a month long across the Pacific and therefore could not be subject to minor engine breakdowns, when he got back to New York he again sent *Dakota* up to Roach, this time to dismantle her leaking boilers and Dickerson's controversial cut-off altogether and to install new iron boilers and a conventional cut-off, in the hope that this handsome steamer could at last produce some revenue.

In April 1871, Webb was back in San Francisco to preside over the inauguration of his new transpacific service. While he was there, he was invited to address a meeting of the San Francisco Mechanics' Institute. The mayor of San Francisco and other local dignitaries showed up at this meeting to honor William H. Webb. In his speech to the Institute, one can get some insight into William Webb's talents as a salesman. He began by telling them that he was "proud to be a mechanic," which earned him a thunder of applause and won him the attention of the audience. Then he told them he was proud to have been the builder of *California*, the first steamship on a regular line between the East and California. Further, he was proud to have been the builder of *Colorado*, the first vessel of a regularly scheduled steamship line between California and China. Each of these references to their home state drew another burst of applause. With the rapport with his audience then well established, Webb added that now he was proud to be the man to establish the first regularly scheduled steamship service from California to New Zealand and Australia (15).

*Nevada*, largest and fastest of the three available steamships, made the first sailing for the United States, New Zealand, and Australia Line, heading out of the Golden Gate on 8 April 1871. *Nebraska* followed on 5 May and *Moses Taylor* on 24 May.

Beginning with the departure of *Moses Taylor* on 24

May, sailings were scheduled every four weeks, on Wednesdays at noon. Since the round trip took just under two months, two steamers might have sufficed in an emergency but would obviously not have allowed time for repairs and reprovisioning in San Francisco between trips. Thus three steamers were required for the service. Webb's intention had been to operate *Moses Taylor* only until *Dakota* was ready to join the line. But rebuilding *Dakota*'s engine took longer than expected, and, on her first trip across the Pacific, *Moses Taylor* proved that her nine and a half knots was not sufficient to maintain the company's schedule. So in June the faithful *Santiago de Cuba* sailed around to San Francisco to join the line until *Dakota* was ready.

For the first several trips, until dockage and coaling arrangements could be worked out in Sydney, Webb Line steamers ran only to New Zealand (with a coaling stop at Honolulu en route), calling at Auckland and Wellington on North Island, and Lyttleton (the port for Christchurch) and Port Chalmers on South Island, and making connections at the last port with a local steamer for Sydney. By mid-summer, however, the steamers were putting in only at Auckland on New Zealand and sailing on to Sydney themselves.

Once his line was in operation, William Webb realized that stopping at Honolulu for coal obliged his steamers to take a much longer route than a more direct line between San Francisco and Auckland would call for. After checking out a variety of possiblilities, Webb decided that the Samoan Islands would make a more logical place to put in, so in June 1871 he dispatched Captain E. Wakeman to the Samoan Islands to investigate whether it would be feasible to establish a coaling station there. As a matter of courtesy, Webb also instructed Wakeman to submit a copy of his report, when it was ready, to Admiral Winslow, the commander of the United States Pacific Squadron based at Pearl Harbor in the Sandwich Islands (Hawaii).

What Wakeman found at Pago Pago was one of the best harbors in the world, so that Webb did indeed establish a coaling station there. From mid-summer 1871, the calls at Honolulu were given up and replaced with coaling stops at Pago Pago, thus shaving several valuable days and expensive miles from the round-trip schedule.

Wakeman also made other discoveries on the islands that were of greater interest to Admiral Winslow. It seems that German naval vessels had recently been calling at Pago Pago, and it was becoming apparent to the native chieftains that the German Empire (which had been

After he had delivered the ram *Dunderberg* to the French Government in 1867, William H. Webb interrupted his busy schedule long enough to have this portrait taken at the studios of A. Liebert & Cie., at 15 Boulevard des Capucines, Paris.

forged by Bismarck earlier that year by uniting a collection of smaller German states and which was therefore some years behind England and France in establishing colonies or even coaling stations in various remote parts of the world) was planning to give the still-unclaimed Samoan Islands the honor of becoming the first German colony in the Pacific. This honor, Captain Wakeman quickly determined, was one that the native Samoan chieftains were prepared to resist. Furthermore, the chieftains informed Wakeman that they would welcome having the United States claim the Samoan Islands as a protectorate, thus to preclude a German takeover.

Once this intelligence was forwarded to Honolulu, Admiral Winslow immediately ordered Captain Richard W. Meade of the cruiser *Narragansett* to sail to Pago Pago and to investigate the situation further. But investigation proved hardly necessary. The Samoan chieftains immediately signed a treaty with Captain Meade "granting to the United States of America the exclusive privilege of establishing in said harbor of Pago Pago, in the island of Tutuila, a naval station for the use of the United States government only" (16). Although an official treaty between Samoa and the United States was not signed until 1878 (when the six-foot "tatooed prince," La Mamea, travelled to Washington to arrange it), and the ultimate agreement with Germany was not signed until after the Spanish-American War of 1898, the steps toward an American protectorate had begun. And in 1872 William Webb received a letter from Secretary of the Navy Robeson thanking him for drawing the Navy's attention to the Samoan Islands (17).

In operation Webb's line to New Zealand and Australia, as any astute reader will already have predicted, encountered many problems. *Moses Taylor*, as already mentioned, could not meet the schedule and so was withdrawn from the service. (She made a few unscheduled freight runs between San Francisco and Hawaii to fulfill contractual obligations made before Honolulu was eliminated as a port of call for the other steamers of the line, often taking ten days for that trip compared with *Nevada*'s seven.)

*Nebraska*, whose average speed on the run was just over ten knots, also had trouble maintaining the schedule, and if she were held up by some unexpected development in port or slowed by a storm at sea, she was never able to make up lost time. *Nevada*, the fastest of the fleet, did maintain her schedules, but she was heavy on fuel and the enormous amounts of coal she had to load for her long voyages across the Pacific left virtually no space in her hold for revenue-producing cargo.

One night in October 1872, *Nevada*, under Captain Blethen, was steaming through the dark on the lonely 1200-mile stretch between Auckland and Sydney, when the small bark *A. H. Badger* suddenly appeared ahead directly in her path. There was no way to avoid a collision, and *Nevada* plowed right into the smaller craft. But *Nevada*, already behind schedule and eager not to lose more time, did not stop to investigate and soon sailed out of sight. The story reported later, though it is hard to credit, was that the pilot of *Nevada* assumed the bark would sink with all hands and that, therefore, out here in the vastness of the Pacific Ocean, no one would ever know it had happened, and there could thus be no charges of negligence or incompetence. Whether or not this version of the story is true, Captain Blethen of *Nevada* (who had

not been on the bridge at the time) did not report the accident on his arrival in Sydney.

Meanwhile, after his craft had been struck by *Nevada*, Captain Leddra, skipper of the bark, ordered all hands, which included his wife and two small children, into the small boat. And, sitting there in the darkness, they watched their ship go under only minutes later. During the following day the group was sighted and picked up by the bark *Alice Cameron*, which subsequently brought the rescued party into Sydney.

Needless to say, by that time *Nevada* was already headed back toward San Francisco. But on her next trip to Sydney two months later, Captain Blethen found himself faced with some very serious charges: not only had *Nevada* struck another ship and left her in sinking condition without stopping to investigate, the captain had also failed to report the matter to the authorities in Australia. In the trial held at the Admiralty Court, Captain Blethen was absolved of any personal responsibility for the accident on the grounds that he had not been in the wheelhouse at the time (though he later admitted seeing the sinking bark astern after the crash). Webb, however, as owner, had to pay a fine of 1600 pounds plus costs. Although pardoned by the Admiralty Court, neither Blethen nor the Webb Line was pardoned by the Australian press, which played up the details of the callous attitude of the American liner's officers after sinking this small local sailing ship. After this incident neither Captain Blethen nor his crew dared leave *Nevada* to walk the streets when their ship was in Sydney (18).

The *Badger* affair had further repercussions for Webb when a group of *Nevada*'s passengers, incensed by the negligence on the part of ship's officers they had witnessed that night, sent a joint letter to authorities in Washington just at the time Congress was discussing the Ship Subsidy Bill. Probably at least partly as a result of this letter from *Nevada*'s passengers, Webb's request for a mail contract from the American government was denied.

It was probably also the *Badger* incident that prompted authorities both in New Zealand and in Australia to begin discovering a series of minor irregularities in the operation of the Webb Line steamers and to demand petty but time-consuming repairs in the name of safety before allowing the steamers to use their docking facilities.

At about the same time, as a result of some event which we can now probably never know about, General Ambrose E. Burnside had developed a personal grudge against William H. Webb. Long after the war was over, General Burnside still harbored this grudge to the extent that he used his considerable prestige and influence to try to put Webb's Australian Line out of business. In the fall of 1872 the general planned to introduce a bill by which Congress would grant him and his partners a subsidy for a line of steamships between San Francisco and Australia. The bill would not only grant Burnside and his partners a far larger subsidy than Webb had requested, it would effectively negate Webb's original request by requiring that any vessels used on the line thus subsidized by Congress be iron propeller steamers. Although this bill, apparently, was never actually introduced, its threat and the enormous influence a former Civil War general then had in the Congress, were other factors persuading Congress to deny Webb's request for a subsidy, without which he could not continue to operate the line.

All of these problems when added together finally made William Webb realize that, once again, this steamship line was costing a great deal more than it was earning and that it was time to end the service. As *Nebraska* was preparing for her departure from San Francisco in April 1873, just two years after the inauguration of the line, Ben Holladay announced that hers would be the company's last sailing.

*Dakota*, meanwhile, with her newly rebuilt engine at last steaming smoothly, had come around to the Pacific earlier that year to replace *Santiago de Cuba*, and, as she had done when Webb was running his line to Aspinwall, she managed to make just one round trip before the company went out of business.

Webb was somewhat more fortunate this time in disposing of his steamers. All four of the ships on the Pacific—*Nebraska*, *Nevada*, *Dakota*, and even *Moses Taylor*—were purchased by Pacific Mail for use on their short-haul lines along the Pacific Coast. *Santiago de Cuba*, which had meanwhile returned to the East Coast, was purchased by Thomas Clyde, who converted her to a propeller steamer and placed her on a line between New York and Cuba, the route for which she had been built originally.

The United States, New Zealand, and Australia Steam Navigation Company represented William H. Webb's last attempt to operate a steamship line. With its failure, one can understand Webb's only slightly hyperbolic statement that all the money he had earned building ships he lost operating them.

# XIV

# Post-Retirement Enterprises, 1872-1889

ALTHOUGH William H. Webb sustained serious losses in his several unsuccessful attempts at running steamship lines on two oceans, other investments he had been making during the more prosperous years of his shipyard left him nevertheless a man of considerable means after his retirement from shipbuilding. These investments, in some instances, particularly in certain marine-oriented businesses, were sufficient to make him an active influence in the companies involved. He often sat on the boards of directors, once his shipyard closed and he was free to do so.

William Webb's involvement in some of these companies actually resulted from a long outdated practice once common in the shipping world. When Webb first began building wooden-hulled packets in the 1830s and 1840s, shipping companies which owned several vessels and sailed to specific ports on regular schedules were rare if they existed at all. More typical at that time were individual ships owned by several small investors, usually including both the captain and the New York agent, among others, and which sailed wherever a lucrative cargo might take them. When William Webb built such vessels, and, as we have seen, he built many of them, it had been his practice to include in his fee a part ownership in the vessel. (We see in his few extant letters references to losses sustained when some ship he had built, perhaps many years before, came to grief in some remote part of the world.) When, later, one of these vessels became part of one of the then-developing shipping lines, William Webb would then, by virtue of the shares in the ship which he already held, become a stockholder in the company.

In later years, when William Webb was building ships on commission from incorporated shipping lines, he tended to continue the practice of including stocks in the company as part of the cost of construction. We have already seen how, in this fashion, Webb became a major stockholder in the prosperous Pacific Mail Steamship Company, and, through this connection, in the Panama Railroad, both of which he sold in the 1870s at great profit. In another example already cited, Webb accepted stock in Jay Gould and Jim Fisk's Narragansett Steamship Company, which operated his *Bristol* and *Providence* on the famous Fall River Line. This stock became even more valuable when, after Fisk was murdered in 1872, Gould sold the company to the Old Colony Railroad, and more valuable still when the Old Colony was absorbed by the New York, New Haven and Hartford Railroad in 1893.

Moreover, from his earliest days as a shipbuilder, William Webb had been involved in the Black Ball Line, whose principal owner was his father's close friend Charles H. Marshall. Marshall, who became a man of considerable means in the shipping business, also included William Webb in several of his other lucrative ventures. It was Charles H. Marshall who in 1858 brought Webb into the profitable American Guano Company. Guano, a highly prized fertilizer at the time, consisted of sea-bird droppings which had accumulated to such depths on certain islands off the coast of Peru that the material could be mined—presumably by workers with de-sensitized olfactory nerves—and loaded onto cargo vessels for transportation to the United States. After his retirement from shipbuilding, William Webb sat for many years on the board of this shipping-related company.

In 1865 William Webb was also appointed a trustee of the Atlantic Mutual Insurance Company, a New York-based concern specializing in marine insurance. The heavy shipping losses during the Civil War had been hard on marine insurers generally, so that when the war ended this company was eager to have the services of such men as William H. Webb, known for their business acumen in the marine field. His association with the company lasted until his death. Another insurance company in which Webb had invested, the Home Insurance Company, had to ask its investors for an assessment when the Chicago Fire of 1870 so depleted its reserves that the available assets were below the amount they were legally obliged to maintain. Each stockholder was assessed 60 per cent of the value of his holdings. As Webb had invested $30,000 in the company, he had to send them another $18,000 in

1871. Even this assessment came to less than Webb had already earned in dividends, and in later years his holdings in this company continued to produce a pleasant profit.

William Webb also bought shares in the Third Avenue Railway Company, New York's first elevated railroad, when the company was formed in the late 1870s and the project was still in its conceptual stages. Because of his sizable investment and also his personal interest and involvement in the project, Webb also sat on the board of this organization, which, after it began operating in 1887, provided him with a considerable income.

In addition to his investments in specific companies, William Webb also owned many pieces of property in New York, in Brooklyn, and elsewhere. On some of these properties, Webb merely collected rents, but on most he made improvements, all thoroughly planned and personally supervised, to increase both value and return. Some of these properties had been put up as security for large loans that were not paid, so that they had devolved to Webb by default. Webb had stumbled onto this form of investment when a piece of property he had been awarded in lieu of a bad debt turned out, given the constantly rising values of real estate in the city of New York, to be more remunerative in the long run than the collection of the debt would have been. Once William Webb began to appreciate the rewards of owning property, he frequently purchased land at sheriff's sales, when it was being auctioned off at a low price for non-payment of debts or taxes.

The start of William Webb's accumulation of lucrative oil-producing properties in Pennsylvania was also the result of a fortuitous accident. It seems that in 1864 a man named Allan K. Williams was employed at William Webb's shipyard as a cashier. One day Williams was visited at the yard by one John R. M. Skinner, an artist, as he called himself, who had once been a friend of Williams' father and whom Williams remembered having seen occasionally at his home when he was a boy. In any event, Skinner, who had by then fallen on hard times, had come to ask Williams to lend him $3500. With this money, Skinner explained, he would buy oil lands in Pennsylvania from which they could both profit.

About a year later Skinner returned to the yard with glowing reports of Pennsylvania oil properties and a request for more money. The part of the story most difficult to credit, but which court testimony tends to substantiate, is that William Webb himself now handed the man a check for $40,000 (without requesting a receipt!) and asked Skinner to purchase Pennsylvania oil properties in

his name also. Several months later, Skinner (who seems to have been aptly named) was back at the shipyard, but this time with a sadder story to tell. He informed Williams and Webb that he had indeed bought property with the money, but when the land turned out to be worthless, he had subsequently sold it to "Jones" for a mere $6000. He then, with due apologies, handed Webb a $2000 down payment, which, he explained, since the land had no value, might well be all he could ever expect.

Skinner apparently had not taken the measure of the man he was dealing with. William Webb himself now travelled to Pennsylvania, where he quickly ascertained, as he already suspected, that Skinner had duped him. Skinner had indeed bought a large piece of property with Webb's $40,000, about 1000 acres of it. But far from being worthless, it turned out to be an excellent site, according to recent surveys, for oil prospecting. Skinner, it seems, after buying this property, had kept a quarter of it in his own name, sold the rest for $30,000, and then returned to Webb with $2000 and the story that the land was worthless.

Webb's first instinct was to sue Skinner. But when the case came to court, Webb discovered that, with no receipt for the $40,000 and no deeds to the property, he was not able to present a very strong case. As a result he had little choice but to accept the compromise solution ordered by the court.

According to this judgment, John Skinner, who had contributed exactly nothing of his own, was allowed to retain the one-quarter of the property already registered in his name; Williams was awarded 7/32 in consideration of his original investment of $3500; and William Webb, who had put up $40,000, was able to claim only the remaining 17/32, or just slightly more than half. What "Jones" got for his troubles, if indeed there ever was such a person, was not divulged in the course of the proceedings.

Not long after, however, local authorities notified William Webb that Skinner, not too surprisingly, had failed to pay his share of the taxes and that, therefore, there was a good chance that the entire property might be sequestered by the sheriff. To avoid losing his own share as well, Webb bought Skinner's share at the sheriff's sale for $4900. And, at about the same time, Webb also persuaded Allan K. Williams to accept $10,000 for his 7/32, so that in the end the entire property came into the possession of William H. Webb.

Once these former farm lands began producing both oil and natural gas, William Webb, often sending his son

Marshall to Pennsylvania as his agent, now started actively seeking properties in this and neighboring areas (all in Venango County), usually, as he had at times done in New York, buying up lands that turned up at sheriff's sales for fairly low prices. By 1875, William Webb owned nearly 3000 acres of oil- or gas-producing lands in Pennsylvania, all of which were ceded to Webb's Academy at his death, although by that time their productivity had been pretty well exhausted.

In a similar manner, William Webb, over the years, accumulated properties in various other areas. He made a large loan, for instance, to Mr. Dudley Sheffield, then an elderly man who was apparently a relative.* As security Webb accepted a mortgage on Sheffield's house in Darien, Connecticut. When Sheffield later died, Webb took possession of the house and subsequently sold it for $3000, a figure which, believe it or not, was in those days a very good price for a house upon a peak in Darien. In similar fashion Webb gained titles to valuable properties in Fishkill, New York; in Plainfield, New Jersey; and even in Knoxville, Tennessee, among others.

Of the many properties owned by William Webb in the cities of New York and Brooklyn, undoubtedly the finest was a large private house standing on the north side of Forty-Second Street, just one lot to the west of Fifth Avenue. By the 1870s, this section along Fifth Avenue had become the most fashionable part of town. When William and Henrietta Webb had first built their home at 415 Fifth Avenue, between Thirty-Seventh and Thirty-Eighth Streets in the 1840s, they were then moving quite far out of town. The better homes of that era, other than the few still remaining in the areas near City Hall, were along Bond Street, Bleeker Street, or Astor Place. But by the 1870s, forty years later, the Webbs, still on Fifth Avenue, found themselves living among some of New York's more fashionable families. Just south of Thirty-Fourth Street, three blocks south of the Webb home but on the west side, where the Empire State Building now stands, were two grand houses belonging to the Astor

family. Across Thirty-Fourth Street to the north of the Astors was the famous mansion of Alexander T. Stewart, proprietor of New York's first large department store (then located in a building on Broadway between Ninth and Tenth, later occupied by Wanamaker's and, since the 1960s, the site of Stewart House apartments). Other residents of lower Fifth Avenue included some of William Webb's friends from the shipping world: Marshall O. Roberts at 107, Moses Taylor at 118, George Law at 259, and Jacob A. Bell at 276. J. Pierpont Morgan's relatively modest brownstone at Thirty-Seventh and Madison was just one block east of the Webbs' home.

Despite living among the members of New York society, the Webbs did not participate in it. William Webb enjoyed the friendship of many of these people, either at business or on his infrequent visits to the Union League Club or the Century Club. But Henrietta Webb apparently preferred a quiet home life and did not much enjoy forays into New York society. One reference in a newspaper, which offers no further explanation, states that the Webbs' elder son, William E. Webb, was "an invalid confined to their home" (1). It is indicative of the private nature of the Webbs' personal lives that no other reference can be discovered to explain in just what manner the younger William Webb was an "invalid," especially considering that we know he was able to make trips to Europe with his family on at least two occasions. Although we know the date of his death (13 January 1919 at the age of seventy-five), not one New York newspaper carried an obituary notice which might have offered a clue to his condition. Whatever its nature, one might fairly assume that one of the reasons for Henrietta Webb's preference for a quiet home life might have been that her elder son required some sort of personal care.

At the time William H. Webb purchased the property nearby on Forty-Second Street near Fifth Avenue, the house faced the great stone wall of the reservoir which then occupied the area along Fifth Avenue between Fortieth and Forty-Second, where the New York Public Library now stands. But Webb, who was then working with the Aqueduct Commission on a plan to close this unsightly facility and bring water into the city from an enlarged, existing reservoir at Croton, in upper Westchester County, realized that the present reservoir was soon to be replaced by a park, thus greatly increasing the value of the property on Forty-Second Street. (No one in the 1870s could have guessed that this land on the outskirts of New York's residential area would one day become prime commercial property.) There was some

---

* The letter from Dudley Sheffield asking for a loan is the only letter among the several in the collection of William H. Webb's correspondence in the Livingston Library, which begins "Dear William." (It hardly bears mentioning that William H. Webb was never in his life referred to as "Bill" or by any name other than "William," even by his parents or other close family members.) All other letters (though admittedly there are no others from family members or close relatives), some from business associates of long standing with whom William Webb was on friendly terms, begin "Mr. William H. Webb," plus an abbreviated address, and then "Dear Sir:"

The Webb family posed for this informal group portrait about 1850 at their home in New York. In the fashion of the times, Henrietta Webb stands at the side of her seated husband, while their two sons, William E. (right) and his younger brother Marshall, are close at hand. This painting is the only know image of the Webb children.

suggestion at the time that William Webb had bought this house on Forty-Second Street either as a new home for himself or possibly for his son Marshall, who was married about that time. Whatever the case, this house was never occupied by a member of the Webb family.

In 1875 William Webb was also able to purchase the lot at the northwest corner of Fifth Avenue and Forty-Second Street, which stood between his original property and Fifth Avenue. Webb and a group of associates then formed a corporation, in which Webb was a major stockholder, and on this beautifully located lot, built the later famous Bristol Hotel. In the construction of this hotel, the house Webb had previously purchased was not dismantled. Rather the architectural firm of Griffith and Thomas, following plans and specifications drawn up for them by William Webb himself, incorporated most of this house, with all of its elegant architectural detail, into the new hotel. The first four floors of the new part of the hotel erected in the adjoining lot were designed to be at the same levels as the four stories of the house. Then three full new stories were added at the top, making seven in all, plus a functioning basement floor.

The popular Hotel Bristol in succeeding years became a valuable source of income to William Webb, yielding approximately $40,000 per annum. In 1902, three years after William Webb's death and just days after the death of Henrietta Webb, the building was sold to Walter J. Solomon, who converted the space into stores and offices; and a new Bristol Hotel was built at the southeast corner of Forty-Second Street and Broadway.

One responsibility which occupied a great deal of William Webb's time and talent during this era of so-called "retirement" gave him no monetary returns at all: this was his position as Chairman of the New York City Council on Political Reform, a position which he held from about 1876 to 1890. As is already well known, the city of New York, in the era of the Civil War and for some years after it, was controlled by a notoriously corrupt group connected with the political club at Tammany Hall. This group, headed by William Marcy "Boss" Tweed, made fortunes milking the city with inflated estimates and large kickbacks as well as by selling offices and influence. But in 1872, the third time that William Webb's friends in the Union League Club suggested (unsuccessfully) that he stand for mayor, the combination of a strong Republican resistance and a series of scathing cartoons by Thomas Nast finally forced the "Tweed Ring" out of office and sent Tweed himself and several of his associates to jail.

During this period, a group of concerned citizens, most of them unabashedly Republican, set up a private, and in theory, politically unaffiliated, committtee to monitor civic affairs, particularly major building programs which might invite graft or a degree of incompetence dangerous to the public welfare. This committtee was later constituted as the Council on Political Reform, and, since William Webb, as a man trained in matters of construction, was a particularly valuable asset to the council, he was eventually asked to serve as its chairman, a position he filled with dedication and competence until he finally begged to resign after fourteen years of unselfish public service.

Although the Council on Political Reform tackled many issues during William Webb's tenure as chairman, the one in which Webb was most personally involved was his well-publicized dispute with the Aqueduct Commission, charged with constructing the aqueduct to carry water from the new Croton Reservoir into the city. When the Commission presented its plan for building the aqueduct system, William Webb studied it and came to the conclusion that the system, as they had designed it, was inadequate for two reasons. First, it provided for only one dam at the mouth of the Croton River, whereas, according to Webb's surveys, three dams would be needed to assure that the water coming into the city was pure enough to be safe for drinking. Also, the cost of the project was so high that it obviously included an unconscionable amount of graft.

When the plan was presented to the city, the assumption of the commission was that their fellow politicians on the city council would approve it as a matter of course. But William Webb, almost single-handedly, fought the plan with such a barrage of reports, statistics, and supporting letters that it was defeated, and a new plan was subsequently submitted which included the three dams Webb had recommended and excluded most of the graft. Thus William Webb not only saved the taxpayers an immense amount of money, he also prevented the construction of an aqueduct system that could have endangered the health of the entire population of the city of New York (2).

William H. Webb's deep concern over civic and business responsibilities led to an invitation to become a member of the Chamber of Commerce of the State of New York in 1851. He remained a member until his death, and served as a Vice-President of the Chamber from 1894 to 1897. Other vice-presidents at that time included Cornelius N. Bliss, who later served as Secretary of the Interior under President McKinley; William E. Dodge, of the firm of Phelps, Dodge & Company; Cornelius Vanderbilt II, a grandson of the Commodore; and J. Pierpont Morgan. John D. Rockefeller and Andrew Carnegie became vice-presidents shortly after William Webb's term ended.**

In 1878, nearly a decade after William H. Webb had given up his shipyard, he received an unexpected honor when King Victor Emmanuel of Italy, by royal decree, bestowed on him the highly coveted Order of Sts. Maurice and Lazarus for his contribution to the Italian Royal Navy in designing and building the ironclad frigates *Re*

d'Italia* and *Re Don Luigi de Portogallo*. A similar honor proposed by the Emperor Napoleon III for building *Dunderberg* (later *Rochambeau*) however, was never granted to William Webb. Napoleon III was deposed in 1870 before the award to Webb had been arranged, and the republican government which succeeded him chose not to honor an award granted by the now thoroughly unpopular former emperor.

One beautiful piece of property, an estate with a rambling house in an interesting mixture of a romanticized mock-Tudor and the misnamed American "Queen Anne" styles, located in Tarrytown, New York, was purchased by William H. Webb in 1882, not as an investment but as a summer home for himself and his family. The house, known as "Waldheim" ("Home in the Woods"), had belonged to the recently deceased John S. Mitchell, who had been a Union general in the Civil War. Tarrytown, in northern Westchester County, an area where several other well-to-do New Yorkers, among whom Jay Gould is perhaps best known, had built summer homes, was about thirty-five miles from the Webb's Fifth Avenue residence and was easily reached by private coach in about four hours or by Mr. Vanderbilt's railroad in about one hour. Although there were other large estates in the immediate neighborhood, Waldheim itself stood on a secluded knoll overlooking the Hudson River. To further guarantee this seclusion, however, William Webb added four more parcels of land to the property.

One can assume that the Webbs followed a pattern similar to that of most wealthy New Yorkers of the day who had summer homes along the Hudson or in the Berkshires farther north. In this case Mrs. Webb and the servants would have moved to Waldheim early in June to avoid the humid New York summer and stayed there until early September. Since both sons were grown by this time, it is unlikely that they would have been there with her all of this time, although the elder son may have been. William Webb, if he followed the typical pattern of New York businessmen, and, given his personality, it is likely that he did, would either have remained in the Fifth Avenue house, perhaps with one servant, or possibly taken a room at one of his clubs during the week, travelling to Waldheim to join Henrietta on weekends.

Unfortunately, William and Henrietta Webb had no grandchildren to bring young voices into their summer home. But both had brothers and sisters and other close relatives living either in New York or in nearby areas of Connecticut. Although we do not know much more about these relatives than their names, and not even that much in some cases, we do get a sense that the Webbs were

---

** A portrait of William Webb painted in 1900—shortly after his death—by Thomas W. Wood, was presented to the Chamber by Mrs. Webb.

close to their families if only from the fact that whenever one of the Webbs' European voyages was reported in the newspapers, the names, otherwise unidentified, of people coming to see them off always included several Webbs and Hiddens.

The most likely visitors at Waldheim would have been William Webb's sister Abigale Henrietta Janes and her family, who lived in nearby Stamford, Connecticut. Noting that Mrs. Janes was the only brother or sister to receive a large legacy in William Webb's will and that Abigale's only son was named William Henry Janes, one can sense that William Webb and his younger sister Abigale must have remained on close terms. Although Webb's will did no more for Abigale's two children, William Henry Janes and Sarah Janes Buckhout, than cancel their debts to him (William Webb's private papers show evidence of many large loans to members of both the Webb and Hidden families), it did leave sizable legacies to Sarah's three daughters, again suggesting that these grandnieces helped fill, to some extent, the role of grandchildren in William Webb's later years.

William Webb sold Waldheim in 1898, probably because he and his wife, both then over eighty, were finding the maintenance of two separate residences more than they wished to handle. The house was sold, through a series of brokers, to Wallace C. Andrews, who in turn sold it two years later to the American Unitarian Association. Apparently this group allowed the house to sit idle for a while, for by the time it was sold to the Hackley School, a boys' preparatory school which had fallen heir to the estate next to Waldheim, the building had been partially destroyed by fire. The school, therefore, demolished the remains of the Webb home and later constructed a new classroom building on the site.

Finally, the most unfortunate event of William Webb's retirement was the unexpected death of his son Marshall, who died of pleural pneumonia at the age of forty-three. Marshall's death took place in the Webb home at 415 Fifth Avenue, on Saturday, 20 April 1889, the day before Easter. Not one word survives to record the reaction of either William or Henrietta Webb to this loss, but Marshall's death was probably the most difficult crisis either William or Henrietta Webb ever had to face. Only the curt notice among the obituaries in the New York papers that the funeral and interment were to be private and that the family requested friends not to send flowers was indicative of the Webbs' wish that they be left to themselves in this time of grief.

The only reference to Marshall Webb's early death in his father's hand came many years later. In 1895 William Webb purchased land at the Woodlawn Cemetery in the Bronx on which he erected a large tomb to serve as a resting place for himself and his family. There is a letter among his papers written at that time requesting, in quite businesslike language, that the remains of his son Marshall be removed from their original burial place to the family tomb.

In the newspaper accounts of Marshall Webb's death, the deceased is referred to as the son of William H. Webb, the "eminent ante-bellum shipbuilder" (3). It is true that William Webb built relatively few ships after the start of the Civil War, and that the great clippers, which were still the main source of his renown, although he had built many other larger vessels, had all been built before the war, which is to say thirty years earlier. In short, by 1889, a time when great 10,000-ton steel-hulled American ships for both coastal and oceanic runs were being produced along the Delaware and few still remembered that there had once been shipyards along Manhattan's East River, William H. Webb, then seventy-three, was already seen by the press as a man of an earlier generation, a man whose life work belonged to an era long past.

Perhaps it was a series of such references in several articles concerning William Webb, usually historically-oriented pieces based on interviews with famous Americans of an earlier time, that inspired William Webb to want to be seen, as he saw himself, not only in the past but also in the present, as a great American shipbuilder. Perhaps partly for this reason, but mostly because of his continuing deep dedication to his profession, William H. Webb in 1888 made the decision to devote the greater part of his fortune to the establishment of an institution for the benefit of past and future shipbuilders. This institution, to be known as Webb's Academy and Home for Shipbuilders, was to have two parts: one a home for elderly or indigent men who had formerly been employed in the shipbuilders' trade, and the other a university-level school dedicated to training young people who wanted to enter the profession, not only in the art of shipbuilding, which had been William Webb's great talent, but also in all of the newest techniques of ship design and construction. In this way William H. Webb, not only in the year 1888, but for many future years to come, would not be relegated to history, but would remain a living part of the world of the shipbuilder, where he knew he still belonged.

# XV

# Webb's Academy and Home for Shipbuilders, 1889-1899

IT is difficult to determine just when William H. Webb decided to devote the bulk of his fortune to establishing Webb's Academy and Home for Shipbuilders. When he first spoke about it in 1888, however, he noted that "the idea for such an institution as this has been in my mind for a long time, but it has been taking shape more especially during the last three years" (1). Questioned further on the subject, Webb went on to explain that there were many shipbuilders who,

after long lives of honest, industrious toil, find themselves, through no fault of their own, reduced to utter destitution in their old age. To provide for such men a happy and comfortable haven is one of my objects. (2)

But providing a home for elderly shipbuilders who had become destitute was only part of William Webb's vision.

Scarcely less important is the other, to educate. . . .
young men who wish to learn the shipbuilder's trade. . . .(3)

Webb's further explanation that "The education I propose to offer will be theoretical, and as practical as it can be made without actual work in a shipyard" (4) reflects his frequently expressed view that the shipbuilder's profession had suffered of late because the apprentice system no longer existed by which a young man could be properly trained in the trade.

Early in the year 1888, William Webb purchased a fine piece of property in the Bronx on a grassy hill overlooking the then picturesque Harlem River. This property of thirteen and a half acres, which lay between 188th Street and 189th Streets on Sedgwick Avenue, had apparently belonged previously to Webb's brother-in-law, Thomas B. Hidden.

In March 1888 William Webb journeyed to Albany to petition the New York State Legislature to pass a bill incorporating "Webb's Academy and Home for Shipbuilders" as a non-profit tax-free institution,

. . . for the purpose of offering free and gratuitous aid, relief, and support to aged, decrepit, invalid, indigent, distressed, or unfortunate men who have been engaged in building hulls of ships or vessels, or marine engines . . . in any section of North America, together with the lawful wives of such persons.

Also further to provide and furnish to any young man, a native of the United States, who may, upon examination, prove himself competent and of good character, and worthy, free and gratuitous education in the art and science of the profession of shipbuilding, both theoretical and practical. . . .(5)

One might note William Webb's emphasis on "good character," a phrase often repeated whenever he speaks of the qualifications for his school and which is also reflected when he makes clear that "decrepit" or "distressed" elderly shipbuilders should bring only "lawful wives" with them to be supported in his institution.

In presenting his bill to the state legislature, William Webb made assurances that he would leave this institution a sufficient endowment for it to be self-sustaining in perpetuity. This endowment was not to be in the form of cash, which could too easily be spent, but in the form of property, to wit: three very large blocks of land situated on the Hudson River and the East River and other valuable pieces of real estate located in New York State and elsewhere. (A great deal more property was bequeathed to the institution in William Webb's will.)

Shortly before this petition was presented to the legislature, a new article was added in William Webb's own handwriting, asserting that:

This institution shall in all religious matters be strictly, positively, and absolutely non-sectarian.

This bill, which clearly reflected no motivation except unselfish and charitable generosity, passed both the Assembly and the Senate with ease, but hit an unexpected snag when Governor David B. Hill refused to sign it. Governor Hill, it seems, was one of the men who had been connected with the Aqueduct Commission, whose schemes for graft had been thwarted by the unflagging opposition of William H. Webb as Chairman of the Council for Political Reform. The governor was now in a position to exact revenge, and he took advantage of it.

The Governor's patently irresponsible position opened him to some strong attacks in the press, however, of which an editorial in the Marine Journal is an example:

Governor Hill refused to sign the act to incorporate Webb's Academy and Home for Shipbuilders. In doing so, he has prostituted his official position to gratify his personal vindictiveness and arbitrarily thwarted the benevolent and philanthropic intentions of one of New York's most distinguished and honored citizens, who has by a lifetime of uprightness, integrity, and distinction as a shipbuilder and owner, reflected credit on the city and country. (6)

Whether moved by such denunciations as these or by pressures within the political system, we cannot know, but eventually Governor Hill did sign the bill, and William Webb's petition became law on 2 April 1889, one full year after Webb had first presented it to the Legislature.

By the time the charter had been run through the Legislature for the second time, it had been augmented to say that the students were to be offered not only a free "theoretical and practical education" but also "board, lodging and necessary implements and materials while obtaining such education" (7).

The charter, repeating verbatim William Webb's original petition, also provided for a Board of Trustees which would include the President of the New York Chamber of Commerce; a member of the General Society of Mechanics and Tradesmen to be designated by the Society; a professor of Applied Mathematical Science in Columbia College to be designated by the trustees thereof (although Webb requested and was subsequently rewarded with the appointment of Professor William Trowbridge); the president of the Society of New York Hospital; five shipbuilders and two marine engineers. The shipbuilders, obviously carefully selected by William Webb, were Richard Poillon, Henry Steers, Andrew Reed, Charles Cramp (of Philadelphia), and, of course, William Webb himself. The two marine engineers chosen were Thomas Rowland and Stevenson Taylor, the latter then a Vice President of W. & A. Fletcher Company in Hoboken, at the time the leading builders of steam engines in America. Albert Bogert, as it turned out, was the trustee designated by the Society of Mechanics and Tradesmen. But when Robert J. Livingston, the recently elected president of the New York Hospital wrote to Webb explaining that he could not serve, Webb penned a note to his friend Stevenson Taylor commenting that "His decision just suited me, for he is no friend of mine—a detestable person." Webb went on to note his pleasure in learning that the Vice President of the Hospital, Mr. Merrit Trimble, "a perfect gentleman and a very old friend," would serve in his stead (8).

Soon after the governor had signed the bill, this Board of Trustees met at Webb's home at 415 Fifth Avenue and immediately appointed a Building Committee of which William Webb was chairman. Arthur P. Jennings was chosen as the architect, and by the fall of 1890, excavations for a building to house Webb's Academy and Home for Shipbuilders had begun. During April of 1891 drafts of two contracts were drawn up, one with a carpenter and one with a mason. When these drafts were returned smothered in inked comments, changes, and specifications, all in the hand of William H. Webb, these carpenters and masons must have questioned whether they wished to work for so demanding a customer. But, in a surprisingly short time, finished contracts, including all of Webb's revisions, were produced, and on 30 April, in another meeting at the Webb home, the plans, specifications, contracts, and revisions were examined and approved by the Board of Trustees (9).

In November 1891, Mr. Richard Poillon, one of the Board members representing shipbuilders, died unexpectedly, whereupon William Webb offered this vacant seat to Thomas S. Marvel, a successful shipbuilder in Newburgh, New York (10).

Subsequently, when the size of the Board was enlarged, a seat was also offered to James O. Poillon, Richard Poillon's son and successor to his business. At the meeting in December 1891 at which these elections were confirmed, Stevenson Taylor also accepted the position of Vice President of Webb's Academy and Home for Shipbuilders; and from this time Mr. Taylor became the recipient of a regular barrage of short notes from William H. Webb detailing each step of the building or the organization of the Academy, or instructing Taylor to attend to some matter on Webb's behalf.

By the Fall of 1893, the Main Hall was finally nearing completion. This was a large building in the highly romantic Romanesque style then popular, consisting mainly of pressed yellow brick with porches, corners, and many other trimmings in New York brownstone. Loyal to the romantic tradition, Mr. Jennings created a building that was totally asymmetric in which unmatched wings consisted of different numbers of stories, each topped with a confusion of towers and turrets. The main or central part of the building was five stories in height, but one large section reached to seven stories. There were towers of differing heights at either end, the larger of which went up to ten stories. Fortunately, the building also had electric elevators. And, although furnished with both gas and electric lighting, the building manufactured its own electricity with a steam-operated generator in the basement.

A classic, nostalgic view of the original Webb Institute, located at Webb and Sedgwick Avenues, in the Bronx. When the Institute moved to Glen Cove in 1947, the property was sold and this lovely example of ornate Victorian architecture was demolished.

Arriving at the Academy on the entrance floor, one encountered a spacious foyer with a grand stairway. Also on this floor was a parlor, a reception room, a suite of rooms for the Matron and another for the Housekeeper, a large dining room, and the office of the Resident Manager. The second, third, and fourth floors consisted mostly of dormitory rooms, with one or two beds to a room, although on the second floor was also found the library and on the third a museum, containing Mr. Webb's wooden half-models, a large model of *Dunderberg*, and several paintings and prints from Mr. Webb's personal collection.

One had to reach the fifth floor before finding anything resembling a classroom or offices for instructors. In that part of the building where the fifth floor was the highest level, there was a large room, extending up to the rafters where was recreated a mold loft. Here students could, as William Webb had done over fifty years before, draw full-scale lines of a ship on the floor.

In the part of the building extending two floors higher, more drafting rooms and a lecture hall were to be found on the sixth floor and physics lab, a gymnasium, and a dark room, for developing photographs taken in shipyard visits, on the seventh.

By the fall of 1893, when his Academy and Home for Shipbuilders was nearing completion and was ready to begin operation, William Webb had become deeply involved in the creation of another organization which would also be of durable benefit to shipbuilders and their profession: The Society of Naval Architects and Marine Engineers. With twelve other men prominent in the shipbuilding world of that day, including Charles H. Cramp and George W. Quintard, William Webb had been a prime mover in obtaining a charter for this organization on 28 April 1893; and William H. Webb's was also the first signature on the charter, which stated the particular business and objects of the Society as:

... the promotion of practical and scientific knowledge in the arts of shipbuilding and marine engineering and the allied professions, and in the furtherance of this object, to hold meetings for social intercourse among its members, and the reading and discussion of professional papers, and to circulate by means of publication the knowledge thus obtained. (11)

With the creation of Webb's Academy and Home for Shipbuilders and the organization of the Society of Naval Architects and Marine Engineers, William Webb, now again, more than twenty years after closing his shipyard, was back where he belonged. He was no longer a man of the past. He was here and now a highly respected senior member of the present and active world of American shipbuilders.

The first meeting of the Society was held in the auditorium of the American Society of Mechanical Engineers at 12 West Thirty-First Street on 19 November 1893. Charles Cramp, who operated a large shipyard building iron- and steel-hulled ships in Philadelphia, gave the major address in which he talked about the duty of this group to return the American merchant marine to its position of leadership on the seas. One wonders what William Webb's reaction might have been, nearly thirty years after he had presided over the production of the 4000-ton *China*, one of the largest wooden-hulled vessels ever constructed, as he listened to Charles Cramp talk about the two ships then being built in his yard, the steel-hulled 11,629 ton, 20,000 horsepower sisterships, *St Paul* and *St. Louis.*

At the end of the talk, the chairman, recognizing the professional stature of one of the older men in the audience, instead of immediately opening the floor to discussion, turned to this man and said, "May I ask Mr. Webb if he will give us a few observations on Mr. Cramp's paper, so as to start these young men in the right direction" (12).

This invitation was followed by a burst of enthusiastic applause, after which William Webb rose and answered:

I consider it a compliment, Mr. President, that you should ask me to say a few words. But it is quite out of the question for me to talk. I am not able to make myself heard; ... (13)

Referring to Mr. Cramp's impassioned plea that this group engage in a fight for the regeneration of an American merchant marine, William Webb went on to comment:

....I agree most fully with all that has been said, and I think as Mr. Cramp says, we are only now beginning to fight. I have done my share. (Applause). I have been engaged these last four years in preparing an institution that may educate young men to begin this fight, and I hope they will be able to continue it. Thanking you for the compliment, I will take my seat. (14)

In the slate of officers elected later in the meeting, William Webb was listed as one of the Vice Presidents. Two years later, he was elected First Vice President and also named the Society's first Honorary Member (15).

At about the same time as the first meeting of the Society of Naval Architects and Marine Engineers, Webb's Academy and Home for Shipbuilders was ready to open its doors. Although workmen continued hammering or plastering in various corners of the building until February 1894, by 20 October 1893, the Trustees enjoyed having their first meeting in the new building, though its vast empty halls must have been somewhat intimidating. Then early in November, when several of the retired shipbuilders and their wives, henceforward known as "guests," began moving in—though the records do not state to what extent they lived up to the requirements for being decrepit, indigent, or distressed—the administration began to function, the kitchen and dining services started operating, and the business of Webb's Academy and Home for Shipbuilders was at last underway. Thus far, providing the grounds, the building, and other necessities had cost William H. Webb something just short of half a million dollars.

In the administration of the institution, William H. Webb served as President for the remainder of his life. Nor was this an honorary position; William Webb devoted full time to his activities as the institution's chief officer, supervising in detail all of its operations. The Vice President, Stevenson Taylor, was also a concerned and active part of the administration, and often had the unenviable job of very subtly taking over certain matters which William Webb, now nearing eighty, might not have been able to manage. The person most responsible for the conduct of daily affairs, however, was the Resident Manager, and Webb persuaded Mr. Andrew Reed, who had formerly served in a managerial position at a shipyard and who was now on the Board of Trustees, to take this position.

The first students, eight of them, did not arrive until mid-January 1894. Each student, before being admitted, had to pass examinations in algebra, plane and solid geometry, and plane trigonometry; and to produce certificates proving a thorough familiarity with arithmetic, the metric system, reading, writing, grammar, spelling, and the history of the United States. He also had to bring proof that his parents were financially unable to educate him elsewhere, as well as two letters attesting to his sound moral character: one from his last teacher and one (or more) from an American citizen in good standing.

The course of study at the start lasted three years and consisted of various aspects of mathematics, ship design, shipbuilding, and marine engineering, all taught by three faculty members. The two intervening summers were spent working in shipyards gaining a more practical knowledge, and thus serving as the "apprenticeship" William Webb so strongly believed to be necessary in shipbuilding and marine engineering.

Apparently some things do not change, for one of the myriad caveats presented to students in the first catalogue warned that "Examinations are held at frequent intervals throughout the year, and students who fail to maintain the required standard of scholarship are summarily dismissed." On the other hand, some things certainly do change, for the same long list of rules and regulations also announced that "Students have until ten o'clock at night to prepare for the succeeding day's work, and at that hour all lights must be extinguished." (A later bulletin changed the hour to 9:30 P.M.)

Although the "guests" had begun arriving in November 1893, and the students in January 1894, the dedication ceremony, postponed until the Episcopal Bishop of New York, Henry C. Potter, one of the principal speakers, returned from Europe, did not take place until 5 May 1894. As might be expected, this was a lengthy affair consisting mainly of long speeches punctuated only by occasional musical renditions by the West Side Vocal Society of New York.

The ceremony began with the presentation of the building, the grounds, and the endowment to the Board of Trustees by William H. Webb and a brief acceptance by the Vice President (obviously the President could not accept the gift from himself) Stevenson Taylor.* After Bishop Potter's speech, which was highly laudatory of

Mr. Webb, there was another by the Rev. Robert Collyer, Rector of the Church of the Messiah on Park Avenue and Thirty-Fourth Street, the parish of which William and Henrietta Webb were members. In the course of his talk, which was in a mercifully lighter vein than some of the others, Mr. Collyer made one particularly interesting and probably accurate observation concerning William H. Webb:

... It is a common stock phrase that every boy in the United States expects or hopes some day to be the President of the United States. Now, I do not believe that this thought ever entered into the mind of William H. Webb. From the day he first began to talk and think, he had but one idea, from the time he became an active boy and was playing about his father's shipyard, and that was to be a great shipbuilder. . . .

After some more singing by the West Side Vocal Society, William Helme, Secretary of the Shipwrights' General Committee, presented a set of embossed and framed resolutions thanking William Webb for his generosity to the members of their profession. And finally the Honorable George F. Roesch, on behalf of the wives and daughters of Mr. Webb's former employees, presented an American flag made by these women. In accepting this flag during his final remarks, William H. Webb noted, with some emotion, that the flag on the opposite tower had been personally sewn and presented by his own sister (presumably Mrs. Janes).

By the time of the dedication, Alexander Orr had replaced Charles Smith on the Board of Trustees as the representative of the New York Chamber of Commerce, and Professor Frederick R. Hutton had replaced Professor William Trowbridge as the delegate of Columbia College. The board had also been augmented by the addition of Herbert Lawrence of a New York shipbuilding firm, and William H. Fletcher of the W. & A. Fletcher Company.

Over the next five years William H. Webb remained active both as Vice President of the Society of Naval Architects and Marine Engineers, thus remaining a respected and participating member of his profession in his last years, and, to an even greater extent, as President of Webb's Academy and Home for Shipbuilders. In this position Webb was kept busy perusing applications, interviewing candidates, hiring faculty and staff, monitoring staff tensions, and fielding student or faculty complaints. On one occasion, for instance, Mr. Reed carried a fish, declared inedible by some students, from Fordham Heights down to 415 Fifth Avenue, so that William Webb could make a judgment on the matter. One hopes, in the

---

* Both of these speeches are reproduced in their entirety in Appendix B.

William H. Webb's obituary as published in the December 1899 issue of *Marine Engineering*.

name of justice, that the weather was not too warm that day.

On 19 June 1896, Andrew Reed declared a holiday called "Founders Day," on which students, guests, and administrators gathered to honor William H. Webb on his eightieth birthday. By this time and for several years to come, the guests outnumbered the students by about three to one, with an average of about forty-five guests at any given time and usually about fifteen students.

The first class of eight students to complete the three-year course of studies at Webb's Academy, graduated on 11 June 1897, with William H. Webb, as president, proudly presiding. By this time both the "Academy" and the "Home" were operating at full capacity, and within a very few years Webb's Academy was recognized nationally as the only institution in the country, other than the Naval Academy, which prepared young men adequately for the professions of Naval Architecture and Marine Engineering. By 1899, J. Irving Chaffee, then Webb's Professor of Mathematics, could write: "The fact that the demand for graduates from Webb's Academy is far in excess of the supply shows the reputation which the institution has already acquired, and vindicates the judgment of the founder."

As time progressed, however, and William H. Webb passed eighty, more and more of his duties were assumed by Stevenson Taylor, assisted by other trustees, especially, of course, by Andrew Reed, who was also Resident Manager. For example, when in 1897 the students presented the trustees with a list of grievances covering many matters but particularly, as usual, the food (in the process revealing also that a serious breach between Mrs. Emma Graf, then the Matron, and Mr. Reed hampered the institution's smooth operation), it was Stevenson Taylor and Herbert Lawrence who took the full responsibliity for handling the matter. Despite such minor problems, as the Nineteenth Century was drawing to a close and the new century was about to begin, Webb's Academy had established itself as an excellent and important educational institution.

On Monday, 30 October 1899, late in the evening, after no particular illness and with no previous warning, William Henry Webb, then eighty-three, died quite suddenly, but peacefully, in his home at 415 Fifth Avenue. In an address to the next meeting of the Society of Naval Architects and Marine Engineers, the president announced:

The Society has suffered in this year the loss of its senior founder, that venerable dean of shipbuilding in this country, Mr. William H. Webb, who has always taken part in our affairs with much interest. His achievements as a master of the art, his kindly presence, his devotions to the welfare of his fellow craftsmen, and the education of their apprentices, will long serve as a noble example to the members of this Society. (16)

But the phrase which would most have pleased William H. Webb was delivered by the Rev. Mr. Collyer at the funeral: "The name of our friend is written on all the oceans by the keels of the ships he built." (17)

# Appendix A

## William H. Webb's Family in America

The earliest known direct ancestor of William H. Webb was a Richard Webb who was born in Gloucestershire in the western part of England in the later 1500s and who was married to Mary Wilson Webb. One of their sons, also named Richard Webb, born in 1611, decided in 1630 at the age of nineteen to indenture himself in order to go to the Massachusetts colony and to make his life in America. Considering that the Plymouth Colony had then been in existence only ten years and that other settlements in Massachusetts were then just starting, it must have taken quite a spirit of adventure not to mention courage to make such a choice at that time.

Richard Webb earned his freedom in just over a year and in 1631, at the age of 20, he became one of the earliest settlers in Cambridge, Massachusetts. But Richard was apparently a wanderer, for in 1638 he joined Thomas Hooker's group, the first settlers at Hartford, Connecticut, and then in 1645, now aged thirty-four, he moved again to Norwalk, Connecticut, where he was among the very first settlers there as well. Richard Webb apparently prospered in Norwalk, for he remained there until his death twenty years later at the age of fifty-four.

It is not known at which of his stopping places Richard Webb married, but he sired five children, among whom, Joseph Webb (b.1640), was the ancestor of William H. Webb. Early in life Joseph Webb and two younger brothers moved to nearby Stamford, Connecticut, where they built and operated a sawmill, and where Joseph married Hannah Schofield in 1672.

Joseph and Hannah Webb also had five children, and again it was from the eldest, also named Joseph (b.1675), that William H. Webb was descended. This second Joseph married Mary Hoyt in 1698, when he was twenty-three, and this couple reared eight children. The third child, Epenetus, (born c.1715) remained in Stamford, worked in the family sawmill, and in the course of his life married three times and fathered a total of nineteen children. The first wife of Epenetus Webb was Elizabeth Lockwood, whom he married in 1739. Elizabeth bore only one child, also named Epenetus, in 1740, and died soon after her son was born.

This second Epenetus was married about 1765 (the year of the infamous Stamp Act), to Mary Leder, and their eldest child, Wilse (or Wilsey) Webb, born on 7 October 1767, we have already encountered. Epenetus and Mary Webb also had two other sons, William, and another Epenetus.

Wilse Webb (who would have been eight when the American Revolution began and sixteen when it ended) married Sarah Jessup in 1792 when he was twenty-five, and this couple were the parents of Mary (b. 1792), Isaac (b. 1794), Abigale (b.1797), and Harriet (b. 1799). Isaac's life has already been detailed in the text and William's aunt Abigale died in New York in 1830 at the age of 33. Of William Webb's other two aunts, Mary and Harriet, no record survives.

# Appendix B

Speeches delivered by William H. Webb and by Stevenson Taylor at the Dedication Ceremony of Webb's Academy and Home for Shipbuilders on 5 May 1894.

## STATEMENT BY THE FOUNDER, MR. WM. H. WEBB.

Ladies and Gentlemen, and Fellow Craftsmen:—You see here an establishment prepared for the reception of old people, decrepit old people, and for the young; but, there is no one who knows how these people are to be provided for in the future, and I thought, on this occasion, it would be proper for me to let you all know just what has been done to provide for the future care of this establishment. Now, for fear that I might say something which would not be prudent, I have transferred to writing what I would like to say, and with your permission I will read it; it is short, and is as follows: You have been invited, in the name of the Trustees of Webb's Academy and Home for Shipbuilders, to be present at the dedication of this building.

In their name I welcome you, and trust you will see enough before leaving the premises to induce you to come again and often, for your presence will always be a solace to the old and a source of pride and encouragement to the young.

This Institution has been erected for the benefit of both the young and the old, as the following extract from its charter will the more fully explain: "Section 3. The object of said corporation shall be to afford free and gratuitous aid, relief and support to the aged, decrepit, invalid, indigent or unfortunate men, who have been engaged in building hulls of ships or vessels, or marine engines for such, in any section of the United States, together with the lawful wives and widows of such persons, and also to provide and furnish to any young man, a native or citizen of the United States, who may, upon examination, prove himself competent, of good character and worthy, free and gratuitous education in the art, science and profession of ship building and marine engine building, both theoretical and practical, together with board, lodging and necessary implements and materials while obtaining such education." The building has now reached a readiness for occupancy after several delays.

I have long since felt a desire to create some such an institution, induced, I suppose, by the fact that I know of those who began life with me, and who have grown old, as I have, now find themselves at the end of their days, after a life of industry, good citizenship and dutiful parentage, in a feeble and next to a helpless condition, and needing aid from some source.

Besides this, I have always had a desire to provide for young men who show an ability and anxiety to become useful, and a talent for mechanics, and whose parents or friends are too poor to pay for a technical education, some means by which their desires could be encouraged, their talents developed, their tastes gratified, to the end that benefit to themselves and usefulness to their country may result.

I well remember the need of some such provision, for when I was a beginner there was none. My own experience was at night, after a hard day's work, upon a small table, with two or three tallow candles (no gas at my home in those days) stuck in wooden blocks for convenience of moving, to find a point or line after once being made.

I could have wished to establish this institution many years since, but while in my early business life and while I conducted affairs alone, I found signal and rapid success, quite the reverse came to me later while engaged with others in legitimate and more extensive business, owing to the bad faith and traitorous conduct of some of my associates, causing very heavy and next to crushing losses, but, thank God, I have been spared to see good fortune come again and the great desire of my heart accomplished.

This building is situated in a park of thirteen and one-third acres, on an elevation on the easterly side of Harlem River, and fronting on Sedgwick Avenue and the old Fordham landing road, affording an extensive and very pleasing outlook. It is a remarkably healthy location. The land forming this park has already been deeded to Webb's Academy and Home for Shipbuilders, and now I give and bequeath to said Trustees, this building in which we now are, together with all other buildings or fixtures on the premises, to be held by them and their successors forever, and for the purposes set forth in the charter under which they are organized. May they accept and receive them in the spirit in which I give them, cherish and maintain this Institution for the benefit of the young and the old, remembering they will always have the poor and needy with them.

To provide an endowment for the support and maintenance for all time of this Institution now founded, I have deeded to the Trustees (my good wife joining in the deeds) three very large blocks of land situated on the Hudson and East Rivers in this city. In addition to the above, I have made title in the name of the Trustees to other very valuable pieces of real estate in this City (my wife also joining me in making those titles). We have also deeded to the Institution several other smaller pieces of real estate in this and other states. All these, it is estimated and believed, will prove fully sufficient to support and maintain Webb's Academy and Home for Shipbuilders, in accordance with its charter, and so long as our country may need ship-builders.

## ACCEPTANCE BY MR. STEVENSON TAYLOR

Mr. Webb, Ladies and Gentlemen: Before responding directly to Mr. Webb, let me, speaking for the Board of Trustees, say a few words to you. Several of us, having been charter members of this Board, have from the beginning been fully aware of all the care and thought that has been bestowed upon this great effort to aid both young and old. All of us have seen it grow to the present glad fruition. The Board, especially the Building Committtee of which our present Resident Manager has been a valuable member, has aided this in a slight degree, but it is just to say that beside the financial requirements of so great a work, the full burden of the many details of the Institution, from the writing and passage of the charter to the search for and purchase of the land; from the directions given to the architects for preparing plans to the awarding of the contracts for the building; from the laying of the foundation to the final completion and equipment of this beautiful Home and Academy, has been borne, of his own choice, by our revered friend, Founder and President, William Henry Webb, who despite his wealth of years, has devoted to it extraordinary energy, attention, care and thoughtfulness, amounting to consecration.

I only touch upon the material side of this great project, leaving its broader and deeper significance to be spoken of by those who follow me. But we, who have been selected (and honored by his selection) to assist in the development of this great beneficence, take advantage of this opportunity to publicly express our appreciation and our willingness to do all we can for the fulfilment of its purpose. And now, Mr. Webb, the Board of Trustees of Webb's Academy and Home for Shipbuilders, accept this magnificent gift of yours with all its responsibilities, asking God in his infinite wisdom to so direct and control our acts that we may completely fulfil the desire of your heart.

# Appendix C

## Transcript of Contract between William H. Webb and C. H. Marshall & Co. for the construction of the packet ship *"Charles H. Marshall"*

Contract with W. H. Webb for building a Ship for C. H. Marshall & Co., the dimensions to be as follows:

| | |
|---|---|
| Length on Deck | 193 feet |
| Breadth of Beam | 40 feet |
| Depth of Hold | 28 feet 6 inches |

With three full decks & long floors, and a swell of about one foot on each side.

Contract includes Hull & Spars, top ironwork including iron boat davits, but excepting strapping blocks & chain for sheets, ties, slings, etc., joiner work to forecastle & houses on deck, but only framing & decking of Cabin aft, which is to be similar in arrangement & size to that on the Ship "Wm. F. Storer."

All of the top timbers, half top timbers, 4th futtocks, apron, knight heads, transoms, inner stern post, upper pieces of deadwood & all of the cants each end to be of live Oak, Locust & red Cedar, frame stanchions of Locust, balance of frame & keelsons of White Oak, Ceiling plank & beams of White Oak, Topsides of pitch pine, all of best quality of seasoned stuff, Deck plank of White pine, Hanging & Lodging Knees of large size & suitable breadths, those of lower deck White Oak others of Hacmatac. Stanchions in the lower hold to be fastened to beams & keelsons.

The lower Masts & Bowsprit to be made spars. House on deck between main & fore hatch, the length & width as may be required, the deck of which will connect with the forecastle, which will extend abaft the bulkhead of forecastle. Cabin to extend forward to the Mizzen Mast, or just far enough to inclose the Mast if required. Contract also includes one Windlass, with patent gear as usual, two Capstans, Steering Wheel as usual, Hawse and scupper leads, Hawse pipes of extra thickness, Carved work on head & Stern, three coats of paint on topsides & two coats on inside of hull & on spars & two coats of verdigris on bottom, Gilding head & Stern & Moulding, Iron Chafing plates on top of hatch combings, bitts, deck rings & eye bolts, iron cleats, chain stoppers on Catheads, One pair of Wooden pumps with Composition chambers & fittings complete & one pair extra pump boxes, Side skids, beds for spare spars, no other plumbing work or side lights or deck lights, includes also the following spare spars: One Topmast, one lower Yard, one topsail Yard and Studdingsail Booms, one topgallast Yard and topgallant Mast.

There will be iron braces 4-1/2 by 5/8 firmly fastened to the inner face of the Ceiling, one between every two beams running from bilge to lower deck lodging Knees & extending from forward of the Fore Mast to Mizzen Mast. Includes also Chain Locker, Deck Ladders & one side Ladder.

W. H. Webb hereby agrees to construct, safely launch & deliver in the East River a ship for the said Messrs. C. H. Marshall & Co. in accordance with the foregoing & in about six months & in consideration of the sum of one dollar now in hand paid, the receipt thereof in hereby acknowledged and the further sum of ($82,800.00) Eighty two thousand Eight hundred dollars to be paid as the work advances, and the said Messrs. C. H. Marshall & Co. hereby agree to pay as herein provided to said W. H. Webb the above mentioned sum upon the faithful performance of the Contract & agreement.

Dated this 12th day of October 1868

(signed) W. H. Webb
(signed) C. H. Marshall & Co.


Author's Note:

In today's business environment, where the construction of a merchant vessel requires the invocation of layer upon layer of rules and regulations as well as the generation of mountains of contract documents by legions of lawyers and accountants, it is difficult to comprehend that the contractual niceties of building *Charles H. Marshall* involved only the document transcribed above—with the possible addition of handshakes and glasses of good port all around.

The vessel was constructed in Mr. Webb's usual manner—to meet the rules of the American Shipmasters' Association (the forerunner of the American Bureau of Shipping)—and was rated "A1" by the Association's Board of Surveyors, although she was not built under their supervision. It will be noted that vessel class is not mentioned in the contract.

It will also be observed that coppering her bottom was not a part of the contract. A recently discovered letter from Rushton Peabody (a son of Captain Enoch W. Peabody, for many years master of the Black Ball liner *Neptune*) to Charles H. Marshall (a grandson of Captain Charles H. Marshall) stated that, on her first voyage, *Charles H. Marshall* was coppered at Liverpool after a crossing from New York of 72 days, and that "when she arrived at Liverpool...it was found that grass nearly a foot long had grown on her bottom, which, of course, somewhat accounted for the prolonged voyage." Contemporary newspapers state that the vessel cleared New York on 14 July 1869 and the Record of the American Shipmasters' Association for 1874 indicates that she was coppered in September of 1869. She did not return to New York until 13 November 1869, and thus the fact that she was coppered on the Mersey is confirmed. Whether this was done for economic reasons, as was frequently the case, or because timely drydocking could not be arranged at New York may never be known. In any event, the use of verdigris as a bottom coating—specified by the contract—was only a temporary measure. Verdigris, a nasty and poisonous copper salt, was not a particularly effective antifouling compound, as may be inferred from the length of grass that grew on the ship's bottom in the relatively short time from her launching to her arrival at Liverpool.

Mr. Peabody's letter, dated 1 March 1927, and Marshall's copy of the construction contract are from the collection of Marshall papers at the New York Historical Society. Copies of these documents were kindly furnished to us by Mrs. Janos Scholz, a great-granddaughter of Captain Charles H. Marshall.

# Bibliographical Comment

In assembling a biography of William H. Webb, one is confronted with an agonizing paucity of original sources. The Livingston Library at Webb Institute of Naval Architecture in Glen Cove, Long Island, possesses the entire collection of William Webb's personal papers. And there are indeed among them a number of letters in William Webb's own hand. But all relate to business. There are no letters there to his wife or his sons or his brother or his sisters. There are no letters which refer even obliquely to William Webb's feelings about any of these people. Only late in his life, at the dedication of Webb's Academy and Home for Shipbuilders on 5 May 1894 does William Webb mention with pride that his wife shares in the donations that made the creation of the Academy possible. And a newspaper account of the occasion mentions that William Webb showed emotion when he told his audience that a flag flying over one of the towers had been personally sewn and donated by his sister.

Almost all extant letters deal with one of two situations: on William Webb's trip to Europe to deliver *Dunderberg* he wrote several letters to the two men, Christian Metzger and Hugh McClellan, who were managing his shipyard in his absence; also a large collection of short notes were written in connection with the creation of Webb's Academy and the construction of the building in the Bronx to house it, most of them to Stevenson Taylor, but some to other members of the original Board of Trustees. The letters from Europe contain detailed instructions for running the yard and only incidental references to family matters, such as that an appointment was postponed as a result of his son's illness. But there are no direct references to his family and certainly none to his relationship with them or his feelings about them. Similarly, the short notes dealing with Webb's Academy, all very formal, concern only specific matters such as construction or organization.

The collection of papers at Webb Institute also includes deeds to various properties, from which some information can be deduced, but very little about Webb's personal involvement with these properties.

Also to be found at the Livingston Library is a collection of notes, known officially as "The Klinger Manuscript," assembled some years ago by Henry C. Klinger. It appears that Mr. Klinger was planning an extensive history of New York shipbuilders and shipyards which he was unable to complete before his death. His wife later donated these notes to the Livingston Library through the offices of Professor Cedric Ridgely-Nevitt, then a Professor of Naval Architecture at Webb Institute. Although these notes contain many fascinating pieces of information not to be found elsewhere, they are not organized and they deal less with William H. Webb himself than with his father Isaac Webb or with Isaac's famous mentor, Henry Eckford.

There are, of course, many volumes on the history of sailing ships which mention William H. Webb and the ships he built. But these are far too numerous and too varied in quality to list all of them here. Most are either superficial or repetitive or both, and, in any event, deal with the ships rather than the man, which is the subject of this section of the study. By far the most informative on the subject of William H. Webb's design innovations is Howard I. Chapelle's *The Search for Speed Under Sail* (New York, 1967), which was written with considerable assistance from Professor Ridgely-Nevitt and which includes many reproductions of William Webb's original drawings of packets and clippers. Also both scholarly and eminently readable is Carl C. Cutler, *Greyhounds of the Sea* (New York and London, 1930). Both Chapelle and Cutler use sound scholarship and a sagacious selection of statistical evidence to stray from the customary popularization of Donald McKay and to show that William H. Webb was the builder of America's most successful sailing ships, including clippers, in the Nineteenth Century. A.B.C. Whipple's, *The Clipper Ships* (Alexandria, Virginia, 1980) gives the clearest short account of Webb's role as a designer of clippers and also contains an interesting and nicely illustrated section showing how clippers were built. Those seeking a more detailed study, particularly of the sailing records of the ships, might consult the seven volumes of William Armstrong Fairburn's *Merchant Sail* (Center Lovell, Maine, 1945-1955), which also tends to support Webb over McKay. Especially informative on the many trans-Atlantic packets produced by Webb early in his career and clearly the definitive work on the subject is Robert Greenhalgh Albion, *Square-Riggers on Schedule: The New York Sailing Packets to England, France, and the Cotton Ports* (Princeton, 1938).

The glamor, deserved or not, associated with the California Gold Rush has invited a plethora of written works on the subject. Though many of these are popularizations and not necessarily reliable, serious studies on the subject carry valuable information both about William Webb's Clipper Ships and about the many large steamships built at his yard for the Pacific Mail Steamship Company. Of particular interest are Victor M. Bertold, *The Pioneer Steamer California, 1848-1849* (Boston and New York, 1932); John M. Pomfret, ed., *California Gold Rush Voyages, 1848-1849* (San Marino, California, 1954); and Duncan S. Somerville, *The Aspinwall Empire* (Mystic, 1983), written sympathetically but objectively by a modern day Aspinwall in-law.

Two books on trans-Pacific liners, Will Lawson, *Pacific Steamers* (Glasgow, 1927), and John Niven, *The American President Lines and its Forebears, 1848-1924* (Newark, Delaware, 1986), contain, in their introductory chapters, information otherwise unobtainable concerning William H. Webb's short-lived steamship line from San Francisco to New Zealand and Australia, as well as some material on the two Webb-built steamships, *Colorado* and *China*, which served on Pacific Mail's route to Japan and China.

The world of New York shipping is covered in many general works, but the two most informative on the subject of William H. Webb and his ships are Robert Greenhalgh Albion's exceptionally scholarly work, *The Rise of New York Port, 1815-1860* (Boston, 1939), and Richard McKay, *South Street: A Maritime History of New York* (New York, 1934), by a nephew of the Boston-based shipbuilder.

For those sufficiently well educated in the field to follow the technical language embedded in his highly florid style, John W. Griffiths' articles in his own periodicals provide the best source for the many innovative concepts of the early designers and builders of Clipper Ships, among whom William H. Webb, Donald McKay, and Griffiths himself were clearly the leaders. Those interested could consult his *Nautical Magazine and Naval Journal* published from 1853 to 1855; *Monthly Nautical Magazine and Quarterly Review* published from 1855 to 1857; or his *Treatise on the Theory and Practice of Shipbuilding*.

Many other periodical articles mention William H. Webb or the ships he built, but most repeat information found more reliably elsewhere and none deals more than superficially with his personal life. Some, however, provide otherwise unavailable data, among them: George W. Grupp, "Charles Henry Marshall: One of America's Foremost Sailing Ship Operators," *Neptune Log*, February, 1930, pp. 6-8, the only source for specific information about one of William H. Webb's closest friends and best customers; an exceptionally interesting and detailed article about the pre-Civil War shipbuilders of New York, with no author's name given, is "The Old Ship-Builders of New York" in *Harper's New Monthly Magazine*, Vol. LXV, No. 386 (July 1882) pp. 223-41; and, dealing with a very specific subject, Cedric Ridgely-Nevitt, "Henry Eckford's United States of 1831," *The American Neptune*, Vol. VIII, No. 1 (January 1948), pp. 7-10.

Information about ships, ship building, and ship launchings, but frustratingly little about William H. Webb personally, can be found in the many newspapers published in the city of New York in this era. The newspaper most likely to carry detailed information about the New York shipping scene was the *New York Herald*. Both the *New York Times* and the *New York Tribune*, however, also carried valuable information and insights, as did the many other local newspapers consulted.

In general, we must sadly conclude that extant detailed information about the life of William H. Webb simply does not exist.

# References

Chapter I— William Webb's Predecessors: New York Ship-builders, 1800-1815

1. John H. Morrison, *History of New York Ship Yards* (New York, 1909), p. 22.
2. *Klinger Manuscript*, Livingston Library, Webb Institute of Naval Architecture, Glen Cove, N. Y.
3. *Ibid.*
4. *Ibid.*
5. Howard I. Chapelle, *The Search for Speed Under Sail* (New York, 1967), pp. 255-256.
6. Morrison, p. 36.
7. *Ibid.*, p. 40.
8. *Klinger Ms.*
9. Morrison, p. 44.
10. Richard C. McKay, *South Street: A Maritime History of New York* (New York, 1934), pp. 68-73.
11. *Klinger Ms.*

Chapter II— Isaac Webb: Shipwright, 1815-1831

1. *Klinger Ms.*
2. *Ibid.*
3. Cf. Cedric Ridgely-Nevitt, *American Steamships on the Atlantic* (Newark, Delaware, 1981), pp. 40-45 for lines drawings, inboard and outboard profiles, and engine renderings of *Chancellor Livingston*.
4. *Ibid.*
5. *Klinger Ms.*
6. Ridgely-Nevitt, *American Steamships*, pp. 52-53.
7. *New York Commercial Advertiser*, 26 April 1820, quoted in Ridgely-Nevitt, *American Steamships*, p. 53.
8. *Klinger Ms.*
9. Morrison, p.36.
10. "The Old Ship-Builders of New York," *Harper's New Monthly Magazine*, Vol. LXV, No. 386, (July, 1882), p. 230.
11. Howard I. Chapelle, *The Search for Speed Under Sail*, 1700-1855 (New York, 1967), p. 257, q.v. for an excellent reproduction of Isaac Webb's lines drawings for *Hercules* (later *Bolivar*).
12. *Harper's*, July 1882, p.229.
13. Morrison, pp. 64-65.
14. Robert Greenhalgh Albion, *The Rise of New York Port, 1815-1860* (New York, 1939), p. 302.
15. *Harper's*, July 1882, p. 230.
16. A.B.C. Whipple, *The Clipper Ships* (Alexandria, Virginia, 1981), p. 52.
17. *Ibid.*
18. Chapelle, *Speed Under Sail*, p. 258.
19. Dr. James E. DeKay, *Sketches of Turkey in 1831 and 1832*, quoted by Cedric Ridgely-Nevitt, "Henry Eckford's United States of 1831," *The American Neptune*, Vol. VIII, No. 1 (January 1948), 7-11.
20. *Harper's*, July 1882, p. 229.
21. *Ibid.*

Chapter III— William H. Webb: Apprentice Shipwright, 1831-1840

1. "The Port of New York," Clipping in Livingston Library, Webb Institute of Naval Architecture, from an unnamed periodical (probably Marine Journal). No author; no date (but clearly after 1890).
2. *Ibid.*
3. *Ibid.*
4. *Harper's*, July 1882, p. 228.
5. Chapelle, *Speed Under Sail*, p. 283.
6. L.H. Boole, *The Shipwright's Handbook and Draughtsman's Guide* (Milwaukee, 1858), quoted in Chapelle, *Speed Under Sail*, p. 283.
7. Chapelle, *Speed Under Sail*, p. 321.
8. Whipple, *Clipper Ships*, p. 19.
9. Chapelle, *Speed Under Sail*, p. 275.
10. Erik Heyl, *Early American Steamers* (Buffalo, 1953), I, p.211.

Chapter IV— William H. Webb: Shipbuilder, 1840-1847

1. *Harper's*, July, 1882, p. 239.
2. Albion, p. 289.
3. *Harper's*, July 1882, p. 289.
4. *Ibid.*, p. 240.
5. *Ibid.*, p. 235.
6. *Ibid.*
7. *Ibid.*
8. Chapelle, *Speed Under Sail*, p. 286; Whipple, *Clipper Ships*, p. 23.
9. Morisson, *New York Shipyards*, p. 143.
10. Whipple, *Clipper Ships*, p. 24.
11. Chapelle, *Speed Under Sail*, p. 276. See also opposite p. 285 for both a lines drawing and an outboard profile of *Yorkshire*.
12. Whipple, *Clipper Ships*, pp. 25-31.
13. Chapelle, *Speed Under Sail*, pp. 289-91; McKay, p. 312.
14. Chapelle, *Speed Under Sail*, p. 327; Whipple, *Clipper Ships*, p. 28.
15. Chapelle, *Speed Under Sail*, p. 322; Whipple, *Clipper Ships*, p. 29.
16. Chapelle, *Speed Under Sail*, pp. 327-329.
17. *Ibid.*, p. 237.

Chapter V— Busy Years in the New York Shipyards, 1847-1848

1. John Haskell Kemble, *The Panama Route* (Berkeley and Los Angeles, 1943), pp. 7-15.
2. Ridgely-Nevitt, *American Steamships*, pp. 128-139.
3. *Ibid.*, p. 141.
4. *Ibid.*
5. Kemble, *Panama Route*, pp. 17-18.
6. *Ibid.*, pp. 20-30.
7. *Ibid.*
8. *Ibid.*

9. Duncan S. Somerville, *The Aspinwall Empire* (Mystic, Connecticut, 1983), p. 27.
10. *Ibid.*, p. 29.
11. *Ibid.*
12. Ridgely-Nevitt, *American Steamships*, pp. 114-121.
13. Kemble, *Panama Route*, p. 27.

Chapter VI— Pacific Mail and the Gold Rush, 1848-1850

1. Kemble, *Panama Route*, p. 33.
2. Niven, pp. 18-19.
3. Kemble, *Panama Route*, pp. 33-35.
4. Somerville, *The Aspinwall Empire*, p. 42.
5. Kemble, *Panama Route*, p. 36.
6. *Ibid.*
7. Ridgely-Nevitt, *American Steamships*; Heyl, I, p.431
8. McKay, *A Maritime History of New York*, p. 313.
9. *Ibid.*
10. Somerville, *The Aspinwall Empire*, pp. 44-46.
11. *Ibid.* p. 46.
12. *Ibid.*
13. Somerville, *Passim.*
14. Ridgely-Nevitt, *American Steamships*, pp. 115-16.
15. Somerville, *The Aspinwall Empire*, p. 54.
16. *San Francisco Herald*, 20 November 1851, as quoted in Kemble, *Panama Route*, p. 40.
17. Kemble, *Panama Route*, p. 40.

Chapter VII— The Clipper Ships, 1850-1856

1. Richard McKay, *A Maritime History of New York*, pp. 312-314.
2. Chapelle, *Speed Under Sail*, p. 327.
3. *Ibid.*
4. Morrison, *New York Ship Yards.*
5. Chapelle, *Speed Under Sail*, p. 359.
6. *Ibid.*, p. 362.
7. *Ibid.*, p. 339.
8. *Ibid.*, pp. 361-362.
9. McKay, *A Maritime History of New York*, p. 317.
10. A. B. C. Whipple, *The Challenge* (New York, 1987), pp. 69-72.
11. Carl C. Cutler, *Greyhounds of the Sea* (New York and London, 1930), p. 184.
12. Chapelle, *Speed Under Sail*, p. 347.
13. McKay, p. 352.
14. Whipple, *Challenge*, p. 51.
15. *Ibid.*, p. 150.
16. *Ibid.*, pp. 182-190.
17. *Ibid.*, p. 201.
18. Cutler, p. 227.
19. McKay, p. 324.
20. Morrison, *New York Ship Yards*, p. 145.
21. Chapelle, *Speed Under Sail*, p. 347.
22. McKay, p. 352.
23. *Ibid.*, p. 351.
24. Whipple, *Clipper Ships*, pp. 110-112.
25. McKay, p. 377.

Chapter VIII— New York Shipyards and the Economic Decline of 1856-1857

1. Heyl, I, pp. 379-380.
2. Kemble, *Panama Route*, pp. 66-71.
3. *Ibid.*
4. Heyl, I, pp. 379-380.
5. Morrison, *New York Shipyards*, p. 106.
6. *Ibid.*
7. Somerville, pp. 43-61.
8. *Ibid.*
9. Kemble, pp. 58-94, *Passim.*
10. Morrison, *New York Shipyards*, pp. 111, 149, 153.
11. Letter from John W. Griffiths to William H. Webb, Private papers of William H. Webb, Livingston Library.
12. Heyl, I, pp. 171-172.

Chapter IX— Webb Rides Out a Depression, 1857-1861

1. *New York Herald*, 8 November 1857.
2. Philip E. Yanaway, "The United States Revenue Cutter Harriet Lane, 1857-1884," *The American Neptune*, Vol. XXXVI, No. 3 (July 1976), p. 174.
3. *Ibid.*
4. *New York Herald*, 8 November 1857.
5. *Harper's*, July 1882, p. 240.
6. *American Neptune*, July 1976, p. 174.
7. *New York Express*, 22 September 1857.
8. *Harper's*, July 1882, p. 240.
9. Undated clipping from the *New York Herald* in the collection of William H. Webb's private papers at the Livingston Library.
10. *Journal of Commerce*, 7 December 1859.
11. *Harper's*, July 1882, p. 240.
12. Kemble, *Panama Route*, p. 82.
13. *Ibid.*, p. 92.
14. Ridgely-Nevitt, *American Steamships*, p. 251.
15. Kemble, *Panama Route*, p. 95.
16. *Ibid.*, p. 92.
17. *Ibid.*, p. 93.
18. *Ibid.*, p. 94
19. Ridgely-Nevitt, *American Steamships*, p. 251.
20. *Ibid.*, p. 252.

Chapter X— The War Years, 1861-1865

1. Letter from William H. Webb to Gideon Welles, Secretary of the Navy, Private Papers of William H. Webb, Livingston Library.
2. *New York Herald*, 1 November 1899.
3. *Ibid.*
4. *New York Times*, 31 October 1899.
5. Private Papers of William H. Webb, Livingston Library.
6. *Neptune Log*, June 1929, p. 4.
7. Will Lawson, *Pacific Steamers* (Glasgow, 1927), pp.25-26.

Chapter XI— Bristol, Providence and China, 1865-1867

1. *New York Evening Post*, 24 July 1865.
2. *Ibid.*

3. Fred Erving Dayton, *Steamboat Days* (New York, 1925), p. 199.
4. John Haskell Kemble, *Side-Wheelers Across the Pacific* (San Francisco, 1942), pp. 6-7.
5. Heyl, I, p. 193.
6. *New York Evening Post*, 22 May 1866.

Chapter XII— The North American Steamship Company, 1867-1868

1. John Haskell Kemble, *Panama Route*, p. 99.
2. *Ibid.*, p. 96.
3. *Ibid.*, p. 100.
4. John Haskell Kemble, *Side-Wheelers Across the Pacific*, pp. 6-7.
5. Heyl, I, 19.
6. *Photographic Portraits of American Ocean Steamships 1850-1870* (Providence, 1986), p. 27.
7. Heyl, I, 382.
8. *New York Evening Post*, 26 January 1866; McKay, p. 314.
9. McKay, p. 314.
10. Letter 4, Paris, 13 August, 1867, from a collection of letters numbered 1-23 from William H. Webb to C. Metzger and H. McClellan, at the Livington Library.
11. Letter 5, Paris, 16 August 1867.
12. Letter 3, Cherbourg, 11 August 1867.
13. Letter 4, Paris, 13 August 1867.
14. Letter 5, Paris, 16 August 1867.
15. Letter 7, Paris, 25 August 1867.
16. Letter 8, Paris, 25 August 1867.
17. *Ibid.*
18. Letter 9, Paris, 29 August 1867.
19. Letter 15, Paris, 19 September 1867.
20. Letter 16, Berlin, 28 September 1867.
21. *Ibid.*
22. Letter 18, Paris, 7 October 1867.
23. Letter 17, Hamburg, 1 October 1867.
24. Letter 20, Paris, 13 October 1867.
25. Letter 22, Paris, 21 October 1867.
26. *Photographic Portraits*, p. 7.
27. "San Francisco News of the World," as quoted in Will Lawson, *Pacific Steamers*, pp. 162-163.
28. *New York Tribune*, 3 October 1868.
29. Kemble, *Panama Route*, p. 199.
30. Private papers of William H. Webb, Livingston Library.

Chapter XIII— Closing Shop, 1868-1872

1. *New York Tribune*, 16 May 1870.
2. *Harper's*, July 1882, p. 241.
3. *New York Times*, 29 January 1867.
4. *Ibid.*
5. Newspaper clipping among William H. Webb's personal papers. No name, no date. Livingston Library.
6. *Ibid.*
7. *Ibid.*
8. *Harper's Weekly Magazine*, 11 November 1866, m.p.
9. Letter from William H. Webb to Christian Metzger and Hugh McClellan, Paris, 21 July, 1867. Private papers of William H. Webb. Livingston Library.
10. *Ibid.*
11. *Ibid.*
12. *New York Herald*, 10 April 1870.
13. Notes in private collection of C. Spanton Ashdown, Member, Steamship Historical Society of America.
14. Letter from William H. Webb to Christian Metzger and Hugh McClellan, San Francisco, 30 November 1870. Private papers of William H. Webb. Livingston Library.
15. *San Francisco Evening Bulletin*, 20 April 1871.
16. *Neptune Log*, June 1929, p. 5.
17. *Ibid.*; Thomas A. Bailey, *A Diplomatic History of the American People* (Englewood Cliffs, 1940), pp. 422-424.
18. Will Lawson, *Pacific Steamers*, pp. 166-167.

Chapter XIV— Post-Retirement Enterprises, 1872-1889

1. *New York Herald*, 1 November 1899.
2. *Neptune Log*, June 1929, p. 5.
3. *New York Herald*, 22 April 1889.

Chapter XV— Webb's Academy and Home for Shipbuilders, 1889-1899

1. Clipping from an unnamed and undated journal in the collection of William H. Webb's private papers at the Livingston Library.
2. *Ibid.*
3. *Ibid.*
4. *Ibid.*
5. From the printed petition for a charter presented by William Webb to the New York State Legislature in April 1888, in the collection of William Webb's private papers, Livingston Library.
6. *Marine Journal*, 7 July 1888.
7. Charter of Webb's Academy and Home for Shipbuilders as approved by the New York State Legislature and signed into law by the Governor on 2 April 1889; from the collection of William H. Webb's private papers, Livingston Library.
8. Letter from William H. Webb to Stevenson Taylor, 24 November 1890, in the collection of William H. Webb's private papers, Livingston Library.
9. Letter from William H. Webb to Stevenson Taylor, 9 May 1890.
10. Letter from William H. Webb to Thomas S. Marvel, 10 December 1891.
11. Original charter of The Society of Naval Architects and Marine Engineers, 28 April 1893.
12. *Transactions*, The Society of Naval Architects and Marine Engineers, Vol. 1, 1893, p. 17.
13. *Ibid*, p. 17.
14. *Ibid*, p. 18.
15. *Transactions*, The Society of Naval Architects and Marine Engineers, Vol. 3, 1895, p. xxiv.
16. *Transactions*, The Society of Naval Architects and Marine Engineers, Vol. 7, 1899, p. xxxii.
17. *New York Herald*, 5 November 1899.

# The Ships of William H. Webb
by
William du Barry Thomas

# SHIPS BUILT BY WILLIAM H. WEBB
## Index by Hull Number

| | | | | | | | | |
|---|---|---|---|---|---|---|---|---|
| 1 | MALEK ADHEL | Brig | 1840 | | 54 | JOSEPH WALKER | Packet | 1850 |
| 2 | JAMES EDWARD | Ship | 1840 | | 55 | UNION | SW Steamship | 1851 |
| 3 | AGNES | Ship | 1841 | | 56 | GOLDEN GATE | SW Steamship | 1851 |
| 4 | HELENA | Pre-Clipper | 1841 | | 57 | SAMUEL M. FOX | Packet | 1850 |
| 5 | LIBERTY | Packet | 1842 | | 58 | ISAAC BELL | Packet | 1851 |
| 6 | WALLABOUT | Stm Ferry | 1842 | | 59 | GAZELLE | Extr Clipper | 1851 |
| 7 | NEW YORK | Stm Ferry | 1842 | | 60 | CHALLENGE | Extr Clipper | 1851 |
| 8 | PRONTA | Sloop | 1842 | | 61 | GREAT WESTERN | Packet | 1851 |
| 9 | VIVA | Sloop | 1842 | | 62 | COMET | Extr Clipper | 1851 |
| 10 | LIGERA | Schooner | 1842 | | 63 | INVINCIBLE | Med Clipper | 1851 |
| 11 | MONTEZUMA | Packet | 1843 | | 64 | SWORDFISH | Med Clipper | 1851 |
| 12 | COHOTA | Pre-Clipper | 1843 | | 65 | JAMES ADGER | SW Steamship | 1852 |
| 13 | YORKSHIRE | Packet | 1844 | | 66 | PLANDOME | Schooner | 1852 |
| 14 | VIGILANT | Sch Rev Ctr | 1843 | | 67 | MANHASSET | Schooner | 1852 |
| 15 | ZURICH | Packet | 1844 | | 68 | ANNAWAN | Ship | 1852 |
| 16 | MONTAUK | Pre-Clipper | 1844 | | 69 | ROBERT MILLS | Bark | 1852 |
| 17 | RAMON DE ZALDO | Brig | 1844 | | 70 | AUSTRALIA | Packet | 1852 |
| 18 | PANAMA | Pre-Clipper | 1844 | | 71 | GEORGE LAW | SW Steamship | 1853 |
| 19 | HAVRE | Packet | 1845 | | 72 | FLYING DUTCHMAN | Med Clipper | 1852 |
| 20 | SILAS HOLMES | Packet | 1845 | | 73 | AUGUSTA | SW Steamship | 1853 |
| 21 | FIDELIA | Packet | 1845 | | 74 | REEMPLAZO | Schooner | 1852 |
| 22 | GENIL | SW Steamer | 1846 | | 75 | VOLANTE | Brig | 1853 |
| 23 | MARMION | Packet | 1846 | | 76 | FANNY | Barkentine | 1853 |
| 24 | COLUMBIA | Packet | 1846 | | 77 | KNOXVILLE | SW Steamship | 1854 |
| 25 | WILLIAMSBURG | Stm Ferry | 1846 | | 78 | YOUNG AMERICA | Extr Clipper | 1853 |
| 26 | BAVARIA | Packet | 1846 | | 79 | FLYAWAY | Med Clipper | 1853 |
| 27 | ADMIRAL | Packet | 1846 | | 80 | SAN FRANCISCO | SW Steamship | 1853 |
| 28 | SIR ROBERT PEEL | Packet | 1846 | | 81 | JOSEPHINE | Brig | 1855 |
| 29 | NEW YORK | Packet | 1847 | | 82 | SNAP DRAGON | Bark | 1853 |
| 30 | ISAAC WRIGHT | Packet | 1847 | | 83 | JOHN BRIGHT | Packet | 1854 |
| 31 | UNITED STATES | SW Steamship | 1848 | | 84 | MILTON | Bark | 1853 |
| 32 | IVANHOE | Packet | 1847 | | 85 | CULTIVATOR | Packet | 1854 |
| 33 | YORKTOWN | Packet | 1847 | | 86 | HARVEST QUEEN | Packet | 1854 |
| 34 | LONDON | Packet | 1848 | | 87 | THORNTON | Packet | 1854 |
| 35 | CALEB GRIMSHAW | Packet | 1848 | | 88 | HOUSTON | Bark | 1854 |
| 36 | AJAX | SW Towboat | 1848 | | 89 | PELAYO | SW Steamer | 1854 |
| 37 | CALIFORNIA | SW Steamship | 1848 | | 90 | GENERAL ADMIRAL | Stm Frigate | 1859 |
| 38 | PANAMA | SW Steamship | 1848 | | 91 | AURORA | Packet | 1854 |
| 39 | CHEROKEE | SW Steamship | 1848 | | 92 | JAMES FOSTER, JR. | Packet | 1855 |
| 40 | TENNESSEE | SW Steamship | 1849 | | 93 | NEW ORLEANS | Packet | 1855 |
| 41 | SAMUEL M. FOX | Schooner | 1849 | | 94 | NEPTUNE | Packet | 1855 |
| 42 | GOLIAH | SW Towboat | 1849 | | 95 | ALAMO | Bark | 1855 |
| 43 | GUY MANNERING | Packet | 1849 | | 96 | ASTORIA | Screw Stmshp | 1855 |
| 44 | GALLIA | Packet | 1849 | | 97 | TEXAS | Bark | 1855 |
| 45 | JAMES DRAKE | Packet | 1849 | | 98 | SABINE | Brig | 1855 |
| 46 | ALBERT GALLATIN | Packet | 1849 | | 99 | AMERICA | SW Steamship | 1855 |
| 47 | CATHARINE | Ship | 1849 | | 100 | SILAS WRIGHT | Packet | 1856 |
| 48 | MANHATTAN | Packet | 1850 | | 101 | FANNY HOLMES | Bark | 1855 |
| 49 | ISAAC WEBB | Packet | 1850 | | 102 | JOHN H. ELLIOTT | Packet | 1856 |
| 50 | VANGUARD | Packet | 1850 | | 103 | ALICE TAINTER | Bark | 1856 |
| 51 | FLORIDA | SW Steamship | 1851 | | 104 | CUBA | SW Steamer | 1856 |
| 52 | ALABAMA | SW Steamship | 1851 | | 105 | INTREPID | Med Clipper | 1856 |
| 53 | CELESTIAL | Extr Clipper | 1850 | | 106 | GUATEMALA | SW Steamer | 1856 |

| | | | |
|---|---|---|---|
| 107 | OCEAN MONARCH | Packet | 1856 |
| 108 | UNCOWAH | Med Clipper | 1856 |
| 109 | WILLIAM H. WEBB | SW Towboat | 1856 |
| 110 | BLACK HAWK | Med Clipper | 1857 |
| 111 | ROGER A. HEIRN | Packet | 1857 |
| 112 | TRIESTE | Bark | 1857 |
| 113 | MOSES TAYLOR | SW Steamship | 1858 |
| 114 | RESOLUTE | Packet | 1857 |
| 115 | HARRIET LANE | Stm Rev Ctr | 1858 |
| 116 | JAPANESE | Stm Corvette | 1858 |
| 117 | MARTINHO DE MELLO | Bark | 1858 |
| 118 | YORKTOWN | SW Steamship | 1859 |
| 119 | HARVEST QUEEN | Bark | 1860 |
| 120 | MISSISSIPPI | SW Steamship | 1861 |
| 121 | ALEXANDER MARSHALL | Packet | 1860 |
| 122 | CONSTITUTION | SW Steamship | 1861 |
| 123 | RE D'ITALIA | Stm Ironclad | 1863 |
| 124 | RE DON LUIGI DI PORTOGALLO | Stm Ironclad | 1864 |
| 125 | GOLDEN CITY | SW Steamship | 1863 |
| 126 | SACRAMENTO | SW Steamship | 1864 |
| 127 | COLORADO | SW Steamship | 1865 |
| 128 | DUNDERBERG | Stm Ironclad | 1867 |
| 129 | HENRY CHAUNCEY | SW Steamship | 1865 |
| 130 | MONTANA | SW Steamship | 1866 |
| 131 | BRISTOL | SW Steamer | 1867 |
| 132 | PROVIDENCE | SW Steamer | 1867 |
| 133 | CHINA | SW Steamship | 1867 |
| 134 | JAMES A. BORLAND | Bark | 1869 |
| 135 | CHARLES H. MARSHALL | Packet | 1869 |

# SHIPS BUILT BY WILLIAM H. WEBB

## Index by Type

| 74 | REEMPLAZO | 1852 |

**Sloops—(2)**

| 8 | PRONTA | 1842 |
| 9 | VIVA | 1842 |

## MERCHANT STEAM VESSELS—(38)

**Sidewheel Steamships—(25)**

| 31 | UNITED STATES | 1848 |
| 37 | CALIFORNIA | 1848 |
| 38 | PANAMA | 1848 |
| 39 | CHEROKEE | 1848 |
| 40 | TENNESSEE | 1849 |
| 51 | FLORIDA | 1851 |
| 52 | ALABAMA | 1851 |
| 55 | UNION | 1851 |
| 56 | GOLDEN GATE | 1851 |
| 65 | JAMES ADGER | 1852 |
| 71 | GEORGE LAW | 1853 |
| 73 | AUGUSTA | 1853 |
| 77 | KNOXVILLE | 1854 |
| 80 | SAN FRANCISCO | 1853 |
| 99 | AMERICA | 1855 |
| 113 | MOSES TAYLOR | 1858 |
| 118 | YORKTOWN | 1859 |
| 120 | MISSISSIPPI | 1861 |
| 122 | CONSTITUTION | 1861 |
| 125 | GOLDEN CITY | 1863 |
| 126 | SACRAMENTO | 1864 |
| 127 | COLORADO | 1865 |
| 129 | HENRY CHAUNCEY | 1865 |
| 130 | MONTANA | 1866 |
| 133 | CHINA | 1867 |

**Screw Steamship—(1)**

| 96 | ASTORIA | 1855 |

**Sidewheel Steamers—(6)**

| 22 | GENIL | 1846 |
| 89 | PELAYO | 1854 |
| 104 | CUBA | 1856 |
| 106 | GUATEMALA | 1856 |
| 131 | BRISTOL | 1867 |
| 132 | PROVIDENCE | 1867 |

**Sidewheel Ferries—(3)**

| 6 | WALLABOUT | 1842 |
| 7 | NEW YORK | 1842 |
| 25 | WILLIAMSBURG | 1846 |

**Sidewheel Towboats—(3)**

| 36 | AJAX | 1848 |
| 42 | GOLIAH | 1849 |
| 109 | WILLIAM H. WEBB | 1856 |

## NAVAL AND OTHER GOVERNMENT VESSELS—(7)

**Steam Ironclad Warships—(4)**

| 90 | GENERAL ADMIRAL | 1859 |
| 123 | RE D'ITALIA | 1863 |
| 124 | RE DON LUIGI DI PORTOGALLO | 1864 |
| 128 | DUNDERBERG | 1867 |

**Steam Corvette—(1)**

| 116 | JAPANESE | 1858 |

**Revenue Cutters—(2)**

| 14 | VIGILANT | 1843 |
| 115 | HARRIET LANE | 1858 |

# The Ships of William H. Webb

From 1840, when William H. Webb assumed the helm of the shipyard of his late father, Isaac Webb, until 1869, when, in the shipbuilding nadir that followed the Civil War, his last packet ship was built, a total of 135 vessels emerged from the modest premises that faced New York's East River at the foot of East 5th to 7th Streets. The yard's output consisted of nearly every type of sail and steam-propelled craft, from record-breaking clippers and immense ocean-going paddle steamships to modest fishing sloops and workaday cargo carriers. Vessels of war were represented, but Mr. Webb focused his attention on these only after a downturn in commercial shipbuilding in the late 1850s.

He is perhaps best remembered for a handful of ships that stole headlines when they were first built—the clippers *Young America*, *Challenge* and *Comet*; the ram *Dunderberg*; the steam frigate *General Admiral*; and the Pacific Mail sidewheelers—but fully one-third of the yard's production was devoted to packet ships, sailing vessels that quietly carried transatlantic passengers and cargo from after the War of 1812 until they were superseded by steamships at about the time of the Civil War. The Black Ball, Swallowtail, Dramatic, Black Star and other packet lines operated their ships on a regular schedule (or as near to one as was possible on the fickle North Atlantic) and, at a time when there was little competition from foreign bottoms, showed the American flag in the many ports of Europe. Westbound, the packets brought hundreds of thousands of European emigres to the United States and delivered the iron rails that made possible the westward expansion of our railroads. On their eastbound passages, they carried the American flour, wheat, corn and naval stores that Europe consumed in prodigious quantities. But the packets were more than simple drudges plodding four times a year back and forth across the angry waters of the North Atlantic. They never received the acclaim of the clippers, but they were capacious, reliable and, in some cases, fast vessels that incorporated the best of the art and science of naval architecture.

Following these introductory paragraphs, short sketches are presented that describe the lives of all of the vessels built by William H. Webb after the death of his father in 1840. The first few vessels, built during the years 1840 through 1843, are sometimes shown as having been built by the firm of Webb & Allen. However, the partnership was in effect a one-sided affair. Mr. Webb was the master carpenter—the shipbuilder—Mr. Allen having for several years been a silent partner, probably not participating in the day-to-day affairs of the shipyard; certainly not as a practical shipbuilder. Three of the vessels contracted for by Mr. Webb (*Martinho de Mello*, *Sacramento* and *Montana*—hull numbers 117, 126 and 130 respectively) were, because of prior commitments in his busy shipyard, subcontracted to his brother, Eckford Webb, whose yard was across the East River in the then independent village of Greenpoint.

About 1895 William H. Webb published a two-volume folio of plans (the "Book of Plates") consisting mainly of lines, sail plans and arrangement drawings of representative vessels built by him between 1840 and 1869. Included are detailed plans of *General Admiral* and *Dunderberg*, two of his most important and interesting accomplishments, as well as a remarkable inboard profile of *Ocean Monarch*, showing in detail the construction of that enormous packet. The plans of nearly all the vessels depicted in the Book of Plates include short paragraphs relating to launching data, speed, sea-kindliness, carrying capacity and other vessel characteristics. It is commonly accepted that these paragraphs were written by Mr. Webb, and that they reflected the criteria by which he judged the success of his own ship designs.

These comments are of interest for the information they present relating to the vessels, such as the degree of completeness of large steam and sailing vessels at the time of launching. But they are of far greater value for the insight they provide, through the medium of his own words, in assessing the stature of Mr. Webb as a naval architect and shipbuilder—whether, for example, a vessel was a "good sea boat" or a "fast sailer." The comments, set in italic type, are printed in their entirety following each of the ship biographies. Bracketed notes have been added where necessary to clarify the text and a few minor changes have been made to Mr. Webb's remarkable and free-wheeling style of punctuation.

The dimensions and tonnage shown below have been taken from a number of sources, and are presented to indicate the approximate size of the individual ships. Because the dimensions used by the admeasurer in determining the tonnage of a ship may not agree with the moulded dimensions to which the shipbuilder has worked (this situation being as true today as it was in the mid-19th century) the reader is cautioned that various sources may quote different figures. Wherever possible, the dimensions shown for a vessel in the Book of Plates have been used. In other cases, the measurements shown in Mr. Webb's book of offsets have been quoted.

Tonnage varied over the lives of many vessels for two reasons: physical alterations to the ship and changes in the tonnage rules. In 1864 the United States adopted the British Moorsom system of admeasurement, and all vessels were readmeasured. Of Mr. Webb's ships, only the last few were admeasured initially under the new rules. In the following text, the tonnage that appeared on a vessel's first register or enrollment is used, except in the case of vessels of war, where no set rules seem to apply.

In general, the tonnage, or burthen, of a vessel admeasured under the old rules approximated her cargo deadweight. However, many factors such as the fineness of her lines (or the sharpness of her model, as Mr. Webb and other ship designers and builders of the era would have expressed it) might cause the vessel to carry more or less than her burthen. Net tonnage was not calculated in the United States until after the passage of the Tonnage Act of 1882.

Fast passages of American sailing vessels have been written of by many authors familiar with packets and clippers, and these have become a part of the marine folklore. The times stated in the following paragraphs have generally been quoted from these reputable historians. Although many of the times have been determined from examination by others of extant logbooks, even these seemingly unimpeachable sources, written by the men who established the records, may sometimes be questioned because of vagaries in the interpretation of starting and finishing points and other more arcane factors. Our purpose in citing fast passages is to establish that a vessel performed well in service in the face of competition, not necessarily to claim that an absolute record was set.

The ownership of many vessels through the 19th century was traditionally by shares, individual owners subscribing a portion of the shares of a ship—the whole usually being divided into 64ths. The principal, or managing, owner (who may not have owned the greatest number of shares) was typically associated with the operation of the vessel. The rest of the shares were taken up by others, such as business associates, family members, friends looking for a well-paying investment, the ship's captain, and, frequently, the shipbuilder and engine builder. Vessels owned in this way were usually chartered to an operating company or individual. Mr. Webb invested in many of the vessels that he built, and was particularly interested in the ships of the Pacific Mail Steamship Company and the Black Ball Line. Both of these operators were valued clients of the Webb shipyard. His share of several of the Black Ball and other packets was 4/64ths, or 1/16th.

In the following short biographical sketches, individual share owners are not mentioned. Rather, it is stated that a ship was built, for example, for the Black Ball Line of C. H. Marshall & Company. It may be assumed that the vessel's owners of record were Charles H. Marshall and a number of other individuals. The last vestige of share ownership appears to have disappeared in the 1960s with the corporate demise of the Dalzell Towing Company, of New York. Their tugs were owned by shares, which, with the passage of time and the division of long-lived assets among successive generations of heirs, were in some cases measured in parts as small as 1024ths, rather than the original 64ths.

These sketches must not be taken to be definitive histories of the vessels. Much has been written of certain of Mr. Webb's ships, such as the clippers and the enormous sidewheelers of the Pacific Mail, but others, like *Agnes*, *Josephine* or the unfortunate, short-lived *Genil*, are so obscure as to have virtually disappeared from view once they were delivered. This is particularly so with those vessels built for foreign owners that left New York upon completion and never returned. But even the careers of some of the more well-known packets have never been properly chronicled. The authors hope that the present work, with all of the gaps that still remain, will permit the reader to gain a bit of insight into the lives of the 135 ships that Mr. Webb designed and constructed.

As one reads these sketches, one common thread runs through many of them. An unusually large number of the ships met violent ends—by foundering, grounding or fire—and with them went untold numbers of men, women and children who had the misfortune to be in the wrong place at the wrong time. This seemed to be an unavoidable fact of life through the end of the 19th century. A seagoing life was indeed a hazardous one, and every voyage in the wooden vessels of that time, which, if they were not frail were firetraps, put the seaman at great risk. Even the seagoing passenger of the day was on his single or occasional voyage exposed to risk of a magnitude that would be totally unacceptable today. That risk was accepted then because there was simply no other way to cross the oceans or to make a trip to a distant coastal destination. But we should not forget that hazard existed in the pioneering days of other forms of transport as well—railroads and aircraft in particular. Perhaps someday we shall look back at the dangers of early space travel in the same way.

In the days of wood and sail, ships suffered from a high mortality rate brought on by the vagaries of wind, sea and fog; the nasty work of marine borers and dry rot; rudimentary (or nonexistent) aids to navigation; lack of ready communication; and the threat of fire on the great wooden tinder boxes that were the ships of the day. But in spite of all of these hazards, a surprising number of the vessels achieved ripe old age. Many lived out their declining years in unglamorous tasks that belied their noble births; others, in final acts of desperation, let themselves become hulks in remote backwaters. That they were by this time badly hogged or sagged, or otherwise overtaken by decrepitude, should not be looked upon as weakness, but rather as part of the natural process of growing old. They had survived a lifetime of pummeling by the elements, and it is a fitting tribute to both their builder and the men who guided them over the world's oceans that they were able to do so.

# The Ships
## From Malek Adhel to Charles H. Marshall

### *MALEK ADHEL*
### Hull 1

This trim and handsome brig was, in 1840, the first vessel built by Mr. Webb, although he had previously supervised the construction of several sailing ships built by his father's firm. *Malek Adhel* measured 80 feet by 27 feet 7 inches by 7 feet 9 inches. Built for Peter Harmony & Company, of New York, she was a vessel of 114 tons, and her first register document was issued on 26 June 1840. The life of *Malek Adhel* is clouded in mystery and spiced with intrigue. A fanciful and entirely unsubstantiated account of the vessel was given by Charles Nordhoff, the author of "Sailor Life on Man of War and Merchant Vessel," who had been told the story by a former crew member of the brig while Nordhoff was aboard *USS Columbus* during the Mexican War. Although built as a commercial trader, *Malek Adhel* seems to have slipped into an exciting, if somewhat shadowy, early career which reportedly encompassed periods as an opium clipper and pirate, the latter after her crew had allegedly murdered the vessel's officers while in the Pacific. *Malek Adhel* rounded Cape Horn with these rogues in charge, continuing her depredations, and was hunted by British and American naval vessels. She was captured in the West Indies and her crew was hanged in Havana. This handsome, but notorious, brig then became a slaver, and was chased and nearly captured by a British vessel, some of the human cargo being thrown overboard in a successful attempt to outrun her pursuer. After two more voyages, she found herself under Brazilian ownership and continued in the slave trade, only to be captured by an American schooner. The vessel was then successively acquired by an American firm, sold to the Mexican government for use in its navy, and captured during the Mexican War by the sloop-of-war *USS Warren* at Acapulco on 6 September 1846. Nothing is known of her subsequent career.

*Built in the year 1840 for the Pacific Ocean trade. Handsome vessel and fast sailer. First vessel built by W. H. Webb as Master Builder for his own account.*

[Note: "For his own account" does not indicate that this vessel was built on speculation or to be owned by the builder, but that she was built under Mr. Webb's name rather than that of his father.]

### *JAMES EDWARD*
### Hull 2

*James Edward* was a ship-rigged vessel of 433 tons, built late in 1840 for James G. Ward for general trading. Her hull dimensions were 120 feet by 28 feet by 19 feet 6 inches. Her launching took place in October of 1840 and she was awarded her first register document on the 10th of December of that year. *James Edward* was char-

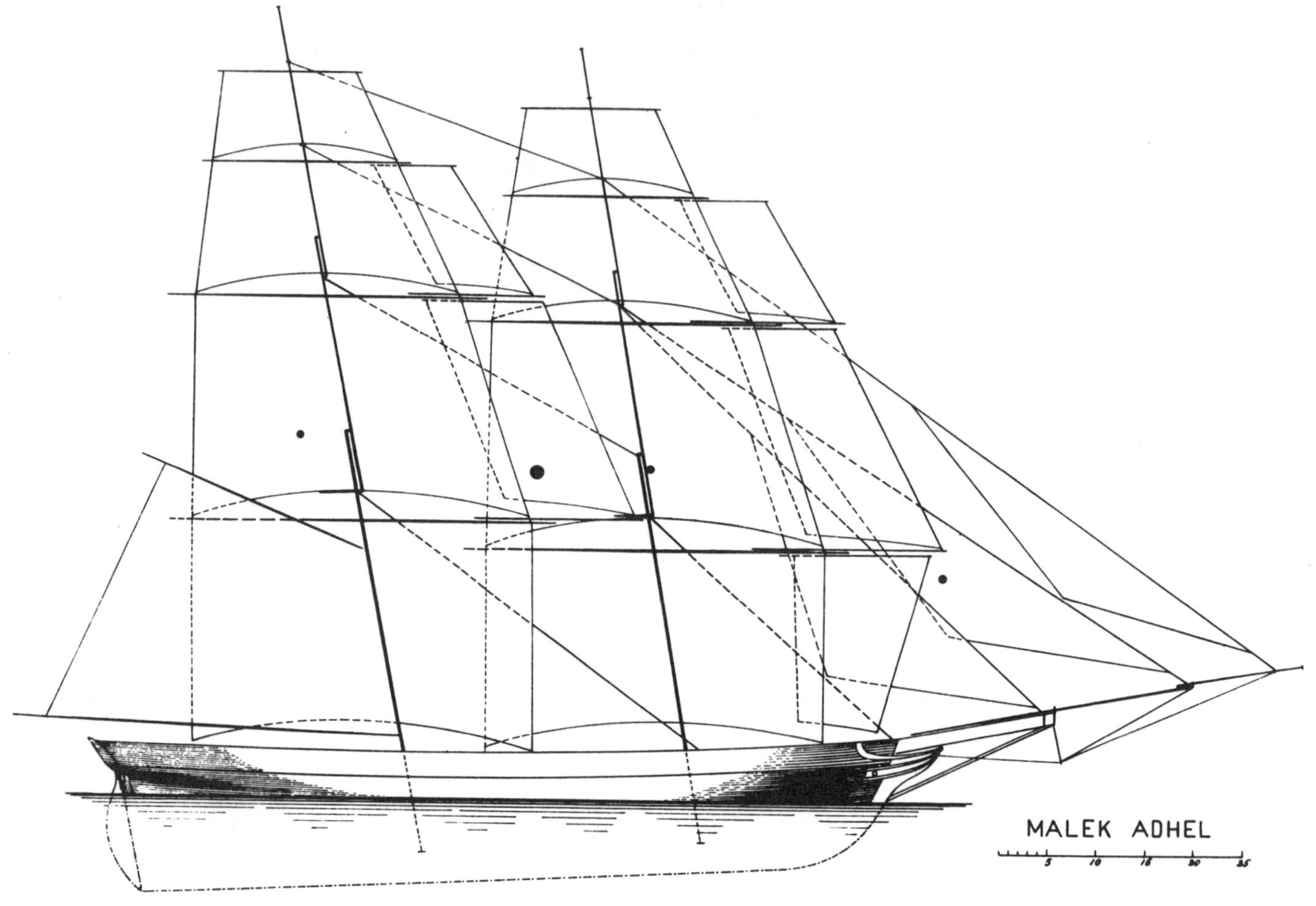


Sail plan of *Malek Adhel,* completed in 1840. This handsome brig was the first vessel built by William H. Webb after he assumed the management of the shipyard of his father, Isaac Webb. This plan, like all the sail plans reproduced here, is from the Book of Plates.

tered to a number of coastwise packet lines early in her career, including Dunham & Dimon's Commercial Line in 1841, John Elwell's Merchants' Line in 1842, Stanton & Frost's Union Line in 1843 and back to the Merchants' Line in 1847.

By the mid-1850s she had become a whaler, operating from New Bedford. Little is known of her whaling days until 10 January 1857, when, under the command of Captain Murray, she put into Port Louis, Mauritius. She had been on an extended cruise which netted the vessel 300 barrels of whale oil, taken in the high latitudes of the Indian Ocean. Her call was made because of a near mutiny by her officers and crew, who refused more duty

whaling under Captain Murray. Protracted negotiations involving the United States Consul followed. On the 29th of January, a heavy gale struck, and *James Edward* slipped her cables and proceeded to sea. She returned on the 1st of February, and, in attempting to recover her anchors and cables two days later, struck hard upon a reef, from which she was released two hours later by a steam tug. But what turned out to be mortal damage had been done. Examination of the vessel on the Port Louis marine railway disclosed that large pieces of her keel, sheathing and copper were missing, and in many places bottom planking had been badly chafed. The damaged areas were covered with felt and the vessel refloated to await instructions from home. By this time she had been abandoned to the underwriters. Her recent catch was loaded upon the British vessel *Bury St. Edmunds*, which sailed for London on the 5th of April. Captain and crew eventually returned to the United States, but the reasons for the crew refusing to work and the outcome of the incident are not known.

### *AGNES*
### Hull 3

The ship *Agnes* was built in 1841 for Frederick A. Wiseman. A vessel of 429 tons and measuring 120 feet by 28 feet by 19 feet, she was nearly identical to her predecessor from the yard, *James Edward*. She was launched early in March of 1841 and delivered later the same month. Her first register document was dated 29 March 1841. No information is known about the early working career of *Agnes* except that for a short period in 1845 she was running for Boyd & Hincken's packet line to Marseilles under the command of a Captain G. W. Wethered (sometimes referred to as Wetherill) of whom even less is known. She appears to have made a number of round voyages between New York and Rio de Janeiro between 1852 and 1855, carrying coffee for the firm of Siffken & Ironsides, but her movements after that period are unknown.

### *HELENA*
### Hull 4

The firm of N. L. & G. Griswold, founded in 1796 by two brothers from Old Lyme, Connecticut, built four outstanding vessels at the Webb yard between 1841 and 1851—*Helena*, *Cohota*, *Panama* and the magnificent clipper *Challenge*. *Helena*, a pre-clipper or clipper-packet, in which the hull form resembled that of a packet but the rig was similar to that used in the true clipper, was launched in August of 1841 and handed over to the Griswolds early in October. She was built for the owner's China trade. A vessel of 597 tons, she measured 135 feet by 31 feet by 20 feet. She departed for Canton via Valparaiso on 30 October 1841 with the redoubtable Captain Deliverance P. Benjamin in command. Abominable conditions prevailed on sailing vessels nearly all of the time at sea. On a hard-driven sailing ship even the relative luxury of the cabin, occupied by the officers and cabin passengers, was not a guarantee of homelike comfort, as

is succinctly and pointedly illustrated by one of Captain Benjamin's log entries during this voyage:

> Fresh breezes with much rain. House all afloat. Not a dry spot in it. I wish the Devil had all such houses.

One can only imagine how those who lived in the forecastle and steerage decks must have been affected by the dank, dismal environment of those areas; forecastle hands lived out entire voyages in a state of perpetual dampness. *Helena* set a number of records for fast passages, and was a ship whose model was one to which newer vessels were frequently compared. Her 51-day crossing from Callao to Hong Kong in 1844 under Captain Benjamin was never bettered and not matched until four years later by Captain Robert H. Waterman, a noted ship driver, in Smith & Dimon's famous *Sea Witch*. *Helena* continued in the China trade for the Griswolds until 1854, when she was sold to Captain Thompson. She left Swatow, China, for Havana on 27 August 1856 with coolies and was never heard from again (although there were reports that she was in Amoy, a short distance up the coast, on the 15th of October.) Captain Thompson was accompanied on this, her last voyage, by his son. The mystery of the ship's disappearance captured the imaginations of many, for three years later, it was reported that a vessel had hailed a China-bound vessel which claimed to have had on board Captain Thompson and his crew. Apparently Mrs. Thompson, who was left in New York with several other children, remained convinced that her husband and son were alive and would return, and, although the rumored news of Thompson was, in the words of an unidentified newspaper, "hailed with joy and public Thanksgiving was given last Sabbath in the Mariners' Church," neither Captain Thompson nor his son nor any of the Chinese passengers was ever seen again. For reasons that are now obscure, the persistent rumor was said to have started in a Scottish newspaper.

*Built in 1841 for the China trade. Proved an excellent sea boat and very fast sailer, made many rapid passages between Ports in the East Indies and New York. Was a very handsome and favorite ship. Draft of water when launched (no extra weight on board). Forward 8 ft., aft 9 ft. 9 in.*

### *LIBERTY*
### Hull 5

*Liberty* was a ship-rigged vessel built in 1842 for James W. Phillips, who was to return to the Webb yard for two later ships—the packet *Vanguard* in 1850 and the clipper *Invincible* in 1851. *Liberty* measured 145 feet by 32 feet by 20 feet and had a tonnage of 689. She was launched early in February of 1842 and received her first register document on the 19th of that month. *Liberty* ran on a number of routes through the close of the 1850s, including Samuel Thompson's Black Star Line to Liverpool (for which the Webb-built packets *Caleb Grimshaw* and *Joseph Walker* also ran) and Stanton & Thompson's Line to New Orleans. She also ran from time to time on other coastwise services between New York and Gulf Coast ports, on which she joined hundreds of vessels transporting the cotton that fed the textile mills of the Northeast. Early in the Civil War, she was running between New York and Liverpool for J. & N. Smith & Company, a firm of New York merchants who were associated with Stanton & Thompson's Line but whose most interesting and controversial activities involved trading with the Confederacy during the Civil War. (This is described later in the history of the bark *Alice Tainter*, hull number 103.) By 1863, *Liberty*'s days as an American vessel had run out. On 29 January 1863, she arrived at New York with coal after a 60-day passage from Liverpool. She was then sold to G. & D. Shaw, of Montreal, and placed under British registry. Flying her new colors, she cleared New York for London on the 2nd of May. Details of her later employment and of her fate are not known.

### *WALLABOUT*
### Hull 6

*Wallabout* was a steam ferryboat built in 1842 with an identical vessel, *New York*, for J. A. Cross for service on the East River between Manhattan and Williamsburg. Steam-propelled ferries had taken over from sailing vessels only twenty years before, and, during the 1840s and 1850s the number of East River ferry routes prolifearated. The line from Manhattan's Houston Street to Grand Street, Williamsburg, which was then a separate riverside community between Brooklyn and Greenpoint, was established about 1842, and it was for this route that *Wallabout* and her sister were built. She was a double-ended vessel of 189 tons measuring 94 feet by 23 feet by 8 feet 9 inches, and was propelled by a beam engine and paddle wheels. The double-ended ferry, in which a wheelhouse and rudder at each end permitted operation over a short route without the need to turn the vessel and also allowed vehicles to be driven on one end of the vessel and off the other, was developed shortly after the introduction of steam propulsion. The basic design of the double-ended ferry has hardly changed since *Wallabout* was built. Only

the materials of construction and the mode of propulsion is different—and, of course, the vehicles carried. *Wallabout* was launched in March of 1842 and was first enrolled on 15 July 1842, entering service on the Williamsburg line at about that time. She was converted to a barge in 1858, in all probability having been found to be too small for the traffic on the line and superseded by a larger vessel. *Wallabout* and her sister were the first steamboats built by Mr. Webb.

### *NEW YORK*
### Hull 7

The steam ferryboat *New York* was a sister vessel to *Wallabout*, constructed in 1842 for J. A. Cross for service on the East River. A double-ended vessel of 189 tons, she measured 94 feet by 23 feet by 8 feet 9 inches. Propulsion was by means of a beam engine. *New York* was launched in May of 1842 and first enrolled on 15 July 1842. She was abandoned in 1854 after spending a mere 12 years on the Williamsburg route. Many early steamboats had short lives, expiring long before their allotted years after the failure of their engines or boilers or deterioration of their hulls, or, as probably was the case with *New York* and her sister *Wallabout*, by being rendered obsolete because of her size. The City of New York and its neighbors across the East River were growing rapidly during this period, and cross-river traffic undoubtedly expanded at the same rate.

### *PRONTA*
### Hull 8

*Pronta* was a sloop-rigged fishing smack built in 1842. She was launched in July of that year, and she received an American register document on the 2nd of August. This formality permitted her to depart from American waters as an American vessel. Upon arrival at her home port, the register was undoubtedly surrendered to the local American Consul, whereupon she ceased to be a vessel of United States registry. *Pronta* measured 44 feet 6 inches by 15 feet by 6 feet, with a tonnage of 36. Little else is known of the vessel, except that her name ("pronta" means "active" in Spanish; the feminine form is properly used in applying the names to a ship) indicates that she may have been built for Central American owners, as indicated by Mr. Webb's comments relating to the similar vessel *Ligera*. Nothing is known of the working career or final disposition of *Pronta*.

## VIVA
### Hull 9

*Viva*, a sister vessel to *Pronta*, was a sloop-rigged fishing smack built in 1842. She was launched in July of that year, and measured 44 feet 6 inches by 15 feet by 6 feet, with a tonnage of 36. Little else is known of *Viva*. Like *Pronta*, she was issued an American register on 2 August 1842, and like her sister vessel, her name ("viva" means "lively" in Spanish) indicates that there may have been a connection to Central American ownership. Nothing is known of her working career or disposition.

## LIGERA
### Hull 10

It is probable that the two-masted schooner *Ligera* was built for the same owners as *Pronta* and *Viva*. She was also described as being a fishing smack, and her name is also an adjectival form denoting fleetness—in the Spanish language, "ligera" means "swift". Like the two sloops, she was launched in July of 1842 and awarded a register document on 2 August 1842. She was slightly larger than the other two vessels, measuring 48 feet by 16 feet by 4 feet 8 inches. Her tonnage was 46. As in the case of the sloops, nothing is known of her career. Unlike the sloops, however, *Ligera* is illustrated in Mr. Webb's Book of Plates by the inclusion of her sail plan. It depicts a handsome and well proportioned little craft, with masts that were stepped with an extreme rake. She looks as swift as her name implies.

*Built in 1842 for a Fishing Smack in Central American Atlantic trade.*

## MONTEZUMA
### Hull 11

With the construction by Mr. Webb in 1843 of the packet *Montezuma* for C. H. Marshall & Company, the owners of the renowned Black Ball Line to Liverpool, there started a business association that was to last until the shipyard closed its gates. From 1843 on, the Marshall firm was to build all of its packets at the Webb yard save one vessel, *Hamilton Fish*, built at Damariscotta, Maine, as *Wm. F. Storer* in 1856 and purchased in 1861. This association was in fact a continuation of one between C. H. Marshall & Company and Isaac Webb, who had built three packets (*Oxford*, *Cambridge* and *New York*) for Marshall's line between 1836 and 1839.

*Montezuma* was a ship-rigged vessel of 924 tons, measuring 162 feet by 35 feet 6 inches by 20 feet 6 inches. She was launched in February of 1843 and took her place on the Black Ball Line in March with Captain Alfred B. Lowber in command. After 11 years on the Liverpool route, *Montezuma* stranded in heavy weather near Jones Beach, New York, on the south shore of Long Island, at about 2:30 on the morning of 18 May 1854. She was carrying about 400 emigrant passengers, all of whom were rescued a day later by the steam tug *Achilles* and two schooners. The vessel's master, Captain William De-Courcey, thought he was off Barnegat as a result of misinterpreting the soundings made just before the ship struck. Lighters removed the cargo, consisting mainly of iron, but *Montezuma* proved to be a total loss, only one of scores of vessels that were lost on the Northeastern coast of the United States in mid-May 1854.

*Built in the year 1843. New York and Liverpool packet. Draft of water when launched with lower masts, topmasts, lower rigging and stays, and 50 tons of ballast on board, was forward 9 ft. 7 in., aft 11 ft. 3 in. Draft loaded 19 ft.*

## COHOTA
### Hull 12

*Cohota* was the second of the clipper-packets built for the firm of N. L. & G. Griswold. Destined for the China tea trade, in which speed had become the criterion by which success was gauged, she was a vessel of 691 tons and measured 145 feet 6 inches by 32 feet 3 inches by 20 feet. *Cohota* was launched in August of 1843 and sailed on her maiden voyage shortly afterwards under the command of W. B. Gerry. Although she and the other clipper-packets were soon to be overshadowed by the true clippers, *Cohota* made some fast runs during her career, but seemed always to be just behind her fleetmate *Helena*. She was sold to George H. and W. P. Lyman, of Boston, in 1847. In the summer of 1850, she went out to San Francisco from Boston in 124 days—only ten vessels made the voyage in less time during that pre-clipper year. *Cohota* nearly came to grief on Toddy's Rocks, Nantasket Roads, Massachusetts, on 19 September 1855. The Boston tug *R. B. Forbes* tried to pull her off, but failed. The next day her masts were cut away and the five feet of water in her holds removed by pumping. The vessel, valued with her cargo at $115,000, was then refloated and towed to Boston, where she was drydocked and repaired. *Cohota* was sold to British owners in 1860, but her career after that time remains a mystery.

## *YORKSHIRE*
### Hull 13

*Yorkshire* was the second packet ship built for the famous Black Ball Line of C. H. Marshall & Company. Launched on 25 October 1843, she was a two-decked vessel of 996 tons and measured 167 feet by 36 feet 6 inches by 21 feet. *Yorkshire* was presented with a set of flags, signals and cabin cutlery by a group of Yorkshiremen living in New York prior to her first voyage early in January of 1844. During her career as a Black Ball liner, she made many fast passages. Her first master was Captain David G. Bailey, who may have contributed to the vessel's early reputation for speed. On the eve of her maiden voyage, the *New York Herald* stated:

> It is a common remark among the sharks when Captain B. goes to sea, that 'it is useless to attempt to follow that ship, for Bailey is on board of her.' Such is his reputation for speed even among the inhabitants of the deep.

In November of 1846 she made the run from Liverpool to Sandy Hook in the then astonishing time of 15-1/2 days. The average year-round westbound time for packets was more than 30 days and many ships spent 40 days or more on the route in the winter months. In what may have been a last hurrah for the pre-war packet trade, *Yorkshire* (then under the command of Captain Edward R. Fairbanks) and two other Webb packets, *Guy Mannering* and *Harvest Queen*, arrived at New York on 6 December 1861 from Liverpool. They had left the Mersey on three successive days (the 7th, 8th and 9th of November respectively), and they brought a total of only 78 passengers. Although the average age of the three vessels was only 12-1/3 years and the Liverpool packet lines were certainly not dead, the small passenger loads suggest that either there was concern among the emigrants over the war that was raging in the United States, or the vessels were only carrying cabin passengers on winter voyages devoted primarily to cargo. If the emigrants were frightened of conditions in their prospective new homeland, their fright was not long lived. By 1862, the packets were disgorging hundreds of steerage passengers at Castle Garden on every voyage. This continued until the business of the packet lines shrank precipitously after about 1867, when all but the Black Ball and Red Swallowtail lines had given up. Nearly all of the packets carried cargo only in the post-war period. The December 1861 arrival of *Yorkshire* proved to be her last one, for on her next voyage, she sailed from New York on 2 February 1862 under Captain Fairbanks, with three passengers and 23 crew members aboard, and was never heard from again. Mr. Webb was of the opinion that *Yorkshire* was a happy combination of speed and cargo-carrying capacity, and her performance confirms that. In some of his later packet ships, many of which could lift more than twice the deadweight of *Yorkshire*, he may not have improved upon the design of that remarkable vessel. One hundred forty-five years after her construction *Yorkshire* again was in the public eye. The United States Postal Service chose a painting of *Yorkshire* by Richard Schlecht to appear on the 28-cent international surface postal card. A ceremony marking the first day of issue of the 'Yorkshire' card was held at Mystic, Connecticut, on 29 June 1988. It may be of interest to Webb Institute and its alumni that *Yorkshire* replaced Donald McKay's *Flying Cloud* on the surface card.

*Built for New York and Liverpool Packet. Proved a fast sailer and excellent sea boat and was a favorite in the trade. Cabin deck from main mast to taff rail. Draft of water when launched with all her spars in and rigging, spare spars, chain cables and hausers [sic] and water, and 84-1/2 tons of ballast, was forward 11 ft. 2 in., aft 12 ft. 6 in. Draft loaded, 18 ft. 3 in.*

## *VIGILANT*
### Hull 14

The two-masted centerboard schooner *Vigilant* was built for the United States Revenue Cutter Service. She measured 56 feet by 18 feet 5 inches by 4 feet 8 inches, was of 50 tons burthen and carried a topsail-schooner rig. The vessel was launched in August of 1843 and shortly afterwards delivered to New Orleans for service out of that port. *Vigilant* was lost at Key West, Florida, in a hurricane on 4-5 October 1844, just over a year after she had been built. All but two of the persons aboard were lost. *Vigilant* was one of only two vessels completed by Mr. Webb for the United States Government during the nearly thirty years in which he operated the shipyard. The other vessel, the steam-propelled *Harriet Lane*, completed in 1858, was also a revenue cutter.

*Built in 1843 for the United States Government. Shoal water service. Draft of water with 6 tons of ballast and everything on board ready for the sea voyage from New York to New Orleans, was 3 feet. Keel 6 inches below timbers.*

## *ZURICH*
### Hull 15

On 22 January 1844 the *New York Herald* declared, "There seems to be a perfect mania in this city for new ships." This newspaper, for many years an avid and reliable chronicler of the marine scene in New York, was referring to the fact that three large vessels were then under construction, one of which, the packet ship *Zurich*, was to be launched that day. *Zurich* was built by Mr. Webb for Fox & Livingston to serve on their line of packets to Le Havre. She was a two-decked vessel of 817 tons, measuring 155 feet by 34 feet 10 inches by 20 feet. She received her first register document on 15 March 1844 and thereupon sailed for Le Havre on her maiden voyage. She was still the property of Fox & Livingston in 1863, when the Le Havre line was abandoned. She was sold in the late spring of that year to D. Hunter, of Ayr, Scotland, but not before Fox & Livingston sent her on a voyage to Buenos Aires and Montevideo, for hides and wool, from which she returned on the 18th of April. Hunter continued to run her under her original name, but she rarely visited the United States while a British ship. Her end came on 13 August 1879, when she stranded on Haisboro Sands, Norfolk, on the east coast of England, while on a voyage from the Tyneside port of South Shields to La Spezia, Italy.

### *MONTAUK*
### Hull 16

The name of William S. Wetmore appears frequently in connection with part ownership of New England-built sailing vessels of the 1840s, but he turned to Mr. Webb's yard in 1844 for the ship *Montauk* for the China trade, where speed of a ship was of great importance. *Montauk* was a clipper-packet of 505 tons, and measured 128 feet by 29 feet 6 inches by 17 feet. She was launched on 7 April 1844 and first went to sea early in May under the command of Captain William McMichael. *Montauk* was a well designed vessel that set a number of records for speed. She sprinted from Macao to New York in 87 days early in 1846, and her time of 31 days in 1848 for the 4,497 miles from Sydney to Hong Kong, under Captain McMichael, was not equalled until Smith & Dimon's clipper *Mandarin* logged the same time in 1855. The 77-day passage of *Montauk* from the U. S. East coast to Anjier was surpassed by Webb's *Helena*, which logged 73 days 20 hours in 1846 (the record passage of 70 days over this route was made by Jacob Bell's *Oriental* in 1850). Anjier was an important landmark for sailing vessels in the Sunda Strait between the East Indian islands of Sumatra and Java; "Anjier for orders" was frequently the destination of sailing vessels voyaging to the Far East.

As a precursor of the clipper ship, *Montauk*, like *Helena*, *Cohota* and *Panama*, represented an important step in the evolution of Mr. Webb's sailing vessel designs.

By the middle 1850s *Montauk* had been acquired by parties at Sag Harbor, New York, and had been relegated to whaling. She arrived at New York early in 1855, at the conclusion of one cruise, carrying whalebone from Honolulu. In the early spring of 1860, she was purchased for $11,000 by a Captain Quayle, rerigged as a bark and fitted out for what appeared another whaling voyage, although, as reported by the *Greenport (N.Y.) Watchman*, "It was commonly thought in Sag Harbor that she was destined for 'black birds' rather than 'black shins.'" That is, she was to become a slaver, a highly risky, but enormously profitable venture. In New York it was also thought that she was to be a slaver, but, upon close inspection of the vessel, nothing tangible was found that would warrant her seizure. After clearing New York, she was last reported at Fayal on the 21st or 22nd of March. Nothing was heard of *Montauk* until the 11th of July, when the fishing smack *Princess* landed a party claiming to be the mate and eleven crew members of the bark *Adela* at Key West, Florida. They stated that their vessel, out of Boston and under the command of a Captain Bowen, was on a voyage from Montevideo to New Orleans with a cargo of hides, when she was destroyed by fire off the East coast of Cuba on the 3rd of July. Their story further claimed that they were taken off and landed at Matanzas, where Captain Bowen chartered a Spanish coaster to take them to Key West. The coaster fell in with *Princess*, which, upon payment of ten dollars a head, agreed to take the crew to Key West. Eventually it was determined that there was no such vessel as *Adela* at Boston, and that the bark that burned was *Montauk*, believed to have just landed 1,200 to 1,300 slaves, probably in Cuba. Later still it was speculated that the vessel was owned by Spanish nationals in Havana. The act of torching a slaving vessel after a successful voyage was a common practice. The profits in the trade were so large that the cost of the vessel could be written off after a single trip, the fire serving to destroy all evidence of the illicit trade under the ruse of an accidental blaze while on a legitimate trading voyage.

### *RAMON DE ZALDO*
### Hull 17

*Ramon de Zaldo* was a small brig built for her namesake in 1844. She measured 90 feet by 24 feet 3 inches by 8 feet 4 inches, and had a tonnage estimated to be 150, carpenter's measure. She was launched in April of 1844 and delivered a short time later. The vessel appears never to have been issued an American enrollment or register, and it is assumed that she traded along the coast of California and Mexico under her Spanish ownership. No record of her service and ultimate disposition have been found.

*Built in 1844, for the Pacific Ocean trade. Proved a good sea boat, fast sailer and satisfactory. Was a handsome vessel.*

### PANAMA
### Hull 18

Like her predecessors from the Webb yard, *Helena* and *Cohota*, *Panama* was a fast packet ship built in the pre-clipper days for the China tea trade. Her tonnage was 612 and she measured 135 feet by 31 feet 8 inches by 19 feet 6 inches. She was launched in August of 1844 and delivered to her owners, the mercantile firm of N. L. & G. Griswold, in October. Her first voyage commenced on 11 October 1844, when she cleared New York for Canton. In 1852, she ran from Shanghai to New York in 91 days, a feat that, up to that time, had only been excelled by a remarkable 88-day voyage of the fleet and famous pre-clipper *HOUQUA* (built by Brown & Bell in 1844) the year before. Two decades after her construction, *Panama* was owned by Hennings & Gosling of New York, for whom she sailed on 7 October 1865 for Galveston under Captain Hansen, with a cargo of hay, salt and lumber. At three o'clock on the morning of the 24th, she was wrecked on Memory Rock, Grand Bahama Island. All but one of the boats were washed away, so the resourceful Captain Hansen ordered the deckhouse cut away and launched into the sea. He, his wife and most of the crew climbed aboard the makeshift raft, which drifted southward past Great Isaac Island, whence it was towed to nearby Bimini by a whaleboat. Two crewmen in the boat landed near the wreck. Captain Hansen was never to forget Memory Rock.

### HAVRE
### Hull 19

As may be inferred from her name, the packet ship *Havre* was built to serve on the line between New York and that French port operated by Fox & Livingston. She was a two-decked vessel of 870 tons, launched in January of 1845 and delivered shortly afterwards. Her hull di-

mensions were 160 feet by 35 feet by 20 feet 6 inches. In November of 1845, she made a fast passage under Captain Allen C. Ainsworth from New York to Le Havre in 15 days, slightly more than the reputed record time of 13 days 20 hours. She remained under the American flag and the property of Fox & Livingston until about 1863, when the firm's interest turned to steamships. At that time, she acquired Norwegian registry under the ownership of M. Thommessen & Son, of Arendal, and was a regular caller at New York on tramping voyages from Northern Europe or the Mediterranean for many years afterwards. During this period she was a member of the flotilla of sailing vessels that maintained the petroleum-and-empty-barrel shuttle across the Atlantic—in the days when the United States was a major exporter of refined petroleum products. From the early 1880s, *Havre* carried a bark rig. She was reportedly broken up about 1908, at the age of 63 years, her last visit to the port of New York having been in June of 1887. If she indeed survived until 1908, she would have been the longest-lasting vessel built by Mr. Webb, barring any that spent their declining years as hulks after being retired from service. *Havre*'s staunch hull of oak and locust outlasted the Long Island Sound steamboat *Providence* by seven years and Mr. Webb himself by nine.

*Built in 1845 for the regular Packet trade, New York and Havre. Proved an excellent sea boat, made rapid passages, was a handsome and very popular ship.*

### SILAS HOLMES
### Hull 20

The ship *Silas Holmes* was built in 1845 for William Nelson & Company's line of New Orleans packets. She was launched in May and commenced her first voyage to New Orleans in mid-June. Named for a one-time New York agent of the line, *Silas Holmes* was a vessel of 644 tons. Her hull dimensions were 190 feet by 32 feet by 19 feet. New Orleans was the predominant Gulf coast destination for packet lines from New York because of the large amount of cotton transshipped at that port from river steamboats. *Silas Holmes* remained on the New Orleans line until she foundered late in 1859. She had left New York for New Orleans under Captain Griffin late in November of that year, and, on the morning of the 2nd of December ran aground on East Key, Dry Tortugas. Salvors removed part of her cargo, and she was refloated. The cargo was reloaded and *Silas Holmes* set out for New Orleans, only to ground again while leaving Garden Key

Roadstead. Again refloated, she left that harbor of refuge on the 11th, after the passengers and crew had signed an undertaking that they were willing to proceed to New Orleans on the vessel. Five days later, on the 16th, *Silas Holmes* foundered. Five women and four men were rescued three days later by the bark *Doretta*, enroute from Bremerhaven to New Orleans, where they arrived on the 27th. The fate of the remaining 32 passengers and crew members is not known.

### *FIDELIA*
### Hull 21

*Fidelia*, named for the wife and daughter of Charles H. Marshall, continued the lineage of Webb-built packets for C. H. Marshall & Company's Black Ball Line to Liverpool. Launched in June of 1845 and completed early in July, she measured 160 feet by 35 feet by 21 feet. Her tonnage was 895. *Fidelia*'s first register document was issued on 15 July 1845, and she cleared New York on her first passage to Liverpool on the same day. She was under the command of Captain William G. Hackstaff, who remained aboard until the packet *Manhattan* was built four years later. Captain Wagstaff then assumed command of the new ship. The vessels of the Black Ball Line displayed a large black circular emblem painted on or sewn to their fore topsails, and a similar device was to be found on the line's house flag, which had a field of red. The line is remembered in the a verse of a traditional sea chantey:

> In the Black Ball Line I served my time,
> To me way-ay-ay, Oh the Black Ball Line.
> In the Black Ball Line it's rise and shine.
> Hurrah for the Black Ball Line.

*Fidelia* remained on the Black Ball Line for nearly 20 years. Her last arrival at New York was on 11 October 1863. The vessel was then under the command of Captain H. Bessling, who moved to the Webb-built packet *Isaac Webb*. *Fidelia* was sold shortly afterwards to George Cairns, of Newcastle-upon-Tyne, for 3,450 Pounds Sterling. Cairns transferred her to British registry. The details of her later career and her ultimate fate are not recorded.

### *GENIL*
### Hull 22

The sidewheel steamer *Genil* was the third steam vessel built by Mr. Webb. Her hull dimensions were 180 feet by 29 feet by 10 feet 9 inches. A vessel of 512 tons, she was launched in September of 1845 for Castellain & Ponvert, of Havana. Shortly after her launching, a survey disclosed that, when completed, she would have excessive trim by the stern. She was then hauled out and additional timbers installed throughout her after body, making her hull form fuller aft and correcting the trim problem. A double-beam engine from T. F. Secor & Company, with cylinders 36 inches in diameter and eight feet stroke, was installed, taking steam from two iron boilers. *Genil*, named for a tributary of the River Guadalquivir in southern Spain, was delivered to Havana in the latter part of January 1846. Her American registry, issued on 16 January 1846 to permit her to make her delivery voyage, was surrendered later that year with the endorsement that she had been given foreign registry. She thereupon entered Cuban coastal service. Her life on the Cuban coast was to be short. In July of 1846, only a few months after her delivery, *Genil* burned while on a passage from St. Jago (now known as Santiago) to Havana. Timbers in the vicinity of the boilers were set alight, and within a short time the unfortunate vessel had burned to the water's edge.

### *MARMION*
### Hull 23

The Liverpool packet *Marmion* was built in 1846 for the firm of Taylor & Merrill, who, five years later, was to order the controversial clipper *Gazelle* from Mr. Webb. She was launched during February of 1846 and cleared New York on her maiden voyage at the end of March in charge of Captain William Edwards. *Marmion*, a two-decked ship of 903 tons which measured 160 feet by 35 feet by 21 feet 10 inches, was typical of the packets of her day. They were workaday ships that could carry large cargoes or hundreds of steerage passengers, but their speed was high enough to permit their assignment to long-haul voyages. *Marmion* made a number of voyages to the Far East for tea and other goods during the Civil War. On one of these, she arrived at New York from Manila (heavily laden with sugar and hemp) on 26 July 1865, after a passage of 124 days. She took 19 days from Manila to Java Head, 43 days thence to the Cape of Good Hope, and 62 days more to New York. Coincidentally, the clipper *Young America* had arrived the day before, also from Manila with the same cargo consigned to the same merchants. Her time to Java Head was an unimpressive 32 days, but she ran to the Cape in 23 days and thence to New York in 45 days, for a complete passage of 100 days.

(The record time of 84 days for this passage was set in 1861 by Samuel Hall's *Wizard*.) Although *Young America* had left Manila a full month before *Marmion*, the packet more than held her own during the first stage of the voyage. But even this did not come close to *Marmion's* 14-day run from Whampoa to Java Head in December of 1863. On 30 December 1867, she cleared New York for San Francisco under charter to William T. Cameron & Company.

*Marmion* was later sold to Wachschlager & Palmer, of New York, and rerigged as a bark to make her more easily handled by a smaller number of men. Her master at this time was Captain William Boyd. In September of 1874, she was reportedly acquired by Captain Boyd and others, her home port having been changed to San Francisco. Some sources list her owners as Eugene Kelly & Company during this period and it is possible that the "others" may have included the Kelly firm. *Marmion* was still running in the autumn of 1879 under the ownership of Gates, Phillips, Sanderson & Brothers, of San Francisco. On 4 November 1879, she left the Pacific Northwestern port of Departure Bay under the command of Captain Jordan with a cargo of 1300 tons of coal for San Francisco. She passed Cape Flattery, Washington, at the entrance to the Strait of Juan de Fuca, at 3:30 in the afternoon of the 7th. That night, at the height of a southeast gale and heavy cross seas in which she labored and strained badly, she sprang a leak, and began to settle rapidly. The next morning, the bark (or barkentine) *Tam O'Shanter*, noting her predicament, hailed her and stood by. At noon, the vessel's entire complement took to the boats and reached *Tam O'Shanter* in safety. She was left to founder at a point 40 miles west of Cape Flattery. Her timbers, weary after 33 years of battling the oceans' might, simply gave up.

### *COLUMBIA*
### Hull 24

C. H. Marshall & Company returned to the Webb yard in 1846 for the packet ship *Columbia*, which was similar in design and of about the same dimensions as her three predecessors. The tonnage of this vessel was 1050, and her dimensions were 170 feet by 36 feet 6 inches by 21 feet 8 inches. *Columbia* was launched on 26 March 1846 and was completed and sailed on her first voyage to Liverpool for the Black Ball Line in April under the command of Captain John Rathbone. His tenure was unfortunately short. On 13 January 1847, while en route from Liverpool to New York, the vessel's after deckhouse was washed away by an enormous wave, and took with it Captain Rathbone, the mate, the second mate, five men and a boy. The body of the sole remaining officer, the third mate, was jammed under the ship's tiller. The crew and passengers managed to get the ship to New York after a stormy and otherwise harrowing passage. After a relatively quiet 23 years on the "North Atlantic ferry," *Columbia* was sold to George Howes & Company about 1869. The new owners, who were active in the San Francisco trade, rerigged her as a bark and ran her in the lumber trade out of her new hailing port of San Francisco. She was later owned by C. L. Dingley and for her last few years by Renton, Holmes & Company. She survived until about 1898 or 1899.

The Liverpool packets carried hundreds of thousands of Irish emigrants to American ports during the late 1840s and early 1850s as these people sought to escape from the effects of the Potato Famine. The American census of 1850 shows large numbers of young Irish persons in practically every city and town in the northeastern states. They built our railroads, manned our canal boats and formed a labor pool that made possible the rapid westward expansion of the United Staes. The regular packets of the Black Ball Line, which left New York and Liverpool twice a month, carried many of these emigrants. In March of 1864, with the Civil War raging, the Black Ball Line alone landed no less than 2,350 steerage passengers from Liverpool, the port at which nearly all Irish emigrants and many from elsewhere in Europe embarked. *Columbia* accounted for 365 of these. Liverpool has retained its affinity to Ireland to this day. A large Irish population resides there, and an unknowing visitor may assume that the frequently-heard Irish accent rather than Liverpudlian is the indigenous speech.

### *WILLIAMSBURG*
### Hull 25

The third and last double-ended steam ferry (and the fourth steamboat) built by Mr. Webb was *Williamsburg*. She was a vessel of 285 tons, launched in June of 1846 for Fleming Duncan and placed in service a few months later. She was larger than her predecessors *Wallabout* and *New York*, measuring 115 feet by 26 feet by 10 feet. Like the earlier vessels, she was built for East River service between Manhattan and Williamsburg. Her propulsion plant consisted of the then conventional beam engine and paddle wheels. *Williamsburg* remained in service until about 1867, when she was dismantled.

### *BAVARIA*
### Hull 26

*Bavaria* was a packet ship built in 1846 for William Whitlock, Jr., to run on the Union Line between New York and Le Havre. The vessel measured 160 feet by 35 feet by 21 feet, and had a tonnage of 908. She was launched in August and received her register document on 16 November 1846. *Bavaria* was the first of a group of three packets for different owners (the others being *Admiral* and *Sir Robert Peel*) completed during November and December of 1846. Minor changes were made from ship to ship, but they were essentially sisters. Although *Bavaria* ran principally on the Le Havre route, she was sent to other destinations from time to time. On 14 February 1853, she cleared New York for Melbourne under Captain Bailey, arriving back at her home port on 26 November of that year. Her return passage was made in 65 days. The record passage between these two ports is not known. Donald McKay's clipper *Lightning* ran from Melbourne to Liverpool in 63 days in 1854, and therefore the voyage of *Bavaria* may be characterized as fast. *Bavaria* was sent to San Francisco and the Far East on at least two voyages during the Civil War—in 1863 under Captain Warren and in May of the following year under Captain Higgins. In 1866 she was to be found on Tupper & Beattie's route to New Orleans from New York, an unusual berth for the vessel in that after the Civil War steamships had nearly ousted sailing vessels from American coastwise routes. *Bavaria* continued to run under the ownership of Whitlock and under the command of Captain Levi Smith until September of 1873, when she was sold to Oulton Brothers, of St. John, New Brunswick, who placed her under British registry and renamed her *Charles H. Oulton*. About a year later she was reportedly owned by A. L. Palmer, of St. John, and shortly afterwards by D. J. McLaughlin, Jr., also of that port. She remained an occasional caller at New York while under Canadian ownership. She was reported by the American Shipmasters Association to have been "broken up" (which, in the case of a wooden sailing ship, might mean that her spars, rigging and other usable gear had been removed and her hull cast aside) in 1876, but was said to still be afloat as late as 1891, probably as a hulk.

### *ADMIRAL*
### Hull 27

Fox & Livingston ordered the packet ship *Admiral* at the Webb yard for delivery in 1846 to serve on the Union Line route between New York and Le Havre. The Union Line included not only the fleet of Fox & Livingston, but also those of Boyd & Hincken and William Whitlock, Jr. *Admiral* was a two-decked ship of 929 tons, measuring 160 feet by 35 feet 6 inches by 21 feet, and, as the second of the three late-1846 packets, differed from *Bavaria* in that her beam was six inches greater. *Admiral* was launched in August of 1846 and her first register document was dated 24 November 1846. Her first master was Captain James A. Wotten, who drove *Admiral* to Le Havre in November of 1847 in 14 days two hours, just two hours more than the best time recorded over the route. He is perhaps better remembered as the master of the New York & Havre Steam Navigation Company's steamship *Franklin* when she ran aground and was lost near Moriches Inlet, Long Island, on the morning of 17 June 1854. With this accident, the steamship company, an enterprise of Fox & Livingston, had lost the second of its two vessels, *Humboldt* having been wrecked near Halifax on the 6th of the previous December. Captain Wotten was succeeded on *Admiral* by Captain Bliffens, who came aboard in 1852 or earlier and remained with the ship until 29 December 1862, the date of what appears to have been her last arrival from Le Havre under the American flag. By this time, Fox & Livingston had all but given up sailing vessels, and, with operation under the Stars and Stripes becoming increasingly hazardous because of the depredations of Confederate commerce raiders, *Admiral* was sold to M. I. Wilson, of Liverpool, and placed under British registry. The details of her later career are not known.

### *SIR ROBERT PEEL*
### Hull 28

Mr. Webb built the third of the late-1846 group of packet ships, *Sir Robert Peel*, for the Red and Blue Swallowtail Line services of Grinnell, Minturn & Company to London and Liverpool respectively. The lines took their names from the shape of their house flags, which were identical except in color—London was red and white; Liverpool, blue and white. House flags, or "private signals," provided a means for identifying a vessel to passing craft, so that a ship's position could be announced upon arrival at the reporting vessel's destination. The packet was named for the British prime minister (1841-46), an advocate of free trade under whose Conservative Party leadership the Corn Laws were repealed and other sweeping reforms carried out, including reorganization of the Metropolitan Police of London. Sir Robert lent his name to the Metropolitan Police Constable, who from that time onward would be known as a 'Bobby.'

*Sir Robert Peel* was launched in November of 1846, at the end of Sir Robert's term as prime minister, and delivered the following month, her first register having been issued on 16 December 1846. Her dimensions were 160 feet by 35 feet 6 inches by 21 feet 6 inches, similar to those of her immediate predecessor, *Admiral*, except that her depth was increased by six inches. Her tonnage was 940. The vessel was originally used on the Blue Swallowtail Line to Liverpool. After a number of years she was moved to the London line. *Sir Robert Peel* was somewhat longer lived than most of the North Atlantic packets, and appears to have been more fortunate than many in that disaster did not visit her. In the summer of 1880, nearly 34 years after she had been built, *Sir Robert Peel* made her last voyage to London for Grinnell, Minturn & Company, departing on the 21st of April and returning to her home port on the 16th of August. Her last voyages were made under Captain Thomas P. Stetson, her longtime master, Captain Nathan F. Larrabee, having left the vessel in May of 1879 after 23 years on her poop deck. Contrary to reports that she was sold to British owners in 1880 (corresponding in time to the abandonment of Swallowtail Line service by Grinnell, Minturn & Company,) *Sir Robert Peel* apparently retained her American registry under the ownership of Nelson Edwards and others, with Philadelphia as her home port. She was removed from the register in 1886-87 for unknown reasons, but probably because she was suffering from 'old age.'

### *NEW YORK*
### Hull 29

*New York* was a two-decked packet ship built to serve on Fox & Livingston's line between New York and Le Havre. She measured 166 feet by 36 feet 6 inches by 21 feet, and her tonnage was 991. *New York* was launched in April of 1847 and took her place on the line about the 20th of May under the command of Captain David Lines. In the late 1840s, the packet trade to Le Havre was shared among three shipowners—Fox & Livingston, Boyd &

Hincken and William Whitlock, Jr. Of these, Fox & Livingston gave up before the Civil War and Whitlock turned his attention to other destinations, but Boyd & Hincken hung on precariously until the mid-1870s, although their sailings were irregular at the end. When *New York* joined the Fox & Livingston fleet, she joined Mr. Webb's *Zurich, Havre,* and *Admiral. Bavaria,* built at the yard in 1846, was a member of the Whitlock fleet. *New York* made two noteworthy eastbound passages, running from New York to Le Havre in 14 days in the autumn of 1848 and in 15 days early in 1850. Both voyages were under Captain Lines. She was on the Le Havre line at least through the summer of 1855, with William C. Thompson in command, but details of her further career are not known.

### *ISAAC WRIGHT*
### Hull 30

*Isaac Wright* was the fifth packet ship built by Mr. Webb for C. H. Marshall & Company. Charles H. Marshall, the principal owner of the mercantile firm that bore his name, started his career at sea, and before he came ashore in 1835 was the master of several Black Ball packets. His first command was *James Cropper* in 1824. *Isaac Wright* was named for one of the founders of the Black Ball Line, inaugurated in 1817 to provide the first scheduled transatlantic service for passengers and freight. She was a vessel of 1129 tons, and was the last of the two-decked packets built for Marshall. Henceforth, all would have three decks. The dimensions of *Isaac Wright* were 175 feet by 37 feet 9 inches by 22 feet 3 inches. She was launched in June of 1847 and made her first voyage on the Black Ball Line, under the command of Captain Edward C. Marshall, shortly after her first register document was issued on the 31st of July. *Isaac Wright* remained on the line until 23 December 1858. At 2:30 that morning fire was discovered in the vessel's forward hold while she was lying at anchor in the River Mersey between Egremont and New Brighton, downstream from Liverpool. With 200 to 300 passengers aboard and a cargo of fine goods and 800 tons of iron in her holds, she was being made ready for departure to New York the following day. All of the passengers were removed safely by ferry steamers and tugs, and departed on *Great Western,* the next scheduled Black Ball liner. The burning vessel was towed to shallow water and anchored well out of the way of river traffic, but little could be done to save *Isaac Wright* and she was soon reduced to charred timbers and memories of a night of near horror for the passengers and crew members who were to sail in her.

### *UNITED STATES*
### Hull 31

Completed in 1848, *United States* was Mr. Webb's first ocean-going steamship. She was the third American vessel designed and built as such, but her hull form, with its counter stern and practically flat deadrise, represented a significant improvement over that of *Washington* and *Hermann* (built at New York by Westervelt & Mackay earlier that year) in which the hull of a sailing packet was adapted to steam propulsion. *United States* had a tonnage of 1857, measured 244 feet 7 inches by 40 feet by 30 feet 10 inches, and was propelled by two side-lever engines built by T. F. Secor & Company, each having a cylinder 80 inches in diameter and a stroke of nine feet. The vessel was launched on 20 August 1847 and ran trials in February of 1848. Built for a group of 33 persons headed by Charles H. Marshall (whose Black Ball Line of packets was an important client of the Webb yard) it was stated that she was to be used for a proposed New York-New Orleans line. As an operator of sailing ships on the Atlantic, Marshall may have used the New Orleans route as a ruse to disguise his true plans for the steamer. He might have been planning to use the steam vessel on the Liverpool line in an attempt to upstage Edward K. Collins, whose Dramatic Line of packets was a fierce competitor of the Black Ball Line, and who, with the assistance of a mail subsidy provided by the United States Postmaster-General, was about to launch his Collins Line, the first important American steamship line plying the Atlantic. The first sailing of the Collins Line, by William H. Brown's steamship *Atlantic*, was on 27 April 1850.

The first voyage of *United States*—to Liverpool—began two years earlier, on 8 April 1848, with Captain William G. Hackstaff in command. She was the first American vessel propelled solely by steam to enter the River Mersey; her three predecessors (*Savannah* in 1819, *Marmora* in 1845 and *Massachusetts* in the same year) was auxiliary steam vessels. There followed two trips to Le Havre, one to New Orleans and a third to Le Havre. She was then sold to the Confederated German States and converted by Mr. Webb into a warship named *Hansa* by the removal of her uppermost deck. Her naval career was short because of the failure of the German states to confederate at that time. She was sold to the firms of W. A. Fritze & Commpany and Karl Lehmkuhl, of Bremen, for whom

she made four voyages on the Bremen-New York route, followed by a charter to the British for Crimean War transport service and finally another voyage to New York. She was, in 1857, offerred to the East India Company an a transport during the Indian Mutiny. In May of 1858, she was acquired by the Atlantic Steam Navigation Company—known as the Galway Line—for which, as *Indian Empire*, she made two transatlantic trips under the British flag. Her first visit to Galway (which was chosen as the line's home port to save a day of steaming time) was marred by a grounding, which delayed the inaugural voyage for several weeks. Once at sea, a piston fractured, and the remainder of the trip had to be made with one engine. The Galway Line lasted until 1863, using an assortment of 16 steamships for a total of 53 westward and 51 eastward voyages, but their terminus at Galway was unpopular. A writer in *The Times* (of London) on 12 November 1863 proclaimed,

> The passengers who go to Galway for convenience are subjected to the greatest possible inconvenience, being without the proper accommodation for themselves and their luggage while waiting to embark. Amid noise, hurry, confusion, wet, dirt and all sorts of discomforts they are taken out a mile or two through the breakers in a small steamer, and perhaps they commence their voyage across the Atlantic thoroughly drenched with sea water.

After several years in layup, *Indian Empire* was badly damaged by fire in the River Thames in 1862. Her wrecked hull finally sank on 4 May 1866 and she was later raised and dismantled. *United States* was an innovative and well-built steamship—one of Mr. Webb's more important, but less appreciated accomplishments, built when he was 32 years of age. She might also have been a commercial success had any of her owners themselves been more successful.

*Draft of water when launched (keel 12 inches below floors) nothing on board was forward 7 ft. 1 in. and aft 8 ft. 4 in. Built in the year 1847 for regular packet between New York and New Orleans but was run as freight steamer between New York and Liverpool. Made good speed with very moderate power, was an excellent sea boat. Carried a large cargo on a very light draft of water. Had two side lever engines 80 inch diameter of cylinder and 9 foot stroke, wheels 35 ft. in diameter. Draft of water when completed and with 1,870,800 lbs. of coal on board was on an even keel 14 ft. 10 in., with full cargo and passengers 18 ft. This ship was afterwards cut down one deck and converted into a vessel of war carrying two large pivot guns forward and one aft and 10 side guns. Draft when fully completed and with 800 tons coal and boiler filled was, forward 16 ft. 1 in., aft 15 ft. 8 in.*

## *IVANHOE*
### Hull 32

The ship *Ivanhoe* was constructed for the New York firm of Taylor & Merrill in 1847 for their line of packets to Liverpool, joining the Webb-built *Marmion* of early 1846. *Ivanhoe* was similar to the Black Ball Line's *Isaac Wright*, measuring 175 feet by 37 feet 9 inches by 22 feet 3 inches. Her tonnage was 1156. The launching of *Ivanhoe* took place in September of 1847 and her register document was dated the 8th of November. The first master of the vessel was Captain William Edwards, who had been in command of *Marmion* on her first voyage. He was succeeded in 1848 by Captain S. C. Knight, a veteran packet mariner who was swept off the vessel's deck by an enormous sea on 29 November 1850. The following March, *Ivanhoe* herself was lost in the vicinity of Montauk Point, Long Island.

## *YORKTOWN*
### Hull 33

Grinnell, Minturn & Company built two identical packets at the Webb yard in late 1847 and early 1848. *Yorktown* was the first vessel. She measured 170 feet by 38 feet 3 inches by 22 feet 3 inches, and had a tonnage of 1150. This two-decked ship was to run on the owner's lines to Liverpool and London, which were referred to as the Blue and Red Swallowtail Lines, respectively. *Yorktown* was launched in October of 1847 and departed on her maiden voyage about 1 November 1847 with veteran shipmaster, Captain William S. Sebor, in command. Captain Sebor had previously commanded four other Swallowtail Line ships from the beginning of the line's service in 1827. At some unknown date (probably in the late 1850s) *Yorktown* was given a third deck. She remained in the service of Grinnell, Minturn & Company for over 21 years, but like so many sailing vessels, her luck finally ran out. With Captain Driver in command, *Yorktown* sailed from London for New York on 6 November 1868. After experiencing extremely heavy weather, in which she lost masts and rigging and began to leak badly, *Yorktown* put into Fayal, in the Azores, on the 24th of December, where she was condemned.

Thousands of wooden sailing vessels were lost over the years because of leaking hulls. Many of the waterlogged hulks that were encountered at sea with frightening regularity had been abandoned by their crews when the hand pumps (with which all vessels were equipped) could not stem the flow of water into the holds. When rising water was discovered, the pumps were manned, and pumping was continued until all was well or the crew gave up from sheer exhaustion. At least one case is recorded where packet ship passengers were hired to man the vessel's pumps, the crew having been depleted by accident or illness.

## *LONDON*
### Hull 34

*London* was the second of the Grinnell, Minturn & Company sister packets built in 1847-48. She followed *Yorktown* by three months. Like her sister, *London* measured 170 feet by 38 feet 3 inches by 22 feet 3 inches. Her tonnage of 1145 was five tons less than *Yorktown*, a difference that can easily be accounted for either by minor differences in the arrangement of the vessels or by two admeasurers interpreting tonnage rules in a slightly different way. *London* was launched in December of 1847 and received her first register document on 18 February 1848, after which she took her place on the Red Swallowtail Line to London under Captain Frederick H. Hebard, whose career as a shipmaster with Grinnell, Minturn & Company, like that of Captain Sebor of the packet *Yorktown*, began in 1827. The vessel was transferred to the Blue Swallowtail Line to Liverpool in 1849, and subsequently could be found on either of the two routes. *London* acquired a third deck at an unknown date, and, like her sister, served the line for many years. She was still running to London in mid-1864, but was later dropped from United States registry, having been sold to British owners and given foreign registry to avoid the depredations of Confederate raiders. While on the Red Swallowtail Line, *London* achieved the dubious honor of making a westbound passage in 85 days, the second worst time recorded over the route.

## *CALEB GRIMSHAW*
### Hull 35

The packet *Caleb Grimshaw* was built in 1848 for Samuel Thompson & Nephew's line of Liverpool packets. Named for the Liverpool agent of the line, she was a vessel of 988 tons and measured 160 feet by 36 feet 8 inches by 21 feet 8 inches. Her short career began when she was launched in February of 1848. Her initial register document was issued on 8 March and she was immediately placed on Thompson's line to Liverpool. Captain William E. Hoxie, whose career in the packets began about 1820, was in command. *Caleb Grimshaw* had a short life, for she burned at sea on 12 November 1849, near Fayal, Azores. The Nova Scotian bark *Sarah* rescued 399 of those aboard, but another 60 or more lost their lives. Fayal was a frequently used harbor of refuge for sailing vessels that encountered difficulties in that part of the Atlantic. Although the destruction of wooden sailing ships by fire could not be blamed on boiler uptakes, which were responsible for many fires that destroyed steamships and steamboats, the hazard of fire at sea was present on sailing vessels. Galley stoves and lightning were occasionally cited as the cause of sailing vessel fires, and spontaneous combustion in cargoes such as cotton, wool and coal was responsible for others, but crew or passenger carelessness may have destroyed more sailing craft than any other cause.

## *AJAX*
### Hull 36

The marked increase in size of sailing vessels during the 1830s and 1840s brought with it the need for towboats to assist them in and out of New York harbor. Most sailing ship captains were unwilling to take their larger and larger vessels into the tricky East River (where most of the clippers and packet ships berthed and where the tides and currents made shiphandling a true art form) without assistance from a steam-powered tug. Although passenger steamboats performed this task at first, sturdy and powerful purpose-built towboats soon made their appearance, ranging the waters beyond Sandy Hook in search of arriving clients and assisting vessels in distress. *Ajax*, the first of three sidewheel towboats built by Mr. Webb, was an early example of the type. She was launched in February of 1848 and completed in December, her original enrollment (signifying that she was in domestic trade) having been issued on the 19th of that month. She was a vessel of 332 tons, and measured 140 feet by 24 feet 10 inches by 10 feet 6 inches. The mercantile firms of New York were the owners of *Ajax* (and many of the other ocean towboats), and the list of her shareholders included such names as Charles H. Marshall, Paul Spofford, Thomas Tileston, Henry and Moses H. Grinnell, Robert B. Minturn and William Whitlock, all of whom had built ships at the Webb yard. Mr. Webb also owned some of the shares of *Ajax*, as he did with many of the vessels he built.

*Ajax* remained in her strenuous role for nearly ten years. Her end came when she stranded on Cape Cod in October of 1858, several months after a false alarm in which it was thought that she had been wrecked on the New Jersey coast. On 3 February 1858, a number of New York newspapers published accounts of the loss of a steam towboat named *Ajax* near Long Branch, New Jersey, while she was towing to safety a bark that had come to grief near Barnegat Inlet. The unfortunate towboat turned out to be *Pilot*, a vessel built by Isaac C. Smith in Hoboken in 1854. The confusion may have been the result of *Pilot* carrying the informal name *Ajax*. In the mid-19th century, it was not unusual for a steamboat to have a name other than her official one painted on the paddlewheel box, provided that the name shown on her register or enrollment document appeared, with her hailing port, on her stern. Several newspapers correctly identified the owner and master of *Pilot*, but they neglected to call the vessel by the name under which she was documented. The practice, which must have been as confusing to those on the waterfront then as it is to marine historians today, was never illegal, but was finally banned by an Act of Congress of 5 May 1864.

## *CALIFORNIA*
### Hull 37

*California* was the first of a trio of sidewheel steamships built at New York in 1848 for the Pacific Mail Steamship Company. Of the other two vessels, *Panama* and *Oregon*, only the former was built by Mr. Webb; *Oregon* was a product of the Smith & Dimon yard. Pacific Mail had been chartered on 12 April 1848 to operate coastwise steamship services from Panama to the Pacific Northwest, an important link in a through route from New York via an overland trip across the Isthmus of Panama. The company later expanded its operations to include both an Atlantic service from New York and a transpacific route, and became a regular customer of the Webb yard. *California* was a vessel of 1057 tons, measuring 200 feet by 33 feet 6 inches by 20 feet. Her propulsion plant consisted of a side-lever engine from the Novelty Iron Works having a 70-inch cylinder with a stroke of eight feet. She was laid down on 4 January 1848, launched in May and delivered in October. Her departure from New York took place on the 5th of October, and she reached San Francisco on 28 February 1849 (as the first American steamer to pass through the Straits of Magellan and the first steamship to enter the Golden Gate) with 365 gold-seeking passengers aboard, about twice her normal capacity. Although she was in later years relegated to secondary routes, *California* remained in the Pacific Mail fleet until

In 1848, William H. Webb built two of the first three steamships ordered by the Pacific Mail Steamship Company—*California* and *Panama*—cementing a business relationship that was to last nearly to Webb's retirement from shipbuilding. In all, he built eleven sidewheel steamships for the Pacific Mail. (Courtesy of the Peabody Museum of Salem)

1870, at which time she was sold to N. Bichard, of San Francisco, who removed her machinery and converted her into a bark. *California* was wrecked on the Peruvian coast in late 1894 as she approached the port of Pacasmayo with a lumber cargo from the Pacific northwest. The crew was rescued, but the old coastwise steamship-turned-sailer had come to the end of the line. She is remembered for having outlived all of her wooden-hulled consorts in the Pacific Mail fleet.

*First vessel built in 1848 for the Pacific Mail Steamship Company for a regular mail packet between Panama and San Francisco. Was the first steamer to enter the Golden*

*Gate. Proved a good sea boat and a very serviceable ship and continued long in the trade. Draft of water when launched, 6 ft. 6 in. forward and 8 ft. aft. When fully completed and boilers filled (no coal) 9 ft. 5 in. Deep loaded draft 15 ft. 8 in.*

## PANAMA
### Hull 38

The second of the three original vessels of the Pacific Mail Steamship Company was *Panama*, a near-sister to *California*. Although the ships were originally to have been identical, a fire in the shipyard in April of 1848 damaged both the forebody of *Panama* and the moulds from which the timbers of the bow were cut and shaped. Upon reconstruction, the lines of the forebody were improved by making them sharper and increasing the flare, and the vessel thus had lines that were lightly different from those of her sister. *Panama* had a tonnage of 1139, but her dimensions were the same as *California*—200 feet by 33 feet 6 inches by 20 feet. *Panama* was launched on 29 July 1848 and delivered to her owner in November. She left New York on the 1st of December and arrived at San Francisco on 4 June 1849, at the height of the California gold rush. *Panama* was relegated to secondary services as traffic increased and larger vessels were built or purchased. In February of 1861, she was one of seven steamships sold by Pacific Mail to the Oregon & California Steamship Company, which had been formed by Holladay & Flint (later Holladay & Brenham) to operate coastwise lines from San Francisco north to the Pacific Northwest and south as far as Mexican ports. The six other steamers (*Columbia, Cortes, Fremont, Oregon, Republic* and *Sierra Nevada*—none built by Mr. Webb) comprised a diverse collection either built for Pacific Mail or acquired by them from erstwhile competitors. The Oregon & California Steamship Company contracted with the Mexican Government to carry the mails on its southern route, and in 1868 presented *Panama* to the Mexican Government, apparently for use as a transport and revenue cutter. She was armed and renamed *Juarez*, but details of her Mexican service are not known.

Ben Holladay, the guiding light of the Oregon & California Steamship Company, is recalled as an entrepreneur who started a line of stagecoaches in the Northwest, and, after selling out to Wells, Fargo & Company, acquired several railroad lines in the area. One of these, the Oregon Central, he renamed Oregon & California—much to the annoyance of Collis P. Huntington, who controlled the similarly named California & Oregon. In 1870, Ben Holladay became a partner of Mr. Webb in the United States, New Zealand and Australia Mail Steamship Company.

## CHEROKEE
### Hull 39

*Cherokee* was a sidewheel steamship built in 1848 for Samuel L. Mitchill's New York & Savannah Steam Navigation Company. She measured 210 feet 8 inches by 35 feet 4 inches by 15 feet 2 inches and had a tonnage of 1244. Her side-lever engine, which had a cylinder measuring 75 inches in diameter and a piston stroke of 96 inches, was built by the Novelty Iron Works. *Cherokee* was launched in June of 1848 and sailed on the first of 31 voyages for the Savannah line on 5 October 1848. In December of 1849 she was purchased by the Pacific Mail Steamship Company for their New York-Chagres route, on which she made 12 trips. From early 1851 onwards, she ran from New York to Havana and New Orleans for the United States Mail Steamship Company. On the evening of 26 August 1853, she was damaged by fire while lying at her North River wharf at the foot of Warren Street, New York, the night before a scheduled departure for New Orleans. Although the vessel's hull and upperworks suffered heavily from the effects of the fire, a survey conducted after she had been raised disclosed that there was little apparent damage to her machinery. As a result, it was thought that she could be rebuilt. The reconstruction of *Cherokee* never took place, and she was laid up—minus her boilers and parts of her engine—at various locations in New York harbor for the next five years. She was sold to W. H. Webster at auction late in 1856, but she remained in layup until August of 1858, when she was scrapped.

## TENNESSEE
### Hull 40

*Tennessee* joined her near-sister *Cherokee* in the fleet of the New York & Savannah Steam Navigation Company. She measured 211 feet 10 inches by 35 feet 8 inches by 22 feet, and had a tonnage of 1275. The differences in tonnage and dimensions between the two vessels was the result of filling out the forebody hull lines of *Tennessee* and the relocating her engine and boilers seven feet forward, all reportedly a result of the shipyard fire of April 1848 and the destruction of the vessel's forebody moulds. The two Pacific Mail liners, *California* and *Panama*, differed in a similar way for the same reason. Mr. Webb later stated that the forebody of *Tennessee* should have

A rare photograph, taken prior to 1870, of Pacific Mail's steamship *California* at a pier in San Francisco. *California*, built in 1848, was the first steamship to enter San Francisco Bay. (Courtesy of The Mariners' Museum, Newport News, Virginia)

been even fuller. The after end lines and the propulsion plants of the two vessels were identical.

The launching of *Tennessee* took place on 25 October 1848 and she was completed during the following March. The first of her 15 voyages to Savannah commenced on 21 March 1849. In November of that year she was sold to the Pacific Mail Steamship Company for their Panama-San Francisco route. She departed from New York on the 5th of December and arrived at Panama on 11 March 1850. The year 1853 proved to be as unfortunate for

*Tennessee* as it was for *Cherokee*. On the morning of 6 March 1853, she stranded about four and a half miles north of the Golden Gate while attempting to enter San Francisco Bay in a dense fog. The sidewheel towboat *Goliah*, built by Mr. Webb in 1849 and sent to San Francisco in the summer of 1850, was dispatched to the scene to rescue the 600 passengers and crew and attempt the salvage of *Tennessee*. In this she was only partly successful; the passengers and crew were saved along with the mail, but the vessel proved to be too securely aground and, lying broadside to the beach and with an increasingly high sea running, she shortly broke her back and became a total loss.

*Built in 1848. First full seagoing Steamer for the regular passenger and freight trade between New York and Savannah, requiring very little draft of water. Draft when launched, forward 5 ft. 10 in., aft 6 ft. 10 in. Keel 10 inches below timbers. Draft with engines and boilers, coal bunkers and boilers filled, all spars in ends, cabins finished, (no coal) [sic], forward 8 ft. 4 in., aft 9 ft. 7 in. With 100 tons coal, water tanks filled and everything on board ready for sea except cabin stores, forward 9 ft. 7 in., aft 9 ft. 10 in. Proved a very successful and popular vessel.*

### SAMUEL M. FOX
### Hull 41

Mr. Webb built two vessels named *Samuel M. Fox* for the firm of Fox & Livingston, a New York mercantile house. They were christened in honor of the senior partner. The first was a three-masted schooner launched in January of 1849 and handed over to Fox & Livingston at about the date on which her first register document was issued, 21 March 1849. She was a vessel of 257 tons, and measured 101 feet by 21 feet by 10 feet 6 inches. No details of her career or disposition have been found, but she was probably built to run on the owner's coastwise lines, which were operated with sailing vessels until steamers were acquired in the 1860s. This schooner appears to have had a short life in the Fox & Livingston fleet, for the second vessel that bore Samuel M. Fox's name, a ship-rigged packet, was constructed during 1850.

### GOLIAH
### Hull 42

*Goliah* was the second sidewheel towboat built by Mr. Webb for the express purpose of towing sailing vessels between New York piers and the high seas beyond Sandy Hook. For this purpose she was fitted with a powerful beam engine, built by T. F. Secor & Company, with a cylinder 50 inches in diameter by eight feet stroke. The vessel's hull measured 145 feet by 25 feet by 11 feet 2 inches, and her tonnage was 333. She was launched in February of 1849 and placed in service in November of that year. Like her predecessor *Ajax*, a list of the share owners of *Goliah* was a roll call of the principals of most of the New York mercantile houses—the shipowners of the city. Included were Charles H. Marshall, Moses H. Grinnell, Robert B. Minturn, Paul Spofford, Thomas Tileston, William Whitlock, Jr., and, later, William Tapscott and James T. Tapscott. William H. Webb's name also

appeared. On 21 May 1850, she departed for San Francisco on a voyage that was to take eight months, with stops for fuel and provisions at several ports between San Juan and Panama. She was used as a towboat and salvage vessel and on various coastal services out of San Francisco (where she was known as *Defender* for a short period) until 1871. Among her owners during this period were James and Peter Donahue, the latter active in railroad construction in the San Francisco Bay area and the founder of the Union Iron Works, later to become the San Francisco shipyard of the Bethlehem Steel Company. As a salvage vessel, *Goliah* came to the assistance of at least two Webb-built steamships: *Tennessee* when she stranded north of the Golden Gate in 1853, and *Golden Gate* when she ran aground at Point Loma in 1854. Her last years at San Francisco, from 1865 to 1871, were under the ownership of the Saucelito [sic] Water and Steam Tug Company. In the latter year she was sold to Cyrus Walker, of Port Gamble, Washington, to run on Puget Sound. From then until she was dismantled and her hull burned in 1899, she served first Walker, then the Puget Sound Commercial Company, of Port Townsend, and finally the Puget Sound Towing Company, who ran her out of her former home port of Port Gamble. Her employment in the state of Washington was mainly connected with the lumber industry.

### GUY MANNERING
### Hull 43

Taylor & Merrill, a New York mercantile house, were the owners of the packet *Guy Mannering*, built in 1849. They also ordered the packets *Marmion* and *Ivanhoe* and the controversial clipper *Gazelle* from Mr. Webb. The vessel was launched in March and took her place on the Taylor & Merrill Line to Liverpool (called the Black Star Line in Liverpool) early in April with Captain William Edwards as master. Her first register document was dated 7 April 1849. *Guy Mannering*, which had a tonnage of 1418 and measured 190 feet by 42 feet 6 inches by 29 feet 8 inches, was the first three-decked ship built in the United States for cargo service. The name of *Guy Mannering* was taken from that of a romance written by Sir Walter Scott in 1815. That the names of *Marmion* and *Ivanhoe* also came from Scott's works indicates admiration for the Scottish author and poet by Robert L. Taylor, his partner Nathaniel W. Merrill, or both.

She remained on the Liverpool line through the end of the Civil War, but her voyages may have been on an irregular schedule in her latter days. In December of 1865,

This undated view of the sidewheel towboat *Goliah*, taken during her long West Coast career, shows the vessel after accommodations for passengers were added. Goliah was built in 1849 and lasted until 1899, when she was dismantled and her hull burned to recover the metal fastenings. Recycling is definitely not a twentieth-century phenomenon! (Courtesy of The Mariners' Museum, Newport News, Virginia)

she departed from New York for the Merseyside port with a cargo of cotton and grain. All went well until she approached Iona, an island off the Western Scottish coast (known most widely and venerated as the place in which St. Columba founded a monastery in 521 A.D. and as the reputed burial place of 48 Scottish, eight Norwegian, four Irish and two French kings.) There she was wrecked on or about the 31st of January of 1866. For several days it was not clear whether or not she had been wrecked; first it was reported that she and another ship, both dismasted, had been seen at anchor in the nearby channel between the islands of Mull and Colonsay; then, she was reported to be safely anchored at Oban and that a ship named *Germania* had been wrecked. Later, a telegram from Oban stated that it was feared that *Guy Mannering* was the unfortunate one, and this proved to be the case. Captain Brown and a part of the crew were saved, but 18 other men were lost. By late January, the surviving crew members were brought to Glasgow, Captain Brown and the mate having remained at Iona. Over 900 bales of cotton were recovered from the vessel in a more or less damaged state; but the grain was ruined by seawater and could not be salvaged.

*Built in the year 1849, for general business, and was the first full three-deck ship built in the United States for a freighter. Draft of water when launched, with all spars (except yards) and lower and topmast rigging on board, was forward 11 ft. 3-1/2 in., aft 11 ft. 4 in. Draft loaded 24 ft.*

## *GALLIA*
### Hull 44

Mr. Webb built the packet ship *Gallia* for Henry Robinson in 1849. She was a vessel of 1190 tons, measuring 171 feet by 41 feet 6 inches by 27 feet 8 inches. Her launching occurred in August of 1849 and she took her place on a line of packets between New York and Le Havre early in October under the command of Captain Addison Richardson. She was the first packet of over a thousand tons to operate between New York and the French port. Organization of the packet lines and their employment of vessels is sometimes confusing to the marine historian. Some of the lines, such as Black Ball, operated their own ships for years at a time—frequently with the same master. Others chartered vessels for a season or single voyage. A single ship might appear on several routes over the course of a few years. In addition, even if the trade names of the lines remained constant over many years, the owners or agents might change, and with each change a new fleet of ships might appear. The service to Le Havre, for which *Gallia* was built, was dominated for many years by three companies which operated their ships independently from time to time (as the Old, First and Second Lines) and sometimes in association with one another as the Union Line. Henry Robinson, the principal owner of *Gallia*, had many years before been a shipmaster for Crassous & Boyd, the founders of the Second Line. The partnership of Crassous & Boyd was succeeded in 1838 by Boyd & Hincken (who continued running sailing packets to Le Havre on a regular schedule until about 1869 and irregularly until the late 1870s,) and it is likely that *Gallia* ran for that firm. Except for her early service to Le Havre and a period starting in 1851 in which she ran from New York to Mobile for Ladd's Empire Line, little is known of the career of *Gallia*. It is stated by some sources that she ceased running as early as 1852. Her movements, all for William Whitlock, Jr., during the latter part of that year were on the New York-Mobile-Le Havre triangular trade that developed because of an imbalance in westbound and eastbound cargoes between New York and Le Havre and the need to transport Southern cotton to Europe. She cleared New York for Mobile on 6 May 1852 to load cotton, after which she proceeded to Le Havre, and then returned to New York, arriving on the 3rd of October with 750 passengers and a mixed cargo. She next departed on the 21st of October for Mobile, probably to repeat the itinerary of the previous voyage. Her movements after that time and her final disposition are not known.

## *JAMES DRAKE*
### Hull 45

*James Drake* was a ship built for Spofford, Tileston & Company in 1849, for service on their line between New York and Havana. The last but one of the small ship-rigged vessels built by Mr. Webb and the only sailing vessel built by him for these owners, she had a tonnage of 482 and measured only 130 feet by 30 feet 6 inches by 20 feet 2 inches. She was launched in May of 1849 and completed late in June, her first register having been dated the 30th of that month. When Spofford, Tileston & Company abandoned sail and started operating steamers, *James Drake* was sold to a Captain Jones and others. A report that she was missing was circulated late in 1853, but it proved to be false. She cleared from New York on 26 November 1853 for Cardiff, Wales, with a typical eastbound cargo—999 barrels of flour, 18,741 bushels of wheat and 637 barrels of rosin, all valued at $30,000. On 10 March 1854, she was posted in New York as missing, but three days later a correction was published stating that she had arrived at Cardiff on 10 January 1854 after what Captain Jones described as a "tempestuous passage of 40 days." She discharged her cargo and sailed on the 24th with a cargo of Welsh coal for Havana. Following her arrival at New York on 20 January 1856 from Trapani, Sicily, with a cargo of salt and her departure for the East Indies on the 12th of April, nothing is known of her movements until she became *Goa*, under the French flag and owned by F. DeConinck & Company, of Le Havre, in the late 1850s. She was still afloat in 1863, but her ultimate fate is not known.

## *ALBERT GALLATIN*
### Hull 46

The packet ship *Albert Gallatin* was built in 1849 for Grinnell, Minturn & Company. She was the fourth of five packets completed for that firm, one of the largest mercantile houses in New York. They employed the vessel on their Blue Swallowtail Line to Liverpool. Named for Thomas Jefferson's Secretary of the Treasury whose death occurred during the year in which the ship was built, *Albert Gallatin* was a three-decked vessel of 1435 tons measuring 190 feet by 42 feet 8 inches by 28 feet 3 inches. She was launched in September of 1849 and cleared New York on her first voyage on the 10th of October, with Captain John A. Delano in command. *Albert Gallatin* remained on the Swallowtail Line through 1868 under Captain Delano, who was relieved for a short time in 1863-64 by Samuel Macoduck, whose unusual name was spelled in a bewildering variety of ways in the arrival and departure columns. The vessel was sold in 1868 to T. Leach & Company, of New Orleans. In 1872, she was sold to owners in Liverpool and transferred to the British flag. Under Captain Groves, *Albert Gallatin* left Antwerp in ballast for Callao on 29 April 1875. She lost her rudder on the 2nd of August in the vicinity of Cape Horn, and, for the next 13 days the crew attempted unsuccessfully to rig a jury steering gear. On the 15th, as the ship drifted close to the rocky shore of Ildefonso Island, Chile, she was abandoned. Captain Groves, his wife and two children, the second mate and four seamen boarded one of the vessel's two boats, and the chief mate and twenty seamen the other. The boats soon became separated and that of the captain landed on Hermit Island the following day. Ten days were spent there, then four in the boat and another two on another island. On 2 September 1875, the captain's boat, with his family and the five crew members, was picked up by the ship *Syren*, which landed them at Honolulu. The hand of Providence must have been with the nine survivors; the probability of a small boat being sighted at sea off Cape Horn in winter was indeed slim. The 21 men in the other boat were never seen again.

### *CATHARINE*
### Hull 47

The 610-ton *Catharine* was, like *James Drake*, a small ship-rigged vessel. She was the last of her type from the Webb yard. *Catharine* was built in 1849 for the firm of Schuchardt & Gebhard, Antwerp agents for E. D. Hurlbut & Company's line of packets between New York and the Belgian port. Launched early in September of 1849 and completed in mid-month, her dimensions were 136 feet by 31 feet 8 inches by 19 feet 6 inches. Later she ran on Hurlbut's New Orleans line and to Mobile, and in 1860 was to be found on Stanton & Thompson's New Orleans line. Her owners in 1863 were Tucker, Cooper & Company, of New York, but nothing is known of her later days.

It is possible that she was sold to foreign registry during the Civil War when operation under the Stars and Stripes became too hazardous in the face of Confederate commerce raiders. These vessels—*CSS Alabama, Florida, Tallahassee, Shenandoah* and others—were responsible for the destruction or capture of upwards of 250 Union vessels. Most of the victims of *Shenandoah* met their ends after the surrender of General Lee at Appomattox Court House, the raider's warlike career having ended in the Pacific Ocean on 2 August 1865, when the British bark

*Barracouta* informed her commander, Lieutenant James Iredell Waddell, CSN, of the capitulation of the Confederacy. Waddell apparently had learned of the war's end on the 23rd of June, but elected to continue his own private war. On *Shenandoah*'s single wartime cruise of 58,000 miles, which netted 36 Union vessels (mainly whalers in the North Pacific,) no less than 20 were destroyed or ransomed after the 23rd of June.

### *MANHATTAN*
### Hull 48

The sixth packet ship built by Mr. Webb for C. H. Marshall & Company's Black Ball Line to Liverpool was *Manhattan*, a vessel of 1299 tons launched in December of 1849. She and *Isaac Webb* were apparently ordered at the same time. The hull dimensions of *Manhattan*, a three-decked vessel (the first of the type on the line,) were 180 feet by 42 feet 6 inches by 27 feet 3 inches. She was completed ready for sea in January of 1850, and her first register document was dated the 12th of that month. She departed on her first voyage a short time afterwards under the command of Captain William G. Hackstaff, who had been the first master of *Fidelia* when she came out of the Webb yard nearly five years before. Hackstaff was relieved by Captain Enoch W. Peabody (who would later spend many years on Webb's Black Ball packet *Neptune*) and he by Captain Thomas Dixon. *Manhattan*'s time on the Black Ball Line was shorter than most of the Webb packets. The vessel's last departure from New York was in late October or early November of 1862, with Captain Dixon in command. She made her scheduled call at Liverpool and left that port to return to New York on the 23rd of December. On 14 March 1863, it was noted that she was over 80 days out of Liverpool, but hope was not given up—an 80-day westward passage of the North Atlantic in winter was not unheard of, and she might have been dismasted or, in distress but still manageable, made her way to Fayal for refuge and repairs. By 21 April 1863, a New York newspaper stated that, since *Manhattan* had been 118 days en route, "much anxiety is felt for her safety." Shortly afterwards, she was posted as missing. *Manhattan* was never heard of again.

### *ISAAC WEBB*
### Hull 49

C. H. Marshall & Company chose the name *Isaac Webb* for its seventh packet ship from the Webb yard in honor of the builder's father. The vessel was launched on 2 February 1850 before a crowd of 5,000 persons. A re-

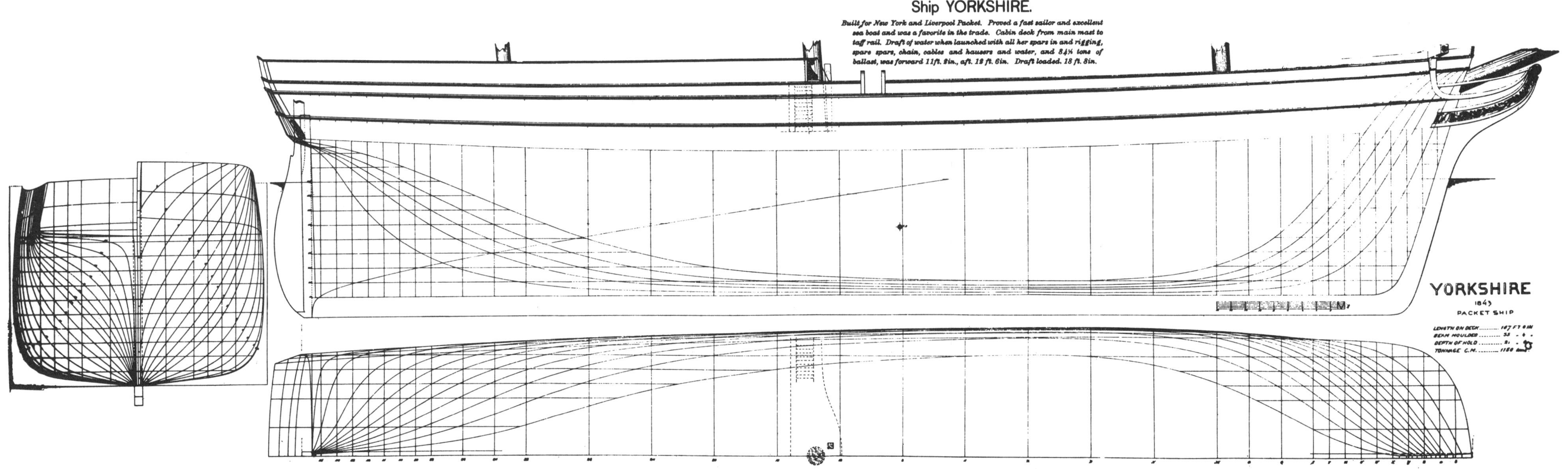


Lines of the packet ship *Yorkshire*, built in 1843 for C. H. Marshall & Company's Black Ball Line. Scale about 1/16 inch to the foot. Compare her lines to those of clipper *Challenge*, reproduced at the same scale.

porter of the *New York Herald*, impressed by the launching and by the ship's size, wrote, "She rested like a swan on her destined element." She was delivered early in March, and her first registry document was dated the 6th of that month. *Isaac Webb* measured 185 feet by 42 feet 6 inches by 27 feet 3 inches. She was five feet longer than *Manhattan*, which had preceded her from the yard by two months. The tonnage of *Isaac Webb* was 1359. She spent many years on the Black Ball Line to Liverpool, making about four round trips per year. These voyages were usually uneventful (except for the North Atlantic weather in winter,) but on 20 June 1863, while under the command of Captain Charles Hutchinson, *Isaac Webb* was captured by the Confederate raider *Florida No. 2* (formerly a Northern bark named *Tacony* that had only been captured by the Confederates on the 12th) and, because she had 11 cabin and 647 steerage passengers aboard, was released by the "rebel pirate" (as Southern raiders were called in the Union press) upon payment of a $40,000 bond. *Isaac Webb* arrived at New York on the 23rd. On the 25th, *Florida No. 2* was burned by her own crew, who transferred to yet another seized vessel when Union naval forces came too close for comfort. The score during her 13-day naval career—15 vessels captured, of which 11 were destroyed.

Later that year, *Isaac Webb* had a minor skirmish with an unidentified British bark off the Grand Banks as she was bound to New York with 460 passengers aboard. Shortly after midnight of 12 October 1863, while on a starboard tack, she was struck on the port side by the other vessel between the fore and main masts, carrying away the bulwark, some of the main rigging and doing some other minor damage. She went to the aid of the British vessel, which not only claimed that she needed no assistance, but also refused to identify herself other than to say she was from Yarmouth bound for Galway. Both continued their voyages. On 29-30 December 1866, with Captain James C. Stowell in command, *Isaac Webb* endured a violent gale while bound to New York. Captain Stowell's report of the voyage as published by the *New York Times* illustrates the environment to which many packets and their passengers and crew were exposed in winter.

> December 29th, had WSW hurricane for 36 hours, sleet and hail, blew away entire suit of sails, staving boats, carrying away spanker boom, springing foremast head and shipped [sic] a sea which shifted everything movable about deck. Barometer 27.50. Owing to intense cold, several of crew were frostbitten and three died.

After this incident, *Isaac Webb* quietly continued her Liverpool trips until November of 1868. She had left Liverpool on the 22nd of September, with 354 passengers aboard, under the command of Captain Stowell, then 34 years old and in the employ of the Black Ball Line for 20 years. On the 10th of October, a terrific gale lashed the ship. The deck was swept by green water, and a companion ladder struck Captain Stowell, hurling him against the bulwark. He was taken below with severe internal injuries, but died seven days later. Mate Daniel Cozzens assumed command. By the 12th of November, provisions had run low, but they were replenished from the passing Nova Scotian bark *Brazil*. As she approached New York, *Isaac Webb* ran aground near Sandy Hook on the night of the 17th. Two days later she was refloated by the New York Submarine Company's salvage steamers *Philip* and *Rescue*, which towed her to her pier at the foot of Beekman Street, East River. She continued her Liverpool voyages for another decade, until the end of the Black Ball Line in 1879. She then made a few tramping trips to London, Le Havre and Antwerp for the Marshall firm, but by late 1880 had reached the end of her career.

The vessel's last voyage began at Antwerp on 2 October 1880, when she cleared for New York under the command of Captain William W. Urquhart. The trip was uneventful until the 24th, when, at midnight, in a southeast gale and heavy and confused cross seas, the ship lurched suddenly to leeward, shifting her cargo and setting her on her beam ends. The main and mizzen masts were cut away to prevent her from capsizing. The crew set to work trimming the ship and pumping. At two o'clock the next afternoon the foremast broke off, taking with it the jibboom. Trimming and pumping continued, but, as Captain Urquhart reported, "The ship was lying over so much and rolling so deeply that it was almost impossible to hold on, let alone to work." A ship's lights were seen and a distress signal given. The Leyland Line steamship *Illyrian*, bound from Liverpool to Boston, saw the signal and hove to. In a rescue mission that was little short of heroic, the 24 men aboard *Isaac Webb* were transferred to a longboat and then to the steamer in high seas and in the black of night. The weary crew landed at Boston on the 27th. As was the custom after daring rescues at sea, Captain Urquhart stated, "All our crew . . . [were] treated in a most kind, hospitable manner by the Captain, officers and all on board [Illyrian], for which we beg here to publicly express our very best thanks." It appears to have been a more civilized age in all but the fury of the full gale. Notwithstanding her unfortunate loss, the three decades in which *Isaac Webb* had battled the North Atlantic was an admirable testimonial to the staunchness of Mr. Webb's ships.

*Built in 1850. Proved a good model and serviceable ship for the [New York and Liverpool] trade. Draft of water, with all spars, anchors and chains and water tanks on board, was 10 feet forward and 10 feet 5 in. aft. Deep loaded draft 23 feet 6 inches.*

## *VANGUARD*
### Hull 50

The name of James W. Phillips appears three times in a list of owners who built ships at Mr. Webb's yard. Phillips' packet ship *Liberty* of 1842 was followed eight years later by another packet, *Vanguard*, and in 1851 by the clipper *Invincible*. *Vanguard* was a vessel 1196 tons, launched early in March of 1850 and completed late in the month. Her first register document was dated 20 March 1850, and it is probable that her maiden voyage commenced shortly afterwards. She measured 175 feet by 38 feet by 22 feet 6 inches, and was a typical two-decked packet. A portion of her early service was in John Elwell's Merchants' Line between New York and Mobile. John Elwell was succeeded by James W. Elwell & Company, which remains active as a ship agent in New York, holding one of the few surviving threads leading back to the days of the packet ship.

*Vanguard* was not so long lived, but managed to last until 1877. In 1866, her rig was changed to that of a bark. She was acquired by Howland & Frothingham, who ran her both on their line to Liverpool and in coastwise trading during and after the Civil War. In the mid-1870s, Captain R. L. Ryfkogle and others purchased her for general trading, Ryfkogle acting as her master. She was driven ashore on 13 October 1877 at Matane, Quebec, after a gale sprang up while she at anchor waiting to load for London. The vessel was condemned, but most of her cargo, consisting of birch and spruce deals loaded at another port, was salvaged and sold; the birch brought seven cents per board foot and the spruce, seven and a half cents.

## *FLORIDA*
### Hull 51

A pair of sidewheel steamships, *Florida* and *Alabama*, were built for Samuel L. Mitchill's New York & Savannah Steam Navigation Company in 1850. The first vessel, *Florida*, was launched in April of that year and departed on her first voyage to Savannah late in October of 1850. The hull of *Florida* measured 214 feet by 35 feet 6 inches by 22 feet, and her tonnage was 1261. The Novelty Iron Works built her side-lever engine, which had a 75-inch cylinder with a piston stroke of eight feet. She remained on Mitchill's line until the Civil War brought an end to coastwise steamship services. The United States Navy purchased the vessel on 5 October 1861 and armed her with eight smooth-bore 32-pounders and one 20-pounder rifle. The name *Florida* was retained. She spent most of the war in the South Atlantic and North Atlantic blockading squadrons, but in 1865 transported Confederate prisoners from the ram *CSS William H. Webb* to New York. The ram was originally the sidewheel towboat *William H. Webb*, built at the Webb yard in 1856. *Florida* was sold at auction on 5 December 1868 to Samuel Ward. It is doubtful that she ever ran again under the American flag, for there followed three quick changes of ownership in June of 1869: to Samuel Pedrick on the 7th, at which time she was renamed *Delphine*; then to Frederick W. Clapp on the same day; and finally, on the 16th, to Haitian owners represented by one I. Annoual. The Haitians were revolutionaries who used *Delphine* for purposes of war during one of the country's frequent 19th-century political upheavals. She was reported to be lying at Port-au-Prince in derelict condition in 1874, but nothing is known of her ultimate fate.

## *ALABAMA*
### Hull 52

The sidewheel steamship *Alabama* was built in 1850 to run on the New York & Savannah Steam Navigation Company's line with her sister, *Florida*. She had a tonnage of 1261, measured 214 feet by 35 feet 6 inches by 22 feet and, like *Florida*, was fitted with a side-lever engine from the Novelty Iron Works. The cylinder of the engine was 75 inches in diameter with a stroke of eight feet. The vessel was launched on 10 June 1850, with the clipper *Celestial*, and made her first trip in January of 1851. She remained on the route between New York and Savannah until the start of the Civil War. The United States Navy acquired her on 1 August 1861, armed her with eight 32-pounders and assigned her to blockading and other duties as *USS Alabama*. She was sold at auction on 10 August 1865 to Samuel C. Cook, who sold her to Henry Steers and his Coastwise Company. She was used on coastal services out of New York for about three years, and in November of 1868 was acquired by the Florida Railroad

Company, which ran her from New York to the then important port of Fernandina to connect with the railroad's line to the Gulf, at which point passengers boarded other steamers for West Gulf ports.

During this period *Alabama* was commanded by Captain Joseph Limeburner, who had been master of Donald McKay's clipper *Great Republic* before the Civil War. Unlike many of the famous clipper ship captains, Limeburner successfully made the transition to steam. Henry Steers again became the owner of *Alabama* late in 1869. Two more changes in her ownership occurred in 1872: she was sold to H. F. Hamill and then to a group headed by George W. Miles. In September of that year her machinery and paddle wheels were removed and she was converted to a schooner, still under the name *Alabama*. Her last document, surrendered on 21 March 1878, states that she had been destroyed by fire, but the date, place and circumstances remain as much of a mystery as the fate of her sister *Florida*.

### *CELESTIAL*
### Hull 53

Mr. Webb built the extreme clipper *Celestial* for Bucklin & Crane, of New York, in 1850. The firm of Bucklin & Crane was active in trading to the Orient during the 1850s. The tonnage of *Celestial* was 860 and she measured 158 feet by 34 feet by 19 feet. She and the steamship *Alabama* were launched on the same day, 10 June 1850. *Celestial* left New York for San Francisco on her first voyage on the 16th of July under the command of E. C. Gardner, who was later to take the clippers *Comet* and *Intrepid* from the Webb yard. Her time on the first voyage, which matched her against *Mandarin* (launched by Smith & Dimon for Goodhue & Company just five days after *Celestial*) was 104 days. Her opponent took 128 days. Subsequent voyages of *Celestial* to San Francisco were made in 1851, 1853 and 1854 under Captain Benjamin F. Palmer, an elder brother of the brilliant shipmaster Nathaniel B. Palmer. These three Cape Horn passages were logged at 108, 119 and 133 days respectively. Three voyages were made in the English tea trade from China to England in 1852, 1853 and 1857, and in 1858 she was sold for service under the Spanish flag. Her owner in 1861 was Bartolomeo Blanco, a New York merchant, for whom Mr. Webb had built a number of vessels. Blanco acted as New York representative for several Spanish shipowners. There is no record of her further career.

The names *Celestial* and *Mandarin* have always been associated with China, as well as with American shipbuilding. In 1921, seventy years after the clipper ship age, the United States Shipping Board built four 10,000 deadweight-ton cargo ships at the Kiangnan Dock & Engineering Works, of Shanghai, China—*Celestial*, *Mandarin*, *Oriental* and *Cathay*—all of which had been names of clippers that ran to the Far East. All four names—this time *Kathay* with a "K"—were repeated in C2 cargo ships built for the United States Maritime Commission in 1944. The clipper *Oriental*, an early vessel of the type, was built by Jacob Bell at New York in 1849; *Kathay* (spelled with a "K" for reasons that are today obscure) was completed by Jacob A. Westervelt in 1853, the peak year for clipper production. About 120 of the vessels emerged from New York and New England yards during that year.

### *JOSEPH WALKER*
### Hull 54

The ship-rigged packet *Joseph Walker* was built by Mr. Webb in 1850 for Samuel Thompson & Nephew's Black Star Line to Liverpool. She was a vessel of 1325 tons and measured 185 feet by 40 feet by 23 feet. She was launched on 22 August 1850 and cleared New York on her first voyage in mid-September with Captain William Hoxie in command. Captain Hoxie had been master of the packet *Caleb Grimshaw* when she burned in 1849. The life of *Joseph Walker* was, like that of *Caleb Grimshaw*, to be short. On the early morning of 27 December 1853, *Joseph Walker* (with Captain Hoxie still aboard) was lying at her East River pier at the foot of Dover Street, near the clippers *Great Republic*, *White Squall*, *Red Rover* and *Whirlwind*. *Great Republic* was Donald McKay's immense new four-master that had arrived from her builder's yard at East Boston on the 28th of November (under tow of the steamer *R. B. Forbes*) and was loading for her first voyage. A million-dollar fire, which started in a building near the waterfront, quickly spread to nearby piers and ships. *Great Republic* and *White Squall* were heavily damaged; both vessels were rebuilt and saw many years of further service. *Red Rover* and *Whirlwind* miraculously escaped the flames, but *Joseph Walker* was not so fortunate. Her masts, sails and rigging were destroyed in less than an hour, and while they were burning, repeated attempts were made to scuttle the vessel. The intense heat of the fire prevented the crew from carrying out the task. Instead, she was almost completely destroyed and eventually sank at her berth.

A number of firms attempted to salvage the ship, but

each was bankrupted in succession. It was not until September of 1855 that she was raised by Dodge, Barnes, Wing & Jones using an effective but expensive method. Eight chains were placed under what remained of the vessel and 40 piles were driven port and starboard. Next, five large trusses were erected through which the chains were led to a number of hydraulic rams, about which the *New York Times* remarked, "the power of each ram being almost incalculable." Whatever calculations were made, were correct, for the vessel was lifted enough so that gangs of men could remove her cargo (and 150 tons of mud) at a succession of low tides, after which she was floated and removed from the slip. Unfortunately, Messrs. Dodge, Barnes, Wing and Jones may not have achieved great wealth from the job, for a week later, in the New York Supreme Court, an injunction was granted in the case of one Janius S. Lewis vs. Walter R. Jones, Charles F. Barnes, The Mayor, Aldermen and Commonalty of the City of New York, Thomas Bell and Alfred G. Benson, by which the defendants were prevented from "receiving the proceeds of Joseph Walker or her cargo or any part thereof, or from meddling with her in any manner, except to raise and properly guard her and her cargo." The Liverpool packets nearly always lifted heavy cargoes on their eastbound passages. *Joseph Walker* had 7500 bushels of wheat, 9000 bushels of corn, 400 barrels of resin and 200 bales of cotton, in all worth $42,000, aboard at the time of the fire. The vessel herself was valued at $90,000.

*Built in 1850 for the Liverpool trade. Was a noble looking ship. Proved a good sea boat. Was a favorite ship. Draft of water when launched, keel 2 ft. 6 in. below timbers, with spars on board, one anchor and chain, forward 10 ft. 8 in., aft 11 ft. Deep load draft 20 ft. 6 in.*

## UNION
### Hull 55

Spofford, Tileston & Company ordered the sidewheel steamship *Union* for one of their coastwise services out of New York. She was a barkentine-rigged vessel of 1200 tons, measuring 215 feet by 34 feet by 22 feet. Two Allaire side-lever engines 60 inches in diameter by seven feet stroke were fitted. *Union* was launched on 9 September 1850 and delivered in March of the following year. After she entered service on the 8th of March, she made nine trips to New Orleans followed by 22 to Charleston. Next came three voyages to Aspinwall in a joint service with Davis, Brooks & Company, then 13 more Charleston

trips. From May of 1854 onwards, she was chartered to the Havre Line, making ten Atlantic voyages. On 15 March 1956, she departed from New York for Trieste, having been sold to Austrian owners for service in the Mediterranean. Renamed *America*, she lasted until 1875, when she was scrapped.

## GOLDEN GATE
### Hull 56

The third sidewheel steamship built by Mr. Webb for the Pacific Mail Steamship Company was *Golden Gate*, a vessel of 2067 tons with a hull measuring 262 feet by 40 feet by 30 feet 6 inches. The size of *Golden Gate* was dictated by the enormous increase in the number of passengers and amount of cargo carried by the line as a result of the discovery of gold in California. She was significantly larger than the first two Pacific Mail liners built by Mr. Webb in 1848, the 200-foot *California* and *Panama*. *Golden Gate* was propelled by two oscillating engines from the Morgan Iron Works, the cylinders of which were 85 inches in diameter with a piston stroke of nine feet. The vessel was launched in an impressive display of shipbuilding skill on 21 January 1851, along with Fox & Livingston's packet *Isaac Bell* and Taylor & Merrill's controversial clipper *Gazelle*. *Golden Gate* received her first register document on 1 August 1851, and departed the following day for Panama. Calls for coal of five and 15 days respectively were made at Rio de Janeiro and Valparaiso. She arrived at Panama on the 16th of October. She soon earned a reputation for speed on the Panama-to-San Francisco run, but with the cost of coal at between 25 and 40 dollars per ton, the company limited her speed. If *Golden Gate* was remembered for anything other than her speed, it was her abysmally bad luck—an outbreak of cholera aboard in 1852 which resulted in 84 deaths, a near collision with Cornelius Vanderbilt's steamer *Sierra Nevada* off the Mexican coast in May of 1853, an engine breakdown and a grounding at Point Loma on one voyage in 1854, a broken shaft in October of 1857, and her loss by fire in 1862.

On what was to be her last voyage, the unfortunate vessel left San Francisco on 24 July 1862 for Panama. On the 27th, when about 15 miles from Manzanillo, Mexico, fire was discovered in the engine room, and the vessel was headed for the shore in order to beach her. Many of the passengers had sought refuge in the stern, but the flames spread in that direction. Boats were launched in the heavy surf, only to have the occupants crushed against

*Union*, completed in March 1851 for Spofford, Tileston & Company, was one of Webb's lesser known steamships. Her career under the American flag was short, for she was sold to Austrian owners for service in the Mediterranean Sea in 1856. (Courtesy of The Mariners' Museum, Newport News, Virginia)

the ship or drowned. The ship broke up, and, when all was again quiet, 175 passengers and crew members had perished. In addition, the baggage, mail and nearly all of the cargo, which included about 1.4 million dollars in gold, were lost. Gold valued at $300,000 was recovered from the wreck and brought to San Francisco by Pacific Mail's steamship *Constitution* the following February. In her fiery end, *Golden Gate* had lived up to the reputation she had acquired in eleven years on the Pacific Mail.

*Three full decks. Built in the year 1851 for the Pacific Mail Steamship Co. Draft of water when launched, with anchors and chain on board, was forward 5 ft. 7 in., aft 7 ft. 7 in. Machinery complete, boilers filled and 50 tons coal on board, was forward 10 ft. 4 in., aft 10 ft. 7 in. With 300 tons coal on board, water tanks filled, ready for sea (except provisions), 13 ft. 8 in. even keel. Proved a fast and excellent sea boat.*

### *SAMUEL M. FOX*
### Hull 57

*Samuel M. Fox* was the first of a pair of packet ships built for Fox & Livingston late in 1850 for the Union

Line to Le Havre. She was a vessel of 1060 tons, and measured 171 feet by 39 feet by 26 feet 6 inches. She was launched in November of 1850 and took her place on the Union Line about the first of the new year under the command of Captain Allen C. Ainsworth. Her register document was issued on 30 December 1850. The vessel saw service from time to time on lines other than that to Le Havre. It was while she was on W. & J. Tapscott & Company's line to Liverpool that her career ended. While leaving Liverpool under tow on 12 November 1856, she was caught in a strong gale and was driven ashore on the West Middle Bank near the mouth of the River Mersey. She later came off and anchored. The next day, her anchor chains parted and she grounded once again, this time on the nearby Great Burbo Bank. Her passengers and crew were landed safely (except for the mate and two crew members who stayed aboard to maintain a watch over the vessel) and two steam towboats from Liverpool attempted unsuccessfully to tow her off, but *Samuel M. Fox* broke up in the storm over the next few days. The packets *Silas Wright*, built by Mr. Webb in 1855, and *Louisiana* were destroyed by the same storm.

### *ISAAC BELL*
### Hull 58

*Isaac Bell* was identical to her sister ship *Samuel M. Fox*, and was built for Fox & Livingston for service on the Union Line of packets between New York and Le Havre. Her dimensions, 171 feet by 39 feet by 26 feet 6 inches, were the same as her sister, but her tonnage was 1072. *Isaac Bell* was launched with *Golden Gate* and *Gazelle* on 21 January 1851, and entered the service of the Union Line early in February. Her first captain was John Johnston. *Isaac Bell* was named for a master mariner turned merchant; Bell was associated with Francis A. Depau, who started the first packet line to Le Havre (the "Old Line") in 1822. Two of Depau's daughters married Samuel M. Fox and Mortimer Livingston, whose partnership—Fox & Livingston—eventually assumed the operation of the Le Havre line. The vessel later ran between New York and New Orleans for E. D. Hurlbut & Company; on Post, Smith & Company's Regular Line to Antwerp; and in 1858 on E. E. Morgan's line to London. In her later years, she was owned by W. S. Drayton, who dispatched her from Cardiff to Shanghai in the summer of 1860 with a cargo of Welsh coal. On the 7th of August she grounded on the North bank of the Yangtzse Kiang as she was approaching the end of the long voyage. Cap-

tain S. M. Warren and his crew were saved, but the vessel, which had become a total wreck within a few days, was sold at auction shortly afterwards.

The most enduring accomplishment of *Isaac Bell* was a 19-day westward crossing to New York, the fifteenth best such passage on record out of the thousands of voyages made by packets. Many shorter passages were made in the eastbound, or "downhill" direction, mainly by clipper ships (these vessels having been used in transatlantic service in addition to their well-known voyages to San Francisco and the Far East,) but also by a number of fast packets. The eastbound record to Liverpool of 13 days, one hour and 25 minutes, never surpassed, was set in January of 1854 by the then-new clipper *Red Jacket*, an outstanding vessel that was designed by Samuel H. Pook, built by Deacon George Thomas at Rockland, Maine, and commanded by the legendary Asa Eldridge. Captain Eldridge is also remembered as the master of Cornelius Vanderbilt's sidewheel steamship *North Star* when she made her grand yachting cruise to Europe in the summer of 1853, and as master of the Collins Line unfortunate steamship *Pacific*, built by Jacob Bell in 1850, when she went missing on a westward crossing of the Atlantic after departing from Liverpool for New York on 23 January 1856.

### *GAZELLE*
### Hull 59

One of Mr. Webb's most controversial ships, *Gazelle* was an extreme clipper launched on 21 January 1851 for Taylor & Merrill, on the same day upon which he launched two other vessels, the steamship *Golden Gate* and the packet *Isaac Bell*. The tonnage of *Gazelle* was 1244, her dimensions were 182 feet by 38 feet by 21 feet, and the date of her first registry was 3 March 1851. Her first captain was Robert Henderson, Jr. She collided with a Spanish ship on 2 October 1852, while on her second voyage, and, on 14 October 1854, while enroute from San Francisco to Hong Kong, she encountered a typhoon in which all three masts and the jibboom were lost over the side and 16 of the 189 Chinese tweendeck passengers drowned. Following a tow to Hong Kong, she was condemned and sold to owners in Peru. Renamed *Cora*, she carried coolies to work in the Peruvian guano deposits. She was later owned by E. Bates & Company, of Liverpool, under the name *Harry Puddemsey*. Her ultimate fate under the British flag is not known.

The controversy over *Gazelle* centered around the

amount of deadrise on her hull. For many years, a debate had been carried on among ship designers and master mariners over the proper shape of the midship section of fast sailing vessels. The designers and builders, such as Mr. Webb, Donald McKay, and John W. Griffiths, generally favored a full midship section with fine entrance and run; many shipmasters were convinced that a midship section with sharp deadrise was required to obtain high speed. When Captains Robert L. Taylor and Nathaniel W. Merrill came to Mr. Webb with their ideas about hull form (to which he was emphatically opposed,) he built the ship in accordance with their wishes. *Gazelle* was probably not as fast as Mr. Webb's other clippers, but the relative performance of vessels with full and sharp midship sections is so influenced by other factors, such as draft, trim, sea and wind conditions, type of rig, condition of the vessel's bottom, and the skill of the master, that it is not possible to reach a general conclusion on the merits of each. However, because clipper ships were built to carry cargo at high speed, the performance of the ships in relation to their carrying ability was of importance. The clipper *Nightingale*, built by Samuel Hanscom, Jr. at Portsmouth, New Hampshire, in 1851, had a midship section similar to that of *Gazelle*. She eventually proved to be a fast vessel, but was not known to be an efficient cargo carrier. If a generalization can be made in relation to the influence of midship section, the vessel with sharp deadrise sailed fairly well in light airs, but, as wind velocity increased, she fell behind the ship having low deadrise.

*Built early in the clipper ship era, expressly for the California trade. This ship was built having a great dead rise of floor, to meet the views of her owners, two retired ship captains, who entertained old received [sic] and erroneous ideas of modelling. She did not prove as fast as other clipper ships having much less dead rise of floor and carrying more cargo. This ship fell in at sea on several occasions with the clipper ships Flying Dutchman of about the same general dimensions having low dead rise of floor, also the Swordfish of much less general dimensions and much less dead rise of floor (all modelled and built by same builder). With light winds and smooth sea Gazelle could barely hold her own, but with stronger winds and heavier sea the others invariably passed the Gazelle. The Flying Dutchman carried about one-quarter more cargo and the Swordfish fully one-third more cargo than the Gazelle. These tests proved what Mr. Webb always maintained that excessive dead rise of floor was not, but rather flat floor was best to secure high speed. Draft of water when launched, keel 44 in. below timbers—with most of spars on board, forward 13 feet 3 in., aft 13 feet 6 in.—deep load draft about 20 feet.*

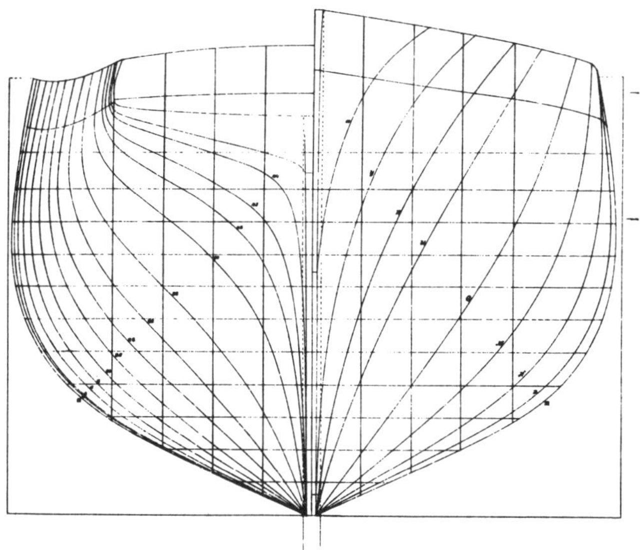

The midship section of the clipper *Gazelle*, constructed in 1851 for Taylor & Merrill, illustrates the controversial high deadrise that was so atypical of William H. Webb's clipper ships. Compare this with the midship sections of the extreme clippers *Young America* and *Challenge*, shown elsewhere in this volume.

## *CHALLENGE*
### Hull 60

One of Mr. Webb's most renowned vessels, the extreme clipper *Challenge* was constructed for the mercantile firm of N. L. & G. Griswold (sometimes irreverently referred to as "No Loss and Great Gain") in 1851 at the peak of the clipper ship craze. In the eyes of a contemporary observer, she was an immense three-decked vessel of 2006 tons, measuring 230 feet by 41 feet 8 inches by 26 feet. At the insistence of her prospective master, hard-driving Robert H. Waterman, she was given an equally immense spread of canvas—too much, in the opinion of Mr. Webb. Her top-heavy rig was cut down several times. When *Challenge* was launched on 24 May 1851, four other ships—the remarkable clippers *Comet*, *Invincible* and *Swordfish* and the packet *Great Western*—were under construction at the Webb yard. On her first voyage, which commenced on the 13th of July, *Challenge* was matched in a race to San Francisco with Donald McKay's *Flying*

*Cloud.* The McKay vessel's time of 89 days completely overwhelmed *Challenge*'s 109 days on a passage that for Captain Waterman was a personal and professional disaster—uncooperative winds, a cruel mate, a poor and inexperienced crew, charges of inhumanity, a shipboard murder and arrest and trial upon arrival at San Francisco—but above all, a lost opportunity to leave *Flying Cloud*, Josiah Creesy and Donald McKay in his wake. But *Challenge* and Captain Waterman went their separate ways to better days, and the ship made many fast passages under Captain John Land and others. Captain Land relieved Waterman in San Francisco.

In the spring of 1852, she ran from Hong Kong to San Francisco in 33 days, a time that was never beaten and tied but once, in 1878, by the ship *Wandering Jew*, built at Camden, Maine, in 1877, long after the day of the clipper ship. Near disaster overtook *Challenge* on at least two occasions. While under the command of Captain John Kenney, she left Le Havre on 29 March 1854 for New York with cargo and 789 passengers on board. On the 17th of April, she was lying to in a western gale and heavy seas about 560 miles southeast of Cape Race, Newfoundland, when her fore and main topmasts, mizzen topgallant mast, the head of the main mast and the fore and main yards were carried away. She arrived at New York on the 10th of May under a jury rig. *Challenge* made a second voyage to China in 1859, under Captain Samuel A. Fabens. She departed from San Francisco on the 10th of August for Hong Kong. On the 17th of September, while off the China coast, she encountered what Captain Fabens stated was the worst typhoon of his life. His log entry for that day was as follows:

> "Laying too [sic] under main Spencer, all other sailes [sic] furled and extra gaskets passed. Typhoon in every sense of the word. At 3 PM the three Royal masts blew over the side at 7 PM all three lower mast heads were twisted off and the ship completely dismasted, the spars chafing badly outside, tearing up the decks, etc., blew away one boat and injured five or six others, some copper torn off. The twisting off [sic] the masts was so sudden that not one Shroud Backstay or Stay gave way, but were obliged to cut away all our rigging to clear the wreck which hung alongside all night. Ends more moderate."

After a period of lay up in Hong Kong, she was acquired by Captain Haskell and placed back in service. In 1861, she was sold to Thomas Hunt & Company, transferred to British registry and renamed *Golden City.* Joseph Wilson, of South Shields, purchased her in 1866, and she spent the next ten years running primarily from British ports to New Orleans or the East Indies. She was lost off the French coast in 1876 after her rudder came adrift and rendered her unmanageable in the English Channel a few days after sailing on a voyage to Calcutta.

*Three decks. Built in the year 1851 for the California and China trade. An extreme clipper. Proved a good sea boat, very fast. She was a very beautiful ship and her reputation was such, that when she was in the London Dry Dock to receive a suit of copper, the British Admiralty made application to take her model, which was granted. Draft of water when launched, with all masts on end and all yards on board, two anchors and chains, was forward, 12 ft. 5 in., aft 12 ft. 11 in. Draft loaded 23 feet. This ship was originally fitted with too much sail at the dictation of her first commander. The lower masts and lower yards were subsequently reduced in lengths [sic].*

### *GREAT WESTERN*
### Hull 61

The Black Ball Line of C. H. Marshall & Company returned to Mr. Webb's yard for the packet ship *Great Western*, launched in May of 1851. Her tonnage was 1443, and she measured 191 feet by 42 feet 8 inches by 28 feet 6 inches, a slight increase in all three dimensions and tonnage over her predecessors, *Manhattan* and *Isaac Webb*, built during the year before. *Great Western* departed from New York on her maiden voyage on 15 July 1851 with Captain David S. Sherman in command. She was extensively rebuilt in 1866 and her bottom was "yellow metalled" (coppered) early in 1873 in preparation for further years of duty. During a downturn in packet-line business that accompanied the financial panic of 1873, she was sent on a voyage to San Francisco under charter to Simonson & Chesebrough, clearing New York on 13 June 1874 with Captain B. L. Simmons in command. Returning, she loaded grain for Liverpool and departed from San Francisco in the late summer of 1875. On the 26th of September, while in the same latitude as Valparaiso, she sprang a leak, but kept on her course. By the 4th of October, the leak had worsened to the point where two pumps could not keep the vessel free of water. Part of her cargo of wheat was thrown overboard, and she was turned about and headed for Valparaiso as a harbor of refuge, where she arrived on the 16th of October. The remainder of the cargo was discharged to shore and *Great Western* was drydocked on the 18th of November, to be repaired, caulked and recoppered. Afloat again on the 1st of December, she was reloaded and departed for Liverpool on the 18th. Her troubles were, however, not quite over. She arrived in the River Mersey on 15 March 1876 and drifted onto Deirls Bank the following morning. She was

towed off without damage and immediately went aground near Egremont. When she finally berthed at Liverpool, the cargo was discharged, her New York cargo was loaded, and she departed for her home port, where she arrived on the 2nd of June, nearly two years after she had departed for San Francisco. She was later sold to the Bellingham Bay Wood & Coal Company. Captain Simmons remained with the vessel after she reached the West Coast for her new owner, and was still in command in 1879, but by 1884 *Great Western* had been dropped from United States registry.

### *COMET*
### Hull 62

*Comet*, an extreme clipper, was built in 1851 for the New York firm of Bucklin & Crane. She measured 228 feet by 40 feet 4 inches by 22 feet, and her tonnage was 1836. Launched on 10 July 1851, she left New York on her first voyage on the first of October under the command of Captain E. C. Gardner. She arrived at San Francisco on 12 January 1852 after a respectable 103-day passage. On her return to New York on 11 August 1852 from Whampoa with teas and silk, she carried her skysails for 85 days (36 of which were in succession) out of the voyage time of 97 days—not a record passage but one on which good weather predominated. *Comet* made a number of fast passages in her early years. In the spring of 1853, while racing Webb's *Flying Dutchman*, she travelled from San Francisco to Sandy Hook in 83 days 18 hours. *Flying Dutchman*'s time was 30 hours longer, and the vessels were within sight of each other several times during the voyage. The following summer, she ran from Liverpool to Hong Kong, carrying coal, in less than 85 days, and in 1857, while returning from the west coast of South America with guano, her time from the Equator to Sandy Hook was 14 days. In 1863, after the age of the clipper ship had begun to wane, she was sold for 8,100 Pounds Sterling to James Baines & Company, of Liverpool. She sailed from New York on 11 March 1863 for delivery to her new owners at London for service on their Australian Black Ball Line. This line bore no relation to the Black Ball Line of C. H. Marshall & Company. *Comet*'s name was changed to a prophetic *Fiery Star* and she soon departed on her first voyage to Australia under the British flag. She left Moreton Bay, Queensland, on 1 April 1865 for London on the homeward leg of her second voyage, but three weeks out her cargo of baled wool was discovered to be afire, presumably as a result of spontaneous combustion. All of her passengers and part of the crew took to the boats, leaving the mate and 17 of the crew aboard to fight the fire. Before *Fiery Star* sank, the British bark *Dauntless* rescued those who stayed behind, but the vessel's boats, in which the 80 occupants had sought safety, were never seen again.

*Comet* was not the only American-built vessel owned by James Baines & Company. Contrary to the usual practice of British shipowners, who preferred ships built in English and Scottish yards, James Baines & Company ordered four clippers from Donald McKay in 1853. These ships—*Champion of the Seas, Lightning, James Baines* and *Donald McKay* (which ended her days on the transatlantic petroleum shuttle under German ownership)— were among the largest and, on the basis of surviving log books, the fastest of clippers. The few log books from that period that are extant tell only part of the story. We shall never know how many outstanding day's runs were made and never reported under conditions rarely seen at sea—a benign sea, a strong wind, an opportune course, a clean bottom and a lightly loaded ship. *Comet* performed outstandingly throughout her life, but her accomplishments were overshadowed by those of vessels that were equal to her from the standpoint of naval architecture, but that might have been fortunate enough to have encountered these conditions at the right time and in the right place.

*Built for the California trade. Had a half poop deck and cabin house extending nearly to main mast; was a most beautiful model, made many very rapid passages to California and East Indies. Made the shortest passage on record (76 days) from San Francisco to New York, proved an excellent sea boat, and a very successful and popular ship. Draft of water when launched, keel 3 feet 4 in. (below timbers), forward 11 feet 8 in., aft 12 feet. Spars all in end and anchors and chain cables on board. Deep loaded, 21 feet 6 in.*

### *INVINCIBLE*
### Hull 63

The clipper ship *Invincible* was built for James W. Phillips in 1851 at a cost of $120,000. She is described by Mr. Webb as a medium clipper, but by others as extreme. She measured 238 feet by 42 feet 10 inches by 25 feet 6 inches, with a tonnage of 1768. *Invincible* was launched on 6 August 1851 and completed early in October, her first register document having been issued on 29 September 1851. She reportedly was built for the Liverpool trade, and Captain Parker P. Norton was to have

commanded her. Unfortunately, Captain Norton died on the day of her launching, and, as a result, James Phillips revised his plans and placed her on the Cape Horn route, requiring that the vessel's rig be altered for the conditions encountered on this route. Captain Henry W. Johnson, whom the *New York Times* described as "recently of the steamer Prometheus and an old East India voyager," was selected as her master. The change in the owner's plans accounts for the long period between her completion and the start of her first voyage, for she did not depart for San Francisco until 20 December 1851, with Captain Johnson in command. She arrived at the Golden Gate on 15 April 1852. From San Francisco, she went to Whampoa, thence to London with tea, and back to New York, where she arrived in February of 1853. On her homeward passage from London, she made the remarkable time of six days and three hours from the Scilly Isles to the Grand Banks. Other voyages to the Far East followed during the next few years, one of which, in 1855, was reportedly marred by a collision with the American-flag ship *A. Chesebrough* near Hong Kong. *Invincible* was beached with water in her hold, but was shortly afterwards raised and repaired. She turned in a respectable 82-day run from Bombay to Liverpool in the summer of 1857. In 1860, she was sold to Spofford, Tileston & Company for the firm's line between New York and Liverpool. Three years later *Invincible* was acquired by a group of investors headed by Henry Hastings, and returned to the long-haul service for which she was built. On 2 May 1863, she cleared New York on the first of a number of voyages to San Francisco, the last of which ended in May of 1867. She was then laid up at Roberts' Wall Street Stores, in Brooklyn. While in layup, *Invincible* was discovered to be afire early in the morning of 11 September 1867. The vessel drifted across the East River to a point near Peck Slip, where she was anchored. The primitive fire equipment of the day was ineffective in fighting the blaze, and she was towed down the river, where several tugs poured streams of water into her. By this time, little could be done to save the unfortunate vessel, and soon *Invincible* had, in the words of the contemporary press, "burned to the water's edge."

*Built in the year 1851 for the China and California trade. Three full decks. A medium clipper. Handsome model. Proved an excellent sea boat and made many rapid passages between New York, San Francisco, Hong Kong, Calcutta and London. Draft of water was, forward, masts all on end, 10 ft. 9 in., aft, chains and anchors on board, 10 ft. Draft deep loaded 23 ft.*

## *SWORDFISH*
## Hull 64

*Swordfish* was a medium-extreme clipper ship built in 1851 for Barclay & Livingston, of New York. David S. Babcock was her first commander. Her tonnage was 1036, and her dimensions, 170 feet by 35 feet 10 inches by 20 feet 2 inches. She was launched on 20 September 1851 and cleared New York for San Francisco on the 12th of November on a voyage in which she raced Donald McKay's *Flying Fish*. Her time was just short of 91 days, the fourth best on record; the Boston ship's time was a slower, but still respectable 100 days. On her second voyage in 1853, with Captain Charles Collins in command, *Swordfish* reached San Francisco in 107 days. She continued westward to Shanghai, and not only was her time of eight days and two hours from San Francisco to Hawaii en route to the Chinese port never equalled, but the 32 days and nine hours for the entire run was a record. Her third voyage around Cape Horn, under Captain H. N. Osgood, was made in 110 days. *Swordfish* was sold to Crocker & Warren in 1854 for $55,000. On the return passage of her voyage to China in 1858-59, she encountered violent westerly gales in which green seas continually broke over the poop of the heavily-laden ship. All of those who worked on deck—master, officers and helmsmen—were lashed to their posts. The vessel's chronically wet decks earned her the irreverent title, "The Diving Bell." In 1859-60, under Captain J. W. Crocker, she made a voyage from Shanghai to New York in 81 days, including five days when she was becalmed in the Atlantic. The record stood for the remainder of the sailing-ship age. *Swordfish* was lost at the mouth of the Yangtze Kiang, China, in July of 1862. She left Shanghai on the 9th of that month for Amoy. While maneuvering on her way downstream, her anchors became fouled and she drifted to the shore, where she grounded and eventually went to pieces. The wreck was sold at auction on the 26th of July.

*Built in 1851 for general trade. Her first voyage made between New York and San Francisco was one of the shortest, 100 days. Carried well. Proved an excellent sea boat and favorite ship. Was a very good model for the Pacific Ocean and East India trade.*

## *JAMES ADGER*
## Hull 65

*James Adger* was a sidewheel steamer built in 1852 for Spofford, Tileston & Company for service between New

York and Charleston. A two-decked, three-masted vessel of 1151 tons and measuring 215 feet by 32 feet 3 inches by 21 feet, she was propelled by a side-lever engine having a cylinder 75 inches in diameter and a piston stroke of 96 inches. She was launched on 10 January 1852 and first enrolled on 1 May 1852. Her regular trips to Charleston were interrupted in January 1857 when she made a single voyage to San Juan del Norte, Nicaragua, with passengers and troops in support of William Walker—the "grey-eyed man of destiny"—an American soldier of fortune who in 1856 had been inaugurated as president of Nicaragua. The passengers were to debark at San Juan in order to transit to the Pacific, but because of the war of insurgency, *James Adger* continued to Aspinwall, whence they crossed the Isthmus of Panama. She returned to the New York-Charleston route upon her arrival in New York. At the outbreak of the Civil War she was chartered to the United States Navy, and was sold to the Navy on 26 July 1861, serving during the Civil War as *USS James Adger*. Following the war, she was sold at auction and ran on the Atlantic coast (mainly between New York and Charleston for Henry R. Morgan & Company) until about 1875, when she was laid up at New York. She was sold to a Boston scrap dealer in November 1877 for $10,500 and dismantled there during the winter of 1877-78. The hulk that had been *James Adger* was beached on Apple Island on 24 May 1878 and burned to recover her metal fastenings.

*Built in 1851 for the passenger and freight trade between New York and Charleston, requiring a light draft of water, was fast and a good sea boat. Proved to be a very satisfactory ship.*

### PLANDOME
### Hull 66

*Plandome* (which Mr. Webb inexplicably refers to in the Book of Plates as "*Plendome*") was a two-masted schooner built in 1852 for Samuel L. Mitchill, of New York, the proprietor of a Savannah line of steamers who favored the Webb yard in building his steamships. The vessel was named for a community on the North Shore of Long Island. Her dimensions were 101 feet by 25 feet 7 inches by 10 feet 9 inches, and her tonnage of 266. *Plandome* was launched in February of 1852 and entered the service of her owner sometime after 13 March 1852, the day on which she was awarded her original register document. She was a typical cargo carrier of the period, and appeared from time to time on several of the estab-

lished coastwise lines. *Plandome* was still the property of Samuel Mitchill through 1863, and, after running throughout the Civil War and beyond, was observed in June of 1868 to have arrived at New York from Havana with a cargo of sugar and molasses consigned to Young, Smith & Company. The circumstances relating to her later career and disposition remain unknown.

*Built for the Atlantic Ocean coasting trade. Proved excellent sea boats, good carriers and were great favorites with their owners.*

### MANHASSET
### Hull 67

*Manhasset*, a two-masted schooner built in 1852, was identical to *Plandome*, built at the same time and also launched in February, perhaps on the same day. However, *Manhasset* appears to have been constructed for one Bernard Flanner of New York. Since she was named for a Long Island community adjoining Plandome, it is surmised that they were ordered (and named) together. She was a vessel of 266 tons, and her dimensions were 101 feet by 25 feet 7 inches by 10 feet 9 inches. According to some sources, *Manhasset* may have been built by Webb & Bell under a subcontract, but this cannot be substantiated. She appears to have been acquired by the New York firm of Benner & Benton prior to 1863, but nothing is known of her later life. By 1867 she was no longer under United States registry, and it is presumed that she was one of the many vessels that were sold or transferred to foreign registry during the Civil War.

*Built for the Atlantic Ocean coasting trade. Proved excellent sea boats, good carriers and were great favorites with their owners.*

### ANNAWAN
### Hull 68

Built in 1852 for Wakeman, Dimon & Company, the 759-ton ship *Annawan* was initially employed on Frost & Hicks' Union Line of New York-New Orleans packets. Her hull dimensions were 150 feet by 32 feet by 19 feet 6 inches. *Annawan* was launched in March of 1852, and was placed in service during April. She remained in coastwise service under the ownership of Wakeman, Dimon & Company at least until July of 1864, when she was reported to have arrived at New York from New Orleans under Captain McNair. Between that time and 1866 she was acquired by R. W. Cameron, of Sydney, New South Wales, Australia, who placed her under the British flag and renamed her *Royal Saxon*. She called at New York as late as December of 1872 under that name, but the details of her career after that visit are not known. She was not shown in register books after 1876.

### ROBERT MILLS
### Hull 69

The 488-ton bark *Robert Mills*, launched in June of 1852, was built for Wakeman, Dimon & Company. Her dimensions were 125 feet by 30 feet by 14 feet. She originally ran in the owners' Star Line between New York and Galveston, but made voyages between many coastal ports and New York for her owners or under charter to others. A typical year was 1855, when she made a number of voyages to New York from Galveston and Savannah with cotton. *Robert Mills* was lost on 28 February 1860 while taking shelter at Holyhead New Harbour, on the island of Anglesey, North Wales. She had sailed from Liverpool for Galveston, but with worsening weather, Captain Reuten (also spelled Reutan and Rauten) attempted to enter the New Harbour, luffed while too close to the breakwater and got her sails aback—she was "in irons"—the worse possible predicament for a sailing vessel in her circumstances. Either there were not enough experienced hands aboard or they were too slow, but she drifted across the harbor with sails aback and grounded at Penrhyn Llanfwrog. Early in April, she was floated off the rocks and towed to the Old Harbour. Later she was shown as a British vessel under the ownership of one "Le Clerk," but it is not known whether she ever ran again.

### AUSTRALIA
### Hull 70

The packet ship *Australia*, built in 1852 for Williams & Guion, ran on the North Atlantic or on the New York-New Orleans packet route through most of her career. She was a vessel of 1447 tons, and measured 190 feet by 42 feet 6 inches by 28 feet. She was launched on 24 July 1852, and her first voyage to Liverpool commenced on the 21st of August under Captain Edwards. Some reputable sources state that *Australia* was wrecked at Akyab, on the Burmese coast, in May of 1864, but this appears to be a ship built at Medford, Massachusetts in 1849 (in addition, the accident occurred at Amherst, near Moulmein, Burma, whence she had sailed on the 6th of July.) As far as Mr. Webb's *Australia*, still owned by Williams & Guion, is concerned, she was on the other side of the world at that time, having cleared New York for Liverpool on 25 June 1864, returning several months later and departing for Philadelphia in December. On this voyage, she lost both anchors and 105 fathoms of chain cable near the Delaware Capes, and had to return to New York for replacements. This was to be an omen of ill luck, for, having finally proceeded to Philadelphia, she loaded for Port Royal and departed late in February. After leaving the Delaware Capes, she encountered heavy weather along the coast and, with Captain Towart unable to determine his position, ran hard aground on Frying Pan Shoal, North Carolina, on the 2nd of March and became a total loss. A sounding of 20 fathoms was obtained just minutes before the vessel struck. The crew left in four boats, two of which (including that containing Captain Towart) were picked up by the coastwise steamer *Charles C. Leary*. The fate of the other two boats is not known.

Writers of clipper ship lore frequently represent *Australia* as a "medium clipper" but she was built to be a packet. The absence of a clear line of distinction between a fast packet and the least extreme of the clippers makes it impossible to categorize many borderline vessels. *Australia* may well have been both.

### GEORGE LAW
### Hull 71

The handsome sidewheel steamship *George Law*, built in 1853, had a short career that ended with her tragic loss. She was constructed for the United States Mail Steamship Company for service between New York and Aspinwall, and was named for the man who was the driving force behind the company. She measured 272 feet by 40 feet by 32 feet, and had a tonnage of 2141. Her machinery consisted of a pair of inclined engines of 65-inch cylinder diameter and 10-foot stroke. The launching of *George Law* took place on 28 October 1852. Nearly a year later, on 20 October 1853, she sailed from New York on her maiden voyage, and monthly trips were made thereafter. Gold from California was the most important commodity carried by ships serving the Isthmian route, and *George Law* brought nearly a million dollars in specie each time she arrived at New York. Her name was changed to *Central America* in 1857 after George Law had resigned from the owning company. On 3 September 1857, she left Aspinwall on her 44th voyage. A routine call was made at Havana, and she departed for New York on the 8th. The next three days brought worsening weather, and by the afternoon of the 11th a full gale was raging and conditions on the ship had so deteriorated that the boiler fires were extinguished. Without steam, the engine could not be run and it became nearly impossible to hold the head of the ship into the wind. *Central America* foundered on the night of the 12th and 423 persons of the 572 aboard were lost, as well as $1,219,189 in gold bullion.

*Afterwards known as the Central American. Side wheels. Built in the year 1853 for the mail service between New York and Aspinwall. Three full decks. Proved fast and an extraordinary [sic] good sea boat. Was a favorite passenger steamer. Draft loaded 20 ft. 3 in.*

### *FLYING DUTCHMAN*
### Hull 72

The extreme clipper *Flying Dutchman*, of 1257 tons, has been described as one of the sharpest vessels ever built at New York. Her dimensions were 190 feet by 37 feet by 21 feet 6 inches. Built for George B. Daniels and others to trade to San Francisco, she was launched on 9 September 1852 and departed from New York on her first voyage to that port on the 15th of October under the command of Captain Ashbel Hubbard. The voyage was made in 104 days. She arrived back at New York on 8 May 1853, the round trip having taken 210 days including port time at San Francisco. Jacob A. Westervelt's clipper ship *Contest*, sailing a month later with Captain William E. Brewster in command, is the only vessel to have bettered that time for the round trip; her voyage was completed in 201 days. On the same return voyage, the time of *Flying Dutchman* from San Francisco to the Equator was a near-record 11 days 10 hours. *Flying Dutchman* continued to make fast passages. Of her later westbound voyages, that of the summer of 1853 was

The steamship *George Law*, carrying a barkentine rig but having two stacks abreast, was one of William Webb's handsomest vessels. She was completed in 1853, but four years later, under the name *Central America*, she foundered in one of the country's worst maritime disasters. (Courtesy of The Mariners' Museum, Newport News, Virginia)

logged at 106 days, and in 1856 her time was 124 days, the third best time of the 25 clipper ships that departed during May and June of 1856. While returning from her fifth voyage to San Francisco (the outward leg of which was made in 102 days), she stranded at Brigantine Beach, New Jersey, at eight o'clock on the evening of 14 February 1858 and soon became a total loss. Captain Hubbard stated that he had been unable to make any observations for three days, and that he mistook Absecon Light for Barnegat Light. By the 18th, *Flying Dutchman* was full of water and, in her exposed location, it was feared that it would be impossible to get her off the beach. On the 22nd, it was reported that she was rapidly breaking up. The ship was valued at $50,000, and her cargo at over $150,000.

*Built in 1852 expressly for the California trade. Has two decks with half poop cabin extending nearly to main mast, proved a very fast sailer and excellent sea boat, successful and popular ship. Draft when launched with all spars on board, forward 11 ft. 6 in., aft 11 ft. 9 in.*

## *AUGUSTA*
### Hull 73

The sidewheel steamship *Augusta* was the fifth vessel built for Samuel Mitchill's line to Savannah. She measured 220 feet 8 inches by 35 feet 4 inches by 21 feet 6 inches, and had a tonnage of 1310. *Augusta* was launched at 11 o'clock on the morning of 30 September 1852, and was then fitted with a single-cylinder oscillating engine from the Novelty Iron Works measuring 85 inches diameter by eight feet stroke. She was completed in December and ran on the Savannah line until the outbreak of the Civil War. She was then sold to the United States Navy for war service as *USS Augusta*. She remained under Navy ownership until sold at auction to C. K. Garrison in December 1868. Garrison renamed her *Magnolia* and placed her on his line from New York to Charleston. He sold her to the Central of Georgia Railway & Banking Company in 1872, which continued to run the ship until they transferred her to a newly organized subsidiary, Ocean Steamship Company, in 1874. *Magnolia* ran to Savannah for about three more years, but on 29 September 1877, while a northbound voyage, she encountered a raging storm which opened the seams of the old vessel. The passengers and crew abandoned the doomed ship on the afternoon of the 30th and were picked up by a sailing vessel. No lives were lost.

## *REEMPLAZO*
### Hull 74

The schooner *Reemplazo* (which means "replacing" in Spanish) was built for J. Carela and M. Garcia in 1852, probably as the replacement for a vessel whose identity is lost in the mists of the past. Her dimensions were 76 feet by 21 feet by 7 feet 10 inches, and her tonnage was about 110. *Reemplazo* was launched in September of 1852 and was completed shortly afterwards. She was apparently built for service in Cuba or Central America, but her career remains a mystery.

## *VOLANTE*
### Hull 75

*Volante* (in Spanish, "flying") was a brig built during the winter of 1852-53 for Schiff Brothers & Company. She was a full-modelled vessel of 307 tons, measuring 112 feet by 26 feet 6 inches by 11 feet 3 inches. She was launched in December of 1852 and her first register document was dated 24 January 1853. On that day she cleared for Havana under the command of Captain Sewall. Ten years after *Volante* was built, L. H. Botsford was listed as her owner and master, and Milford, Delaware, as her hailing port, while by 1873, she had become the property of Freyer Brothers and was flying the Argentinian flag out of Buenos Aires. Little has been found to record her career after 1875. On the 30th of April of that year she called at New York with a cargo of sugar from Pernambuco, Brazil, clearing for Montevideo on the 19th of June after having been surveyed by the American Shipmasters Association. This appears to have been her last call at New York. She was last listed by A.S.A. in the record book for 1891, still under the ownership of Freyer Brothers, but the vessel may have been lost or broken up at any time during the preceding 16 years.

*Built in 1853 for the Mediterranean trade, was a handsome vessel and proved very suitable for the trade.*

## *FANNY*
### Hull 76

The two-masted topsail schooner *Fanny* was built in 1853 as a general cargo carrier for charter to the mercantile house of Schiff Brothers & Company. This small vessel was typical of the many hundreds of cargo-carrying schooners that plied the offshore and coastal waters of the United States through the end of the 19th century. Their cargoes consisted of coal, brick, stone, salt, ice, hay, fish, lumber, potatoes, fruit, sugar, guano and a myriad of other commodities. *Fanny* measured 85 feet 6 inches by 21 feet 2 inches by 9 feet, and had a tonnage of 150. She was launched on 27 January 1853 and cleared New York on her first voyage on the 26th of February under the command of Captain Jamison. Notwithstanding the vessel's size, this voyage was to have taken her to Melbourne, Australia, with passengers and a cargo of flour. Fate soon intervened, for *Fanny* was dismasted shortly after the first of March. Her wreck was encountered at latitude 35-00 north longitude 62-30 west (about 205 miles northeast-by-north of Bermuda) by the ship *Mary* on the 25th of March. On 7 April 1853, the bark *Io* happened upon *Fanny* at a position 310 miles northeast of the previous sighting and reported that the mainmast of the "new and beautiful clipper vessel" had been cut away, her foremast was broken off at the deck and the wreck looked as if it had been boarded and stripped. It was assumed by the master of *Io* that the passengers and crew had been taken off by an outbound vessel. Finally, on 2 November 1853, more than seven months after the first sighting, the German vessel *Marie Otten* encountered the wreck 200 miles east southeast of the place at which *Io* had found her. As far as is known, those aboard *Fanny* were never heard from. Included in the vessel's complement was Edward Laight. Laight and John Wendell, of New York, were the owners of *Fanny*.

*Built in 1853 for general business. Draft of water when launched with masts in end and one anchor and chain on board, was forward 5 ft. 6 in., aft 5 ft. 5 in. Proved a good sea boat, and was a fast and handsome vessel.*

### *KNOXVILLE*
### Hull 77

*Knoxville*, a sister vessel to *Augusta*, was built at a cost of $200,000 for Samuel L. Mitchill's New York & Savannah Steam Navigation Company. She measured 220 feet by 35 feet 6 inches by 22 feet, and her tonnage was 1240. Unlike *Augusta*, which was fitted with an oscillating engine, she was propelled by a side-lever engine. The engine, built by the Novelty Iron Works, had a cylinder 85 inches in diameter and a piston stroke of 8 feet. *Knoxville* was launched at ten o'clock on the morning of Saturday, 20 August 1853 and completed in March of 1854.

A trial trip was made on the 13th of that month and she sailed to Savannah for the first time on the 15th. She proved to be a fast vessel, equal to or better than *Augusta*, and fell easily into the routine of the line's twice-weekly schedule requiring each ship to complete a round trip every week. Ships of the coastwise lines to southern ports usually departed from New York on Saturday afternoons if weekly service was maintained; departures on twice-a-week services usually took place on Wednesdays and Saturdays. Then as now, Sunday was a day of rest in the harbor.

On 1 December 1856, *Knoxville* arrived at New York on schedule, but her departure two days later was cancelled due to unspecified difficulties with her machinery. Repairs continued for three weeks while the vessel lay at the line's terminal at Pier 4, North River. Shortly before eight o'clock on the evening of the 22nd of December, a small fire was discovered in the tween deck near the boilers by the Second Steward, one of two crew members aboard. After the alarm was given, the first persons on the scene were local police who attempted to use a fire hose on the pier, but were unable to connect it to the hydrant. City fire apparatus, consisting mainly of hand-powered pumping wagons, responded within twenty minutes, but found that the flames had spread thirty feet forward and aft of the starting point. By nine o'clock, the entire upper portion of the vessel, with the exception of about 20 feet near the bow, was ablaze, and an attempt was made to scuttle *Knoxville*, but without cargo, coal or ballast aboard, she only heeled over to starboard. The mainmast fell at half past ten, and by morning the ship had been completely gutted by the fire. The *New York Times* stated that the fire was probably caused by careless workmen and dutifully reported that, in addition to the virtual destruction of the hull and upperworks of *Knoxville*, "all the plate and furniture were destroyed."

### *YOUNG AMERICA*
### Hull 78

*Young America*, an extreme clipper, was Mr. Webb's most renowned sailing vessel. Built in 1853 for George B. Daniels, of New York, she measured 243 feet by 43 feet 2 inches by 26 feet 9 inches. Her tonnage was 1961. She was launched on 30 April 1853, at which time Mr. Webb stated:

This is my conception of what a well-designed and built sailing ship should be. She will be fast and handy, amply stiff and carry well. The ship of the future should be like this one, or possibly even a little fuller, but with no loftier spars or bigger sail spread.

The first of nine masters of *Young America* was Captain David S. Babcock, who had commanded *Swordfish* when she was delivered from the Webb yard two years before. The vessel cleared New York on her maiden voyage to San Francisco on 10 June 1853. Her time enroute was 110 days, not a fast passage, but a commendable time during the height of the southern winter. Of 31 clippers that sailed to San Francisco from Boston, New York and Philadelphia during May, June and July of 1853, only three completed their voyages in the same or less time — *Invincible* (a two-year old Webb ship) and *Contest* (built by Jacob A. Westervelt in 1852) logged 110 days, and Webb's *Flying Dutchman*, 106 days. *Young America*'s times in 1854 and 1855 were 111 and 107 days respectively, but on her fourth voyage, under the command of Captain Nathaniel Brown, Jr. in 1859, she was dismasted in the South Atlantic and took 174 days outbound. After having been fitted with new spars and rigging at Rio de Janeiro, she ran from that port to San Francisco in 69 days, the second fastest time on record. In 1855 she engaged in a little-appreciated informal contest with another Webb clipper, *Challenge*. *Young America* arrived at New York from Manila on the 31st December with a cargo of sugar and hemp after a passage of 100 days (not a record time by any means, for Samuel Hall's *Wizard* was to make the passage in 1861 in 84 days.) For several days she sailed in company with *Challenge*, which was carrying hemp, indigo and other cargo to New York from the same port. *Challenge* did not arrive at her destination until 15 January 1856, her time en route being no less than 114 days.

*Young America* continued to make fast passages throughout most of her career. As late as 1877, her time from San Francisco to New York was 92 days, well above the record of 76 days set by *Comet* in 1853, but nonetheless an admirable performance by a 24-year old ship. She also fell victim from time to time to near catastrophe. On 18 May 1856, her jibboom broke into three pieces while she was lying to in the vicinity of Cape Horn, but fortunately this did not lead to further failures of rigging and spars. On 3 December 1868, while off the River Plate on a voyage from New York to San Francisco carrying rails for the Pacific railroad, she was struck by a whirlwind, which set her over on her beam ends. Captain George Cumming (who commanded the vessel from 1864 to 1874) ordered shrouds and stays cut away, but the gale carried away the fore royal and main topgallant masts and the mizzen mast down to the lower mast head. Four hours later the gale had subsided enough to permit the ship to be pumped free of water. Ten days later a jury mizzen mast had been rigged, and the main top and topgallant sails set. She arrived at San Francisco 117 days after leaving New York. Captain Cumming was presented with a thousand dollars in gold by the underwriters. A third close call occurred at five o'clock on the morning of 2 October 1870, when she struck a reef near Cape St. Roque, on the Brazilian coast (where Mr. Webb's packet *John Bright* would be lost four years later). A portion of her cargo was jettisoned and the ship was floated off without serious damage.

The vessel was sold to Abram Bell's Sons in 1860, then to Captain Robert L. Taylor (the man who had insisted upon extreme deadrise when *Gazelle* was built in 1851,) and, about 1870, to George Howes & Company. Her last assignment under the American flag ended on 6 October 1883, when she arrived at New York under Captain Charles Matthews after a 13-month voyage that had taken her to Portland, Oregon, and San Francisco. She put into Rio de Janeiro on her way home to repair leaks that had developed in her aging hull in her last battle with Cape Horn. *Young America* was sold after her arrival for $13,500 to Austrian owners. As *Miroslav* under the Austrian flag, she was a regular visitor to New York and the Delaware River, frequently transporting petroleum in wooden barrels eastbound and returning with empty barrels. In the days prior to the development of the bulk tankship, sailing vessels were widely used to carry both petroleum in barrels and case oil. The absence of fires aboard sailing vessels (other than in the galley) made them more attractive for the service than steamships. The career of *Miroslav*, unusually long for a clipper, ended some time after 17 February 1886, when the venerable ship passed the Delaware Breakwater outbound from Philadelphia, in command of Captain Vlassich and carrying 407,306 gallons of crude oil valued at $26,965 (equivalent to about $2.78 per barrel in 1886 dollars). She was never heard from again, and it was not until the 27th of May that the press stated, "Anxiety is felt as to the safety of the Austrian ship Miroslav"; by the end of June she was officially posted as missing. *Young America*, Mr. Webb's masterpiece in sail, exemplified the country's spirit — in name and in deed — at a time when showing the American flag in foreign ports on an outstanding merchant ship was a symbol of national pride and honor.

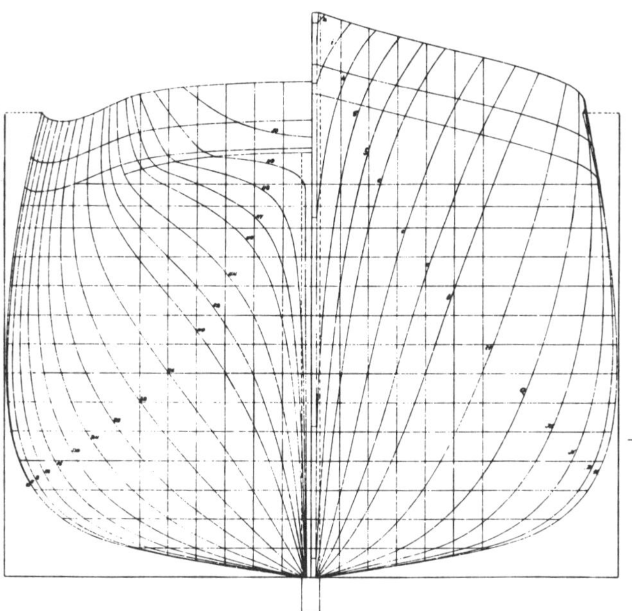

The midship section of the extreme clipper *Young America* clearly illustrates William H. Webb's philosophy in the design of fast, cargo-efficient sailing vessels—low deadrise combined with extremely fine entrance and run. *Young America*'s successful career was an eloquent testimonial to Webb's masterful design.

*Built expressly for the California and East India trade, was a most beautiful and successful ship and made many very rapid passages under different commanders. Among others from New York to San Francisco 103, 107, 110, 112, 117 and 116 days. San Francisco to New York 92, 97, 85, 101, 103 and 83 days. The latter being the shortest record for loaded ship. San Francisco to Liverpool 103 and 106 days, and Liverpool to San Francisco, 117, 111 and 99 days. The latter being the shortest on record. New York to Portland, Oregon 127 days. Portland to San Francisco 7 days. San Francisco to Antwerp 118 days. New York to Liverpool 18 days. Liverpool to Melbourne 81 days. Callao to Queenstown 74 days. Glasgow to Otago, New Zealand 88 days. Otago to Callao 36 days. San Francisco to Hong Kong 47 days. Manilla [sic] to New York 88 days. This ship after 30 years of continuous service, with only slight repairs, was sold to Austrian parties, who changed her name. No other sailing ship has made such a record.*

*Liverpool to San Francisco (13800 miles) in 99 days. Wonderful passage and shortest on record. This ship beat, in*

*a race from San Francisco to New York, the noted English Clipper Ship Escosesa, leaving port on the same tide, 5 days; and the still more noted Clipper Ship David Crockett, sailing about the same time, 11 days. She sailed on a voyage from New York to San Francisco from 50 degrees of latitude in the Atlantic Ocean to 50 degrees of latitude on the Pacific Ocean in six consecutive days.*

## *FLYAWAY*
### Hull 79

The medium clipper *Flyaway* was built in 1853 for Schiff Brothers & Company, of New York. Her tonnage was 1274, and she measured 190 feet by 38 feet 3 inches by 21 feet 6 inches. She was launched on 23 June 1853 and departed for San Francisco under Captain Merrill Sewall on the 20th of August, nine days after her register document had been issued. Her name was chosen under the assumption that she would be a swift vessel, and it appears that she was, but she never made headlines in the way of some of the other Webb clippers. Her voyage to San Francisco in 1854, made under Captain Sewall, was logged at 110 days. Of the 63 clippers that made the New York-San Francisco run that year, only three (*Archer*, *Witchcraft* and *Flying Cloud*) bettered her time. Her time in 1855 was 106 days, and in 1857, 125 days. She was sold to other American owners about 1858 for $50,000, and in March of 1859 to Spanish owners, Galwey, Casado & Teller, by whom she was later renamed *Concepcion*. By 1875 the vessel had been rerigged as a bark and had acquired British registry under the name *Bothalwood*. She stranded and became a total loss on 20 January 1881 at St. Ouen's Bay, on the western coast of the island of Jersey, Channel Islands, while on a voyage from Cartagena to Leith with a cargo of esparto grass.

*Built in 1853 for the California and Pacific Ocean trade. Was a good sea boat, a handsome and successful vessel. Very suitable model for the trade for which she was designed.*

## *SAN FRANCISCO*
### Hull 80

*San Francisco* was a sidewheel steamship of 2272 tons built in 1853 for the Pacific Mail Steamship Company at a reported cost of $350,000. Her dimensions were 276 feet by 39 feet 10 inches by 23 feet 6 inches. Her paddle wheels were driven at 22 revolutions per minute by a two-cylinder oscillating engine, built by the Morgan Iron Works and having cylinders measuring 65 inches in diameter by 8 feet stroke. The vessel was launched on 9

June 1853, and her first registry document was dated 23 November 1853. *San Francisco* had an abbreviated career that came to a tragic end on her maiden voyage.

Under the command of Captain James T. Watkins, *San Francisco* cleared New York on the 20th of December bound for San Francisco via the Straits of Magellan, with stops for coal and stores at Rio de Janeiro, Valparaiso and Acapulco. Aboard were 700 passengers and crew, including eight companies of the Third Regiment, United States Artillery, en route to duty in the new state of California. On the 24th, when the ship was at a position approximately latitude 38 degrees north and longitude 69 degrees west, a fierce gale struck, disabling the vessel and sweeping the deck of cabins, spars and fittings. The brigs *Napoleon* and *Maria*, the bark *Kilby* and the ships *Antarctic* and *Three Bells* rendered assistance over the next 12 days. The first two vessels became separated from the steamer and elected to make for port with news of the accident; the other three were able to stand by and saved many of those aboard. *San Francisco* was abandoned on 5 January 1854 and foundered shortly afterwards. Estimates of the number of lives lost in the disaster vary from 164 to 246, most of whom were washed overboard by the heavy seas that swept her deck on the 24th. Some were stated to have fallen victim to an outbreak of cholera. The survivors on *Kilby*, which was bound from New Orleans to Boston, had a further harrowing experience when she was partially dismasted by the gale. They were transferred to the westbound packet ship *Lucy Thompson* and arrived at New York on the 14th of January.

The wreck triggered one of the earliest organized search-and-rescue missions. When the disaster was announced by *Napoleon* and *Maria*, a number of ships of the Revenue Cutter Service (later the Coast Guard), two pilot boats and the steamships *Union* and *Savannah* (chartered by Pacific Mail for the occasion) were dispatched to the approximate position at which *San Francisco* had been reported (the coordinates reported by *Maria* and *Napoleon* were about 28 miles apart). None of the searching vessels found her. Using his extensive knowledge of ocean winds and currents, Navy oceanographer Lieutenant Matthew Fontaine Maury plotted the probable course of the drifting vessel, and it was at a point close to that estimated by Maury that she was found by *Three Bells*. She had drifted about 220 miles from her original position. On 1 February 1854, the master of the ship *Amelia* reported in a letter to his ship's owners that, "in latitude 30-12 north and longitude 50-5 west I picked up a case marked 'Lieutenant L. Loeser, U. S. Army, Str. San Fran-

cisco,' containing a dozen cane-bottomed chairs." First Lieutenant Lucien Loeser, who, with his wife and his wife's sister, survived the wreck of *San Francisco*, only to board *Kilby* and, finally, *Lucy Thompson*, on their way to safety ashore. Loeser was being posted to his second California assignment, the first (1846-1848) having ended when he was selected to carry important dispatches to Washington announcing the discovery of gold at Sutter's Mill. His journey, via Peru, Panama, Jamaica and New Orleans, took from August to December of 1848. A narrative of the wreck of *San Francisco*, written in 1854 by Miss Lucia Eaton, Loeser's wife's sister, survives as a priceless document in American maritime history.

*Side wheel steamer built in the year 1853 for the Pacific Mail S. S. Co for service on the Pacific Ocean. Draft of water when launched, one anchor and chain on board, was forward 6 ft. 8-1/2 in., aft 6 ft. 9-1/2 in. Was a beautiful and very fast ship.*

### JOSEPHINE
### Hull 81

The brig *Josephine* was unusual in that she was laid down as a sidewheel steamer, but completed as a sailing vessel. Her hull measured 130 feet by 21 feet 5 inches by 9 feet 6 inches, dimensions that more closely match those of a typical steamboat for inland waters than of a sailing vessel. Her tonnage was 268. Most of Mr. Webb's vessels were completed in approximately the same order as the sequence of hull-numbers. *Josephine*, completed in 1855, is one of the few exceptions; her completion date is about two years later that those of the vessels ahead of and behind her, *San Francisco* and *Snap Dragon*. It is presumed that her construction as a steamer had commenced in 1853, and that the contract was later cancelled. Mr. Webb then put her up for sale as a schooner or brig, eliminating the need to invest in an engine and boiler on speculation. The date of her launching is not known, but on 15 September 1854, the *New York Times*, in describing recent activities of local shipbuilders, stated that Mr. Webb had during the year built "a schooner of 350 tons," and that she was "built for sale." Of the career and disposition of *Josephine*, little is known. Her first registry document was issued on 19 July 1855. She was owned by Macondry & Company, of San Francisco, in 1863, but appears to have been dropped from American registry by 1867.

There is no doubt that the unusual hull of *Josephine* was derived from her original design as a steamboat,

| VESSEL | LENGTH | BEAM | DEPTH | TONS | L/B | L/D |
|---|---|---|---|---|---|---|
| JOSEPHINE | 130.00' | 21.42' | 9.50' | 268 | 6.07 | 13.68 |
| SNAP DRAGON [130'] | 130.00' | 27.23' | 16.71' | 494 | 4.77 | 7.78 |
| BLACK HAWK [130'] | 130.00' | 28.11' | 17.45' | 458 | 4.62 | 7.45 |

suitable for operation at moderate speed in protected waters where exposure to the stresses imposed by wave action would not be encountered. This may best be observed by comparing the proportions of her hull to those of *Snap Dragon*, a slightly larger sailing vessel of more conventional dimensions: 140 feet by 29 feet 4 inches by 18 feet, with a tonnage of 628. Further comparison can be made to the medium clipper *Black Hawk*, which was 178 feet by 38.5 feet by 23.9 feet. The table above shows the comparative dimensions and the tonnage of the three vessels, those of *Snap Dragon* and *Black Hawk* having been converted to correspond to a common length of 130 feet.

The proportions of length, beam and depth are entirely different for sailing and powered ships. In general, sailing vessels require relatively greater beam because of the dynamics stability while under sail, but, naval architecture having been described as a collection of compromises, a long, slim hull form is desirable for high speed. Similarly, they require greater depth so that the lee side of the weather deck is not submerged when under way. Two criteria may be used to compare the two types of craft: the ratios of length to beam (L/B) and length to depth (L/D). These ratios are compared in the table above, and the marked difference between *Josephine* and the other two vessels (which are of quite similar proportions despite their entirely different configurations) may readily be seen.

The rig of *Josephine*, consisting of only two masts on an exceptionally long hull, made her appear almost as if she was a three-master with the mainmast missing, but this rig permitted the use of immense main staysails. Her slim hull form and unusual rig combined to make her the "extraordinarily fast sailer" as reported by Mr. Webb, but her shallow depth must have made her difficult to sail close-hauled without completely submerging her lee deck, even though he describes her as a "good sea boat." If *Josephine* had any shortcoming, it must have been that her cargo deadweight was deficient. It is said that old-measurement tonnages approximated the burthen, or deadweight, of a vessel, and a comparison of the tonnages of *Josephine* and the 130-foot version of *Snap Dragon* and *Black Hawk* (268, 494 and 458 respectively) confirms

*Josephine*'s lack of cubic capacity. A 130-foot *Black Hawk* could be expected to have a lower tonnage than *Snap Dragon* because of her finer hull.

*Built in 1855. This vessel was originally laid down for a side-wheel steamer but before completion was converted into a brig. She proved not only a good sea boat but an extraordinarily fast sailer, always working to the windward of other vessel. She had great renown in the Pacific Ocean.*

### *SNAP DRAGON*
### Hull 82

Mr. Webb built the bark *Snap Dragon* in 1853 as a cargo carrier for the firm of Wakeman, Dimon & Company, of New York. She was a vessel of 618 tons, measuring 140 feet by 29 feet 4 inches by 18 feet. Launched on 1 October 1853, she entered service late in November, her first register document having been issued on the 21st of that month. Before the Civil War, she ran on Wakeman, Dimon & Company's line between New York and Galveston. Her later career remains a mystery, although it is probable that, like the other members of the Wakeman, Dimon & Company fleet, she delivered wartime supplies to Union-controlled Southern ports. Wakeman, Dimon & Company was shown as her owner in 1863, but she was no longer listed as having United States registry after the Civil War.

*Built for the general freighting business. Proved a good sea boat and good sailer. Draft when launched with all spars on board, one anchor and chain, forward 8 ft., aft 9 ft. 3 in. Deep loaded 17 ft.*

### *JOHN BRIGHT*
### Hull 83

*John Bright* was a ship-rigged packet named for a British merchant and statesman who, with Richard Cobden, successfully opposed the notorious Corn Laws. This vessel was built in 1853 for the firm of Williams & Guion for service between New York and Liverpool. Her dimensions were 190 feet by 42 feet 6 inches by 28 feet 4 inches, and her tonnage was 1444. She was launched on the morning of 29 December 1853 (the day after the Webb-built packet *Joseph Walker* burned in the fire that nearly destroyed

Donald McKay's clipper *Great Republic*) and departed from New York on her first voyage on 30 January 1854 under the command of Captain Edwards. Fifteen years later *John Bright* was still running as a Liverpool packet for Williams & Guion under the command of Captain J. R. Dewar, who had come aboard early in the Civil War. On 3 August 1869, during the shipping doldrums that followed the war, she departed for San Francisco under charter to Platt & Newton. In April of 1872, she was recoppered and major repairs were made to her hull. Following her rebuild, she continued running to Europe as a tramp under Captain Joseph E. Hadley.

On 3 November 1874 she was again sent on the long voyage to San Francisco with Captain Hadley in command and a carrying mixed general cargo, valued at $70,772 and consisting of a bewildering array of items from canned goods and croquet sets to whiskey and wheelbarrows. At 10:30 on the night of the 10th of December, she struck a reef at Capo de Sao Roque (Cape St. Roque,) on the coast of Brazil 140 miles north of Natal. Although the ship was rolling heavily, the entire crew abandoned her calmly and with precision: at midnight, the three boats were provisioned and watered; at 3:30 a.m., with the ship's foredeck awash, the boats were manned. At four o'clock that afternoon, one of the boats sprang a leak, and the occupants boarded the other two. The following morning, the 12th, the boats lost sight of one another. That of the chief mate arrived at Natal on the 13th, and he and his crew boarded the Brazilian coastal steamer *Inajuca* for Pernambuco, the point of rendezvous selected before the boats separated. *Inajuca* picked up Captain Hadley and the remainder of the crew on the same evening and landed *John Bright*'s entire complement at Pernambuco on the 15th. One of a mariner's greatest fears is that of inaccurate charts; Captain Hadley later claimed that the St. Roque reef was five miles north and four miles west of the position shown on his charts.

## *MILTON*
### Hull 84

The bark *Milton*, built in 1853 for Wakeman, Dimon & Company, was a lengthened version of *Robert Mills*, constructed for the same owners the year before. *Milton* had a tonnage of 536, and measured 132 feet 6 inches by 30 feet by 14 feet. She was launched in December of 1853 and took her place on her owner's Star Line between New York and Galveston on the 30th of that month, when she cleared for the Gulf port under Captain W. Bradford.

She remained in Wakeman, Dimon & Company's service on coastal routes for at least ten years with Captain Bradford still aboard. During the early part of the Civil War, it is probable that many of her voyages were made to Union-held Southern ports in support of Union armed forces. In the summer of 1864 she was sold to Tupper & Beattie, and her later career was spent in general trading. Her first voyage for this firm—to Cadiz for salt, wine and generral cargo commenced in August of that year. In June of 1868, she arrived at New York on a typical voyage—carrying a cargo of railroad iron from Cardiff, Wales. "Railroad iron" was the contemporary term used to describe rail, much of which was imported into the United States from Great Britain prior to the 1880s to supplement the meager output of our own mills. Nearly all of this rail was transported in sailing vessels. *Milton* appears to have been sold to Norwegian owners in the early 1870s, and was afterwards an occasional visitor to the port of New York. Her movements after 21 April 1877, when she arrived at New York from Rotterdam, are unrecorded.

## *CULTIVATOR*
### Hull 85

The 1448-ton packet ship *Cultivator* was launched early in March of 1854 for James C. Ward. A three-decked vessel, she was identical in dimensions to *John Bright*, measuring 190 feet by 42 feet 6 inches by 28 feet 4 inches. The two vessels that followed her, *Harvest Queen* and *Thornton*, had approximately the same measurements. With these vessels, Mr. Webb achieved a high point in the design of packets. All of the packets he built after this time were of nearly the same dimensions and rig. She cleared from New York on her first voyage, to Liverpool, on the 30th of that month, and returned to New York on 30 June 1854 with cargo and 807 passengers, after a 33-day passage. In April of 1856, *Cultivator* received some notoriety in that her master, Captain George B. Austin, was pursued by United States Marshals after complaints by crew members accusing him of assault with a deadly weapon. There is no further record of the affair or of the weapon alleged to have been used. Later that year, she was running to Liverpool for Williams & Guion, under the command of Captain W. H. Russell. In 1869, she was still a Williams & Guion vessel, even though the firm had turned its attention to steamships. Later she was acquired by Paul Upton, of Salem, Massachusetts, who was also her master. About 1879 she was purchased by C. T.

Russell & Company, and W. H. Russell came back aboard as her master. Her home port during this phase of her career was Norfolk, Virginia.

The demise of *Cultivator* came on a westward voyage in 1880, not too long after she was bought by Russell. She left Liverpool for Baltimore on 8 April 1880 with a cargo of iron ore. All went well until the morning of the 14th, when, at a point about 600 miles west-southwest of Fastnet Light, Ireland, a southwest gale and very heavy cross seas were encountered. The ship labored heavily and sprang a leak that night. Two pumps were rigged and manned continuously, but by the following day the crew was exhausted. It was found that water was entering at the bow. A stage was rigged and, with a heavy sea running, several men tried unsuccessfully to install a patch of felt, canvas and wood battens. On the 16th, the pumps were choked with cargo and the ship was headed east-southeast in an attempt to run for a port of refuge. At one o'clock that afternoon, the Norwegian bark *Winslow* sighted *Cultivator*'s distress signal and picked up the entire crew. By this time, the 26-year veteran of the Atlantic, about half full of water, was settling fast. The rescued men were transferred to the British steamship *Victoria* the following day, and were landed at Liverpool on the 27th of April.

### *HARVEST QUEEN*
### Hull 86

*Harvest Queen* was built in 1854 for C. H. Marshall & Company to run on the Black Ball Line to Liverpool. With a tonnage of 1383 and measuring 188 feet by 42 feet 6 inches by 28 feet 6 inches, she was nearly identical to the packets *Cultivator* and *Thornton*, which were constructed in the Webb yard at the same time. *Harvest Queen* was launched in May of 1854. She cleared New York for Liverpool on 16 June 1854 under the command of Captain Edward Young, joining the Black Ball fleet in maintaining the line's twice monthly sailing schedule, for which eight ships were required. After twenty years on the Liverpool route, *Harvest Queen* was dispatched on a long voyage to San Francisco on 29 September 1874, from which she never returned to New York. Captain Henry Janssen, the vessel's master since 1870, was in command. As the vessel was nearing the end of a leg of the voyage that was to have taken her from San Francisco to Liverpool, she was run down and sunk by the White Star liner *Adriatic* in the Irish Channel on the night of 30 December 1875. All hands perished in the disaster. The owners of *Harvest Queen* sued White Star before the United States District Court to recover $220,000 in damages for the loss of the ship and cargo. Litigation over the accident dragged on for nearly four years. After a long trial, the case was dismissed. In an appeal before the United States Supreme Court, the case was again dismissed on 3 October 1879. Scraps of rigging and timber, gathered not only from the waters around the site of the collision, but also from the deck of *Adriatic* herself, were introduced as evidence during the trial. Among these exhibits was one black-painted timber bearing the letters "EEN" in gilt.

The evidence presented during the trial was sufficient to cause a controversial verdict to be reached in which blame for the accident was placed solely upon the packet. Not only was it freely admitted by White Star that *Adriatic* indeed sank *Harvest Queen*, but diagrams illustrating the testimony of crew members of the steamship and describing the movements of that vessel and *Harvest Queen* in the moments prior to the collision were contradictory. In addition, the loss of all hands on *Harvest Queen* made it impossible for the court to hear the other side of the argument. In his opinion, Chief Justice Morrison R. Waite of the Supreme Court wrote the following comments:

> Much of what was done at the time of this disastrous collision will never be known. It is impossible to tell with any certainty how the vessels came together. The ship [*Harvest Queen*] came and went in a comparatively short time, and necessarily in the midst of great excitement. . . . It seems incredible that so large and apparently so strong a ship could be broken into and sunk in so short a time without inflicting serious injury on the Adriatic. . . . It is a very remarkable fact, under the circumstances which we know to have existed, that no hail was heard from the deck of the ship before or after the collision. This seems entirely inconsistent with the idea that she could have had her proper complement of men standing watch or attending to their duties at the time. . . . But it is not my intention to look for the faults of the ship. I place the decision entirely on the ground that upon the case as it stands the steamer is free from blame.

There is no doubt that the combined effect of the loss of *Neptune* in April of 1876, and the financial burden placed upon C. H. Marshall & Company by the unfavorable and, in retrospect, astonishing legal action relating to the sinking of *Harvest Queen* contributed directly to the demise of the firm. Several years of declining cargoes on the Black Ball Line's Liverpool route (which may have prompted Marshall to send *Harvest Queen* on her last voyage to San Francisco and beyond) led to the abandonment of regular service at about the time that Chief

Justice Waite rendered his decision in 1879. Within two years, *Isaac Webb* and *Alexander Marshall* had been abandoned at sea, and service had deteriorated to the point where the few remaining ships only made occasional tramp voyages to European ports. It was then that the once-proud ships of the Black Ball Line, flying C. H. Marshall & Company's distinctive house flag—a black ball on a red field—were no longer to be seen on the oceans of the world.

### *THORNTON*
### Hull 87

The third of three nearly identical packet ships that occupied the Webb yard in the spring of 1854, *Thornton* was built for Williams & Guion's Black Star Line to Liverpool. She measured 190 feet by 42 feet 6 inches by 28 feet 4 inches, and her tonnage was 1422. She was launched early in June of 1854 and departed on her first voyage on 23rd of that month with Captain Charles Collins in command. She was still on Williams & Guion's New York-Liverpool route and Captain Collins was still aboard in July of 1862, but her command passed to Captain Wells later that year. On 2 November 1867, *Thornton* cleared New York for Liverpool on what appeared to be a routine winter crossing with Captain Hatten in command. On the 5th of December, as she neared the end of her voyage, she ran aground on one of the treacherous shoals in the approaches to the River Mersey. Because of the exposed location, the unfortunate vessel was subjected to the full force of the sea. On the 7th she had 15 feet of water in her holds, and at that time it was decided to take off the crew. Two days later, tugs were alongside removing sails, rigging, cargo and loose fittings. By the 11th, six days after having stranded, she had completely broken up and was declared a total loss. The ship's timbers and bales of cotton (there were 1187 aboard as part of her cargo) washed ashore along the north coast of Wales as far away as Prestatyn for days afterwards.

### *HOUSTON*
### Hull 88

*Houston* was another of the group of vessels built in 1854-55 for Wakeman, Dimon & Company's Star Line between New York and Galveston. She was a 518-ton bark, measuring 132 feet 6 inches by 30 feet by 14 feet, and was of the same design as *Milton*, built six months before. *Houston* was launched early in June of 1854 and completed shortly afterwards. She left New York on her maiden voyage to Galveston on the 16th of June under the command of Captain W. A. MacGill. She was next back in New York on 3 January 1855 from Liverpool with 22 passengers and a cargo of coal and salt, after a run of 44 days. During the voyage, one of her crew, John Kennedy, fell from the jib boom, was run over by the vessel and came to the surface, but the difficulty of rendering assistance resulted in his death. Few men who fell from sailing vessels were ever rescued; the difficulty of stopping and returning to the point at which the unfortunate crewman fell made the odds against survival almost unsurmountable. If the accident happened in winter, as in the case of poor Kennedy and countless others, the chance of being picked up alive approached zero. *Houston* ran to Gulf ports during the pre-Civil War period, and spent the war years on coastwise supply voyages for the Union forces. Port Royal, Washington and Key West were typical destinations for her—and other vessels of the fleet of Wakemen, Dimon & Company—during this time. Although *Houston* was still afloat in the mid-1870s, details of her post-war activities are not known.

### *PELAYO*
### Hull 89

In 1854, the Webb shipyard built the sidewheel steamship *Pelayo* for Cuban owners represented by Bartolomeo Blanco, a merchant in New York. In a ship register of 1863, the owner is identified only as "Pelayo," but this may have been Pelayo, Pordo & Company. *Pelayo* measured 200 feet by 30 feet by 14 feet, and had a measurement tonnage of 850. She was fitted with two masts and was rigged as a topsail schooner. Her draft was ten feet. She was described by a surveyor employed by the Atlantic Mutual Insurance Company as having "cabins on deck with a promenade deck fore and aft." He also stated that she "appears well-built and fastened" and was a "good-looking boat." Otherwise little is known of the appearance of the vessel. Pease & Murphy, proprietors of the Fulton Iron Works, supplied her beam engine, the cylinder of which measured 64 inches in diameter and 8 feet stroke. Launched in July of 1854, she was completed and delivered two months later. As was generally the custom for foreign-flag vessels built in the United States, an American registry document was issued on 13 September 1854 to permit her to be delivered to Cuba. The papers were surrendered to an American consular official upon arrival at her Cuban home port. *Pelayo* was used mainly in Cuban coastal passenger and cargo service.

*Built in 1854 for the Cuban passenger and freight trade. Proved a good sea boat and fast and satisfactory vessel.*

### GENERAL ADMIRAL
### Hull 90

A 74-gun steam frigate completed in 1859 for the Imperial Russian Government, *General Admiral* was the first of four major war vessels built by Mr. Webb. Significantly, none was built to the order of the United States Government. She measured 313 feet 7 inches by 54 feet 6 inches by 33 feet 7 inches. The dimensions alone of *General Admiral* attest to the extraordinary size of this vessel. In the United States Navy, only *USS Niagara* (constructed at about the same time in the New York Navy Yard by George Steers) was of equal size and power. Her propulsion machinery consisted of two back-acting engines, built by the Novelty Iron Works, on her single propeller shaft. Each engine had a single cylinder 84 inches in diameter and 45 inches stroke of piston. Six boilers supplied steam. Her propeller, which was rigged so that it could be lifted clear of the water when the ship was under sail alone, was a then astonishing 19 feet in diameter. *General Admiral*, named for Grand Duke Constantine Nikolayevitch, General Admiral of the Imperial Russian Navy and brother of Tsar Alexander II, was ordered late in 1853. For reasons related to international politics, her keel was not laid until 1857—on the 21st of September, the birthday of the Grand Duke. She was launched exactly one year later, on 21 September 1858, with her decks crammed with an estimated 3,000 persons and many thousands more in the shipyard, on the adjoining properties and aboard a fleet of harbor vessels. This assemblage of New Yorkers responded with "one long loud huzza" and Dodsworth's band struck up the Russian national anthem when the towering bulk of *General Admiral* began her natal slide down Mr. Webb's launching ways. The vessel was delivered in June of 1859, first stopping at Kronstadt, where her guns (except for two 10-inch Dahlgren pivot guns built by the West Point Foundry) were to be fitted. Then, as now, the construction of large naval vessels consumed vast amounts of time and money. Little is known of the career of *General Admiral* in the Imperial Russian fleet. In the 1870s she was replaced by a more modern ironclad vessel of the same name.

*Of 74 broad-side guns, built for the Imperial Russian Government, draft when launched with two anchors and two chains, propeller bearings, propeller gear, after portion of main shaft and other machinery on board was forward 11 ft., aft 15 ft., with six boilers, propellers and other machinery on board was forward 12 ft. 9 in., aft 16 ft. 8 in., with all spars, machinery complete, tanks, etc. was forward 15 ft., aft 18 ft. 4 in., with boilers filled with water, all anchors, chains and hausers [sic] on board was forward 16 ft. 6 in., aft 18 ft. 6 in., with crew of 750 men, stores and provisions and coal for five months steaming, gun carriages (but without guns) on board was 22 ft. 9 in. aft. This ship made the passage from New York to Cherbourg in 11 days and 8 hours mostly under steam but at times under sail alone with moderate winds and no sails set above the topsails.*

### AURORA
### Hull 91

The packet ship *Aurora* was constructed for the New York firm of Grinnell, Minturn & Company in 1854. She was launched on the 23rd of September before a crowd of 2,000 persons, of which the *New York Times* reported that "the air resounded with their cheers as the noble vessel glided into the water." The same newspaper also stated,

> The vessel has been built with the best regard to ventilation, while the rig in her fore and main masts is done in a novel manner, so as to combine economy in managing with greater speed in sailing.

"The best regard to ventilation" was a response to the appalling overcrowding of steerage passengers and the frequent occurrence of outbreaks among confined passengers of what was called "ship fever"—typhoid fever. By regulation, the number of steerage passengers that could legally be carried on a packet was determined by dividing the usable tween deck area by fourteen, under the assumption that each emigrant required a space measuring all of two feet by seven feet. The sanitary conditions and ventilation in the tween deck spaces of these vessels must have been intolerable, particularly when hatch covers were in place and battened down in heavy weather. In 1853, deaths among steerage passengers reached ten or more percent on many passages, the worst probably being the arrival in December of that year of the 1568-ton packet *Constellation*, after 100 of her 922 passengers had been buried at sea. *Aurora* was probably certified to carry upwards of 800 persons in steerage. The sail plan of *Aurora* has not survived, and so the attributes of her rig that combined "economy in managing with greater speed in sailing" must remain a mystery. These may have been the personal preferences of her first commander, Captain Richard L. Bunting, who, in keeping with the practice of the day, was undoubtedly assigned by Grinnell, Minturn & Company to superintend her construction.

The first register document of *Aurora* was issued on 4 December 1854 and she departed for Liverpool a few days later under Captain Richard L. Bunting. *Aurora* was a vessel of 1639 tons, and measured 200 feet by 43 feet 7 inches by 21 feet. She was a typical full-modelled packet of the type that carried most of the passengers and freight during the early-to-mid 19th century. Rarely glamorous, these ships went about their business with little fanfare, except notice in the arrival and departure columns and, when misfortune struck, in the casualty reports. Misfortune was sometimes caused by a lack of experienced seamen, for, despite their human cargoes, packets frequently sailed with incompetents or the dregs of the waterfront in the forecastle. Upon arriving at New York on 9 December 1855, Captain Bunting reported,

> Had a very inefficient crew; not one half knew the ropes; was obliged to hire ten of the passengers to assist in working the ship.

*Aurora* was acquired prior to 1861 by Howland & Frothingham, of New York, for whom she continued to run as a Liverpool packet through the end of the Civil War under the command of Captain A. R. Barker. In the summer of 1867, as business for the packet lines was winding down, she sailed on the first of several voyages to San Francisco under charter to firms that operated regular lines to that port—Sutton & Company, Simonson & Chesebrough and others. On her return passages she usually carried wheat from San Francisco or guano from Callao. Captain Galen O. Norton relieved Barker around 1873, and occasional tramping voyages were made to Europe during this period. On 9 January 1878, at the completion of another Pacific voyage, she arrived at New York with a cargo of guano from Callao, her passage around Cape Horn having taken 128 days. By that time, she had been acquired by Captain Robert L. Merriman, who was in command of her on that trip. Captain Merriman was in charge when she made a voyage to Calcutta in 1882, departing from New York on the 2nd of August under charter to James W. Elwell & Company. She carried a cargo of oil products, loaded at either Pratt's refinery on Newtown Creek or the Standard plant at Constable Hook, Bayonne.

The vessel came to a spectacular end in February of 1884. She sailed from New York on her second voyage to Calcutta on 25 September 1883, again under charter to James W. Elwell & Company. Her cargo consisted of 59,332 cases of kerosene, or "case oil." Case oil derived its name not from chemistry but from packaging—two five-gallon tins encased in a wooden crate. Most of the petroleum sold in the Far East was case oil, used almost exclusively for lighting. The tins, emptied and flattened, were as valuable in the Orient as building material as their contents were as fuel. *Aurora* arrived at Calcutta on 5 February 1884 and anchored in Garden Reach, on the Hooghly River below the city, so that her cargo could be lightered ashore in small craft. Early in the morning of the 20th, fire was discovered among some loose waste in the forecastle. The crew barely had time to jump from the ship and swim the short distance to shore before the fire reached the cargo. The vessel then burned furiously all day, and by late afternoon she filled with water, broke in two and sank. Over 28,000 cases of her cargo were still aboard. An unidentified correspondent wrote of the fire:

> It was feared that the wind and tide would drive the burning oil to the crowded part of the river. But fortunately the wind shifted and the flaming liquid was driven upon the beach. The scene was wonderful beyond description. The river seemed to be on fire. The cases exploded with a sound like volleys of musketry. Dense black smoke enshrouded the town. The escape of the shipping and the town from a disastrous conflagration was narrow.

The loss of *Aurora* led to changes in the way kerosene cargoes were handled at Calcutta. Prior to the accident, it had been proposed that kerosene-laden vessels should only be allowed to discharge at Port Canning, about 25 miles distant on the Mutlah River, well away from the crowded city. The spectacle of the Hooghly afire on that February day quickly brought renewed interest to the proposal.

> *Built for the New York, Liverpool and London trade, was considered an excellent type of ship for that business, proved a fair sailer and a good sea boat. Gave general satisfaction. Draft of water when launched, forward 9 ft. 7 in., aft 10 ft. with all masts standing, yards on deck, one anchor and chain on board.*

### *JAMES FOSTER, JR.*
### Hull 92

In 1854, C. H. Marshall & Company once again turned to Mr. Webb for a packet ship for the Black Ball Line. *James Foster, Jr.*, a three-decked vessel of 1410 tons, was launched in December of that year and received her first register document on 3 January 1855. She departed for Liverpool under Captain James M. Porter on the same day. The ship's dimensions were 190 feet by 39 feet 9 inches by 27 feet 6 inches. As was the case with all of the large North Atlantic packets, accommodations were provided for a few cabin passengers, but many hundreds of emigrants were carried in steerage. We do not know the maximum number of steerage passengers allowed to be carried by *James Foster, Jr.*, but she had 750 aboard on one voyage in 1855 and similar vessels sailing from British and continental ports are known to have carried as many as 1,029. The concerns over passenger safety expressed today (and in 1912 following the loss of *Titanic*) pale into insignificance when compared to those that might have been voiced about conditions on the packet— a vessel less than 200 feet in length having hundreds of persons aboard; with no means of communications in case of emergency; and having little, if any, provision for the survival of the passengers (other than a handful of small boats) in case of the loss of the ship.

*James Foster, Jr.*, plied the route to Liverpool without reportable incident until 7 March 1869, when she arrived at New York after a 76-day passage with many of the 146 passengers ill with typhus and allegations of brutal treatment of passengers and some of the crew by the ship's officers and senior crew members. An investigation of the incident brought to light both questionable practices on the part of the packet lines and design deficiencies on vessels, all affecting the welfare of steerage passengers. At a subsequent trial, the charges of cruelty were proven and the third mate, carpenter and boatswain were found guilty. Captain Andrew Armstrong and the chief mate had meanwhile died of typhus. Long jail terms were imposed on the three surviving culprits. The special committee formed to investigate the affair made the following statement in its report:

> The principal security against recurrence of such evils, is to be found perhaps, in the rapid discontinuance of sailing vessels in the carrying of immigrants to this port, and the institution of steam vessels in their stead.

By this time, few emigrants were carried on packets, but this disastrous voyage, on which severe weather and the outbreak of fever among passengers and crew alike contributed to the appalling conditions on the ship, may have led directly to the demise as passenger carriers of those packet lines that remained in business. *James Foster, Jr.* had earned her place in history.

The vessel continued to run for C. H. Marshall & Company until the firm gave up ship operation in 1881. Her last voyage for Marshall was to Antwerp, her departure from New York taking place on 2 May 1881. On her return passage, she was consigned to Snow & Burgess, who had taken over the operation of what remained of the former Black Ball fleet. *James Foster, Jr.* made another eastbound voyage to Antwerp, clearing New York on 12 August 1881, and this proved to be her last under the American flag. She was sold at that time to owners in Bremen and, renamed *Hudson*, continued to cross the Atlantic until about 1890, when she was dropped from registry.

### *NEW ORLEANS*
### Hull 93

The firm of Stanton & Thompson, operators of one of the many lines of packets between New York and New Orleans, was the owner of the packet *New Orleans*, completed at the Webb yard early in 1855. She was a ship-rigged vessel of 924 tons, measuring 166 feet by 35 feet 6 inches by 21 feet 6 inches. These dimensions suggest that she was a two-decked vessel. *New Orleans* was launched on 4 January 1855, received her first documents on the 14th of February and departed for New Orleans on the following day. She was commanded by Captain John L. Rich, who, in accordance with customary practice, had supervised her construction. By 1860 *New Orleans* was owned by William T. Frost and running for Frost & Hicks' packet line—still to New Orleans and with Captain Rich still in command. He was relieved by Captain Bell and, later, by Captain F. L. Hewitt. In the summer of 1862, *New Orleans* made at a voyage to Liverpool, returning to New York with coal. Next came round voyages to Milford Haven under Captain Russell, and Newport, Wales, with Captain Hewitt back on the poop deck. On 5 February 1864 she sailed from New York for Glasgow with a cargo of 2,580 barrels of flour, 34,219 bushels of wheat, 24,803 pounds of lard, 343 bags of animal matter and 7,200 barrel staves. By the end of May, she had not arrived in the River Clyde, and was presumed lost with all hands. *New Orleans* was insured for $8,500 and her cargo for $10,450.

### *NEPTUNE*
### Hull 94

Four months after the completion of the three-decked packet ship *James Foster, Jr.*, a similar vessel, *Neptune*, joined her in the Black Ball Line fleet of C. H. Marshall & Company. *Neptune* was of about the same length and depth as the earlier ship, but was given nearly five feet more beam. Her dimensions were 190 feet by 42 feet 6 inches by 28 feet 3 inches, and her tonnage was 1406. *Neptune* was launched on 3 March 1855 and took her place on the line to Liverpool early in May under the command of Captain Enoch W. Peabody, who had previously been master of Marshall's packet *Manhattan*, built by Mr. Webb in 1850. Captain Peabody, a son-in-law of Alexander C. Marshall, remained aboard until about 1870, and *Neptune* survived as a Black Baller until 1876. On one harrowing trip, *Neptune* departed from Liverpool on 19 August 1862 with a mixed cargo and 376 passengers. All went well for a month, until the 19th of September, when the ship was about 150 miles east-southeast of Montauk Light, Long Island. There she was struck by a ferocious hurricane, which swept away all of her boats, stove in her bulwarks and snapped off the jibboom and all three masts at the deck. Five seamen were in the rigging at the time and were lost, and several passengers received severe injuries. The next day, two packets—Williams & Guion's *Chancellor*, Captain A. G. Spencer, and Spofford, Tileston & Company's *Ellen Austin*, Captain W. H. Garrick—happened upon the unfortunate ship. The former vessel furnished a small boat and the latter transferred eight men to *Neptune* to help in constructing a jury rig from the tangle of spars and rigging still attached to the vessel. By evening, rudimentary fore and mizzen masts had been rigged and later a mainmast was added. *Neptune* arrived at New York on the 25th, with Captain Peabody thankful for the help he had received and the 376 passengers equally thankful that they had survived the ordeal.

While returning from Liverpool in April of 1876 under the command of Captain J. H. Spencer, she ran aground on Sable Island, off the southern tip of Nova Scotia, in dense fog and heavy seas on the night of the 12th of that month. Within three hours the vessel was full of water. Three boats were launched, into one of which Captain Spencer, his wife and their four children embarked. The boats remained within sight of *Neptune* until daybreak. The captain's boat then went to the north side of the island in an attempt to make an easier landing, but was driven out to sea and drifted for three days. On the fourth day, a landing was made, and the boat's party walked seven miles to the island's lighthouse, where they were reunited with the remainder of the crew from the other boats. The schooner *Vanilla* was chartered locally to take the survivors (one man having died from exposure) to Halifax, where they arrived on the 2nd of May. *Neptune* soon broke up where she had gone aground.

## *ALAMO*
### Hull 95

*Alamo* was the next of the large group of vessels built by Mr. Webb for Wakeman, Dimon & Company, of New York. She was a 507-ton bark of the same dimensions as *Houston*, 132 feet 6 inches by 30 feet by 14 feet. Her launching took place on 17 February 1855 and she was completed during the following month. She cleared New York on her first voyage to Galveston, on Wakeman, Dimon & Company's Star Line, on the 20th of March under Captain Frederick Sherwood. *Alamo* was involved in an unusual incident in February of 1858. On the 17th of that month, she arrived at Galveston in charge of the first mate, who stated that she had been aground on Timbalier Island, Louisiana. While she was aground, Captain Muggelt, his wife, two passengers and part of the crew set off in the vessel's boat to summon assistance. After their departure, the wind came around to the north, releasing *Alamo* from the beach. The mate sailed her to Galveston and there awaited the arrival of the master, who arrived later on a steamer from New Orleans. In 1867, *Alamo* was acquired by the Bremen firm of Siedenburg & Wendt, and ran for at least 15 years under German colors, visiting New York frequently during this period. In late 1883, she was purchased by George E. Forsyth, of Halifax, Nova Scotia, and placed under British registry. She was still afloat in 1890, but details of her career beyond that year are not known. In at least 35 years of operation under three flags, *Alamo* retained her proud Texan name.

## *ASTORIA*
### Hull 96

*Astoria* was a screw steamship built by Mr. Webb in 1855 for service in Alaskan waters. Her connection with the "Russian American fur trade" (see below) is indicated by her name, which derives from the German-American fur trader, Colonel John Jacob Astor. She was a vessel of 428 tons, measured 160 feet by 24 feet 6 inches by 12 feet, and was rigged as a bark. Her engine had two cylinders measuring 26 inches by 26 inches, and turned a propeller that could be unshipped and hoisted from the water when the vessel was under sail. A single boiler supplied steam. The engine of *Astoria* was unique among the steam vessels built by Mr. Webb. It was the only one in which the cylinders were inverted and above the crankshaft—a configuration that was to survive for a century in the steam-reciprocating engine. The other screw steamers, all of which were naval vessels, were propelled by horizontal or back-acting engines. The sidewheelers, which predominated in the yard's output of steamers, were equipped with a variety of vertical beam, oscillating, inclined or side-lever engines.

*Astoria* was launched on 26 April 1855 (having been

coppered on the stocks) and received her first registry document on the 30th of June of that year at New York. Later, she was to have San Francisco as her home port, with B. C. Saunders as her managing owner. She was transferred to Russian registry in November of 1856 and renamed *Alexander II* after the then reigning tsar. Under the Alaska Purchase Treaty, she became an American-flag vessel again on 23 April 1868 as *Alexander*. During this period, she was owned by Hutchinson, Kohl & Company, and her home port was again San Francisco. She was later transferred back to Russian registry under the ownership of Phillipeus & Company, of Vladivostok. The date of this change is not entirely clear. Her American registry document is endorsed that it was surrendered in 1876, but American Shipmasters Association record books, which state that the vessel was rebuilt in 1879, imply that the change in ownership occurred at that time. Whatever the true date of her sale, her name reverted to *Alexander II*, and she survived until about 1896. During her second Russian phase, the vessel appears to have been used in long-distance trading, calls at Philadelphia in June of 1883 and San Francisco in May of 1888 having been recorded.

*Built in 1855 for the Russian American fur trade. Draft of water when launched, with all masts and yards, anchors and chains, engine nearly completed, (no boilers), was,*

This photograph of *Alexander II*, built in 1855 as *Astoria*, was taken in the 1880s, near the end of her long career and after extensive reconstruction. (Courtesy of The Mariners' Museum, Newport News, Virginia)

*forward 4 feet 10 inches, aft 8 ft. 5 in., mean 6 ft. 7-1/2 in., drag of keel 12 in. Draft loaded aft 11 ft. 9 in.*

## TEXAS
### Hull 97

The next of the Wakeman, Dimon & Company vessels was *Texas*, a 554-ton bark of the same dimensions as *Houston* and *Alamo* (132 feet 6 inches by 30 feet by 14 feet), launched on 23 July 1855 and not completed until September. Although *Texas* was built for New York-Galveston service, her first voyage, which commenced on 10 September 1855, appears to have been to Melbourne, Australia, under charter to George B. Daniels, the owner of the Webb-built clippers *Flying Dutchman* and *Young America*. Upon her return, she entered the coastwise trade and continued to make voyages on the East and Gulf coasts, carrying cotton for Wakeman, Dimon & Company, until the Civil War. During the early part of the Civil War, she carried supplies from New York to southern ports under Union control. On 2 December 1862, she cleared New York Alexandria, Virginia, in company with another vessel of Wakeman, Dimon & Company's fleet, the brig *Empire*. Neither vessel was ever seen again. On 28 February 1863, the press at New York reported that the two vessels had not arrived, but it was not until the 11th of July that they were declared missing.

## SABINE
### Hull 98

*Sabine* was the next-to-last of the fleet of vessels built by Mr. Webb for Wakeman, Dimon & Company in the mid-1850s. Her service was to be primarily on the New York-Galveston route. *Sabine* was rigged as a brig and, with a tonnage of 399 and measuring 120 feet by 28 feet by 12 feet 6 inches, was somewhat smaller than the other vessels. She was launched on 16 August 1855, and cleared New York on her first voyage to Galveston on the 11th of September, one day after the departure of *Texas*. Captain Abraham H. Trask was in command. She was back in New York late in January of 1856 with cotton and resin from St. Marks, a small port south of Tallahassee, Florida. The working life of *Sabine* was unfortunately short, for she burned at sea on 2 October 1859. She had left New York for Galveston on the 3rd of September with a full mixed cargo and one passenger. After an extremely slow passage, fire suddenly erupted around her mainmast on the 2nd of October while she was about 15 miles south-southeast of Galveston. Those aboard had barely time to launch two boats and, after two days,

landed at Quintana, Texas, near the mouth of the Brazos River. Here they embarked upon the steamboat *Alice* for Galveston, but she ran aground while in the canal of the Galveston and Brazos Navigation Company, a waterway completed in April 1854 with a nominal depth of all of three feet! The shipwrecked party was then forced, as reported by the New York marine press, to "procure another conveyance to Galveston." The fire had apparently started in some chemicals in the tween deck, and spread so rapidly that the passenger and crew members were lucky to escape with their lives; even the gold watch of the captain was left hanging in his cabin.

## AMERICA
### Hull 99

The Imperial Russian Government was the client of Mr. Webb when he built the bark-rigged sidewheel steamship *America* in 1855. This vessel, which had a measurement tonnage of 544, was launched on 15 September 1855 and received an American register document on the 18th of December for her delivery voyage, which commenced on the 24th when she sailed for San Francisco via the Straits of Magellan or Cape Horn (which is not really a cape but the southern tip of a small island, Isla Hornos) with stops en route for coal at, perhaps, Rio de Janeiro, Valparaiso and Panama. Captain Hudson was on the bridge for this long passage. In 1857, the vessel's American register was surrendered with the endorsement that she had been transferred to foreign registry. Assuming that the voyage to a port at which she could be handed over to Russian authorities took five months (four months to San Francisco and an additional month to refit and proceed to the nearest Russian port of Sitka), the reason for the lapse of at least a half year is not clear. *America* measured 170 feet 6 inches by 27 feet 6 inches by 12 feet 2 inches, and was of handsome appearance. She was propelled by an oscillating engine with two cylinders measuring 45 inches in diameter by 60 inches stroke. The purpose for which the vessel was built and the details of her career are not known, although her name indicates that she was used in Alaskan waters.

*Built in 1855 for the Imperial Russian Government. Was in continual service during 30 years in the North Pacific Ocean. Was a very beautiful and popular vessel. Draft of water when launched, without anything on board, forward 4 ft. 6 in., aft 5 ft. 10 in. When completed and boiler filled, forward 7 ft. 11 in, aft 8 ft. When ready for sea, coaled for 30 days and extra stores and much duplicate machinery on board, was forward 10 ft. 8 in. and aft 10 ft. 6 in. Total weight of engines, boilers, wheels, coal bunkers was 150 tons.*

### SILAS WRIGHT
### Hull 100

Unfortunately, *Silas Wright* had a short career as a Liverpool packet on the Black Star Line. She was built late in 1855 for the firm of Williams & Guion, the line's proprietor and New York agent. A vessel of 1340 tons, her dimensions were 190 feet by 42 feet 6 inches by 28 feet 3 inches. She was launched on 22 December 1855 and was placed in service early in the new year. She cleared New York for the first time on 16 January 1856 under Captain T. W. Freeman. *Silas Wright* lasted less than ten months. She was lost on 12 November 1856 during her fourth voyage to Liverpool, in a fierce gale that also claimed the Webb-built packet *Samuel M. Fox* and the New Orleans-based ship *Louisiana*. As the three vessels were being towed down the River Mersey to open water in preparation for their departures for their American destinations, all three were driven aground by the unusually high winds. *Silas Wright*, drawing 23 feet, grounded on the Mersey's West Middle Ground. Although the vessel remained upright, her bottom planking had been badly damaged when she struck and from the pounding she was subjected to as the storm continued to rage. She was abandoned and declared a total loss after tidal waters, for which the Mersey is noted, filled her holds and rose to a height of three feet above her upper deck at high tide.

### FANNY HOLMES
### Hull 101

Many merchants and shipowners competed for the lucrative coastwise routes between New York and Gulf Coast ports. The long established firm of E. D. Hurlbut & Company had run sailing vessels to Apalachicola, Mobile, New Orleans and other points, all of which served as entrepots for the cotton trade to the Northeast and Europe. After experiencing financial difficulties in 1855, Hurlbut's business was assumed by Post, Smith & Company, who turned to Mr. Webb for the construction of the barks *Fanny Holmes* and *Alice Tainter*. *Fanny Holmes* was a vessel of 700 tons, and her hull dimensions were 140 feet by 32 feet by 17 feet 6 inches. She was launched in November of 1855 and departed from New York for Mobile on the 5th of December with Captain G. H. Smith in command. Captain Smith was her master for the entire time *Fanny Holmes* was in service. While loading cotton for Antwerp at West Pass Anchorage, at Apalachicola, Florida, on 3 April 1860, she burned to the water's edge and was a total loss, although a part of the 871 bales was salvaged. It was supposed that the fire was caused by carelessness on the part of a longshoreman. One of the vessel's crew saw a longshoreman with a pipe in his hand late in the evening, and, as a newspaper account of the fire said, "as [he] has not been seen since, he must have fallen asleep with the pipe in his mouth and has probably burnt up in the vessel." A clipping containing this news story was found in one of Atlantic Mutual's Vessel Disaster Books; a terse handwritten comment in the margin states, "Hope he has."

### JOHN H. ELLIOTT
### Hull 102

The firm of Post, Smith & Company, operators of the Regular Line of packets between New York and Antwerp in addition to their coastwise routes, built the ship *John H. Elliott* at the Webb yard in 1856. She had a tonnage of 1077 and her hull measured 172 feet by 36 feet by 23 feet. This vessel was launched in February of 1856 and cleared New York for Antwerp about the 15th of the following month under Captain J. H. Tucker. She later ran for Tapscott's Line to Liverpool, and in mid-1862 was to be found on J. & N. Smith & Company's (the successor to Post, Smith & Company) line to Liverpool. Her last voyage began in New York on 26 November 1862, when she cleared for Liverpool with Captain Somers in command. In January of 1863, she left Liverpool to return to New York, but grounded a short distance from the Merseyside port on one of the treacherous shoals that abound in the estuary of the River Mersey. At first it appeared that she could be refloated, but by the next morning she was reported to be high and dry, with her back broken and her stern post gone, probably as a result of the extreme tidal range present in the region. At that time the pathetic wreck that had been *John H. Elliott* was abandoned and shortly afterwards was seen to be afire.

### ALICE TAINTER
### Hull 103

The bark *Alice Tainter*, a sister vessel to *Fanny Holmes*, was built for Post, Smith & Company as a cargo carrier in 1856. She was a vessel of 667 tons, and measured 140 feet by 31 feet 5 inches by 17 feet 8 inches. She was launched in February of 1856. Her first register document was issued on 2 April 1856, and on the following day she departed for Antwerp on her first voyage with Captain Spencer in command. During the Civil War, *Alice Tainter*, then under the ownership of J. & N. Smith & Company (successors to Post, Smith & Company,) made one interesting voyage—conducted quite openly under the noses of the Union navy, Union customs officials and others in high places. Under charter to Charles Stillman, a Brownsville, Texas, steamboat owner and entrepreneur (and business partner of Richard King and Mifflin Kenedy, founders of the famous King Ranch,) she left Matamoros, Mexico, on 25 June 1862 for New York, where she arrived on the 16th of July. *Alice Tainter* was one of a number of vessels that regularly made trips carrying cotton from Matamoros (across the Rio Grande from Brownsville) to New York and New England. There was no doubt in the north that the cotton loaded at Matamoros was grown and compressed in Texas, the Mexican connection being the most transparent of ruses. But the trade was carried on without interference. In this way, the mills of New England were kept running and many Union soldiers were clothed in Confederate cotton throughout the war. The extent of the traffic from Matamoros to the north was such that, on 6 April 1864, the ship *Banshee*, the bark *Manhattan* and seven brigs were loading at the Mexican port—all for New York or Boston. On their return voyages to Matamoros, these vessels carried such articles of value to the Confederacy as cavalry boots, coffee, powder and soap.

After the Matamoros voyage, *Alice Tainter* traveled legitimately from New York to New Orleans and Matanzas, Cuba, and back to New York. She then made the long voyage to Shanghai in 1863, clearing New York on the 23rd of April, and returning from Liverpool on 25 July 1864. It was during this voyage that the vessel was transferred to Bermudian registry under the ownership of Pendergast Brothers. She subsequently traded between South America and ports in the United States under this mid-nineteenth century flag of convenience. On 6 December 1874 she arrived at New York with coffee from Rio de Janeiro after a 47-day passage. Little is known of her movements after this, but she appears to have been removed from the register sometime about 1876.

*Built in 1855. Draft of water when launched with all masts in end, all yards on deck, water tanks, two chains and two anchors on board, forward 6 ft. 6 in., aft 6 ft. 9 in. Deep load draft, 15 ft. 4 in. Proved a good sea boat. Good carrier.*

## *CUBA*
## Hull 104

The sidewheel passenger and cargo steamship *Cuba* was constructed in 1856 for owners represented in New York by Bartolomeo Blanco, a Spanish merchant and importer in the South Street area. She was a vessel of 820 tons, measuring 200 feet by 30 feet 6 inches by 13 feet. There were several deck houses on her main deck and a light hurricane or promenade deck above. Her beam engine, which had a cylinder 66 inches in diameter by 9 feet stroke, was built by Pease & Murphy's Fulton Iron Works. *Cuba* was launched on 7 April 1856, a few minutes before the Collins Line steamship *Adriatic* was launched at the adjoining shipyard of George Steers. She was delivered late in June. The vessel received an American register document on the 26th of June, and cleared for Havana on the 1st of July with a Captain Adams in command. She was transferred to Cuban registry upon her arrival at her home port. In 1863 she appears to have been owned by Pelavo, Pordo & Company, of Havana, and was probably used in Cuban coastal service but little else is known of her career.

*Was built in 1856 for the Cuban trade. Draft of water when launched, with main parts of engine and boilers on board was, forward 4 ft. 9 in., aft 5 ft. 6 in. Proved a fast vessel. Load draft 11 feet.*

## *INTREPID*
## Hull 105

*Intrepid* was a short-lived medium clipper built for Bucklin & Crane in 1856 to trade to California and China. She was a vessel of 1173 tons, and measured 180 feet by 37 feet by 23 feet. Her launching took place in June of 1856. She cleared New York on her first voyage on 1 July 1856 under the command of Captain E. C. Gardner, and was 145 days enroute to San Francisco. From there she went to Shanghai before returning to New York. It was Captain Gardner's ill fortune on this voyage to have been beaten to San Francisco by *Neptune's Car*, an extreme clipper of 1616 tons built by Page & Allen, at Portsmouth, Virginia, in 1855. She was one of the few clippers built south of New York. Captain Gardner's chagrin was not so much brought about by the other vessel's time of 134 days, but by the fact that, for most of the trip, she had been in charge of a 19-year old woman. With Captain Joshua A. Patten lying ill with "brain fever" and the mate under arrest for neglect of duty and insubordination, it was the responsibility of the master's wife,

Mary A. Patten, an experienced navigator who first went to sea on *Neptune's Car* as a 16-year old bride, to guide the vessel to her destination. Both Mary Patten and E. C. Gardner were honored by the United States Maritime Commission during World War II (as was William H. Webb) by having ships named for them—a Liberty ship for Mrs. Patten and an N-3 coastal cargo ship for Captain Gardner.

The vagaries of passages 'round the Horn are illustrated by her second voyage in 1858, upon which she departed from New York on the 20th of October in company with the clipper *Crest of the Wave*. *Intrepid* had a 132-day passage to San Francisco, but *Crest of the Wave* took 163 days, reporting that she had been 36 days off Cape Horn. On her third voyage, *Intrepid* left Macao for New York on 17 March 1860 with a cargo of tea, silk, general Oriental cargo and firecrackers reportedly worth over a half million dollars. The "general Oriental cargo" probably included fine Chinese lacquered furniture and ceramics; much of the valuable antique Chinese ware traded today were brought to this country in the tea and silk clippers, sometimes as the personal property of the vessels' masters. The voyage of *Intrepid* was to take her on the customary route followed by the tea clippers through the Sunda Strait between Sumatra and Java, past the port of Anjier, then across the Indian Ocean to the Cape of Good Hope and on to New York.

Unfortunately, *Intrepid* never made it to Anjier, but ran hard aground on 31 March 1860 at Belvedere Reef, in the Gaspar Strait, about 200 miles to the north between the islands of Bangka and Bilitung. Some of the cargo was jettisoned in an attempt to get off and the vessel's masts were cut away in an attempt to prevent discovery by Malay pirates, but on the following day the pirates arrived and a fierce battle took place in which *Intrepid*'s crew lived up to the ship's name by holding the attackers off with a pair of nine-pounder cannon, small arms, boarding pikes and buckets of boiling water. The ship's complement, less one man killed in the battle, took to their boats after setting their once proud ship afire, and, shortly afterwards, were picked up and taken to Anjier by a French clipper. The pirates returned to *Intrepid* and removed more of the cargo, but were eventually driven away by Dutch gunboats. Some accounts of the incident state that the wrecked hulk of *Intrepid* was later refloated and towed to Singapore, where what little remained of the cargo was landed and sold for $2,000.

## *GUATEMALA*
### Hull 106

*Guatemala* was a sidewheel steamer built in 1856 for persons represented in New York by Bartolomeo Blanco. She was constructed alongside the larger *Cuba*. Rigged as a topsail-schooner, she was a single-decked vessel of 217 tons, and her hull dimensions were 127 feet by 22 feet by 8 feet 3 inches. The beam-to-depth ratio indicates that she was probably intended for use on coastal or short ocean passages rather than longer deep sea routes. She was launched in April of 1856 and delivered in June after her machinery, a beam engine with a 30-inch cylinder and having a piston stroke of six feet, had been installed at the Fulton Iron Works, and her hull had been caulked and coppered. She was given an American register (which was to be surrendered upon arrival at her home port) on 26 June 1856 and departed the following day for Belize, under the command of a Captain Simmons. Unfortunately, an unspecified defect in her machinery brought her back to New York for repairs. She resumed the voyage after repairs and adjustments were made. *Guatemala* appears to have remained afloat at least through 1863, but nothing else is known of her career.

## *OCEAN MONARCH*
### Hull 107

When the packet *Ocean Monarch* was built for Stanton & Frost in 1856, she was the largest vessel of the type ever built. Measuring 240 feet by 46 feet by 30 feet 3 inches and having a tonnage of 2145, she was used primarily on the owner's Liverpool route. *Ocean Monarch* was launched in September of 1856 and entered service on the 8th of October, when she cleared for Liverpool with Captain Pitkin Page in command. The Stanton of Stanton & Frost had previously been a partner in the firm of Stanton & Thompson, owners of the Webb-built packet *New Orleans*. In May of 1857 *Ocean Monarch* arrived at New York with 954 emigrants in steerage, in addition to a handful of cabin passengers. A prodigious cargo carrier in her short career, she loaded 120,000 bushels of wheat and corn, 150 tierces of rice and 150 tons of fustic (a tropical wood yielding a yellow dye) for her first voyage. The *New York Times* remarked, "She still has capacity in her upper deck for 50,000 more bushels, but owing to her heavy draft of water (being 23 feet) her owner has deemed it prudent to dispatch her with her main deck empty." Still under the command of Captain Page, *Ocean Monarch* sailed from New York for Liverpool on 5 March 1862 with her cavernous cargo holds filled with 10,811 barrels of flour; 10,892 bushels of wheat; 33,774 bushels of corn; 193 tierces of beef; 76 barrels of pork; 62,020 pounds of tallow; 298,643 pounds of lard; 739,800 pounds of bacon; 7,000 pounds of shoulders and 10 tierces of tongues (for a total of a then astonishing 2,700 tons of cargo.) On the 9th, while at 38 degrees north latitude and 60 degrees west longitude, *Ocean Monarch* was struck by a violent southwest gale, causing the cargo to shift and the vessel to ship a large amount of water. The boats were quickly swung out, and the second mate and 21 crew members boarded one, which soon became separated from the ship, leaving Captain Page and the rest of the crew on the deck of *Ocean Monarch.* The 22 survivors were picked up by the schooner *Oliver H. Booth* on the 14th, and later transferred to the ship *James R. Keeler*, which landed them at New York on the 24th of March. Four days later, the British ship *Timour* arrived at New York and her master reported that on the 15th they had sighted "a great number of flour casks and other ship wreck, in fact the sea seemed all strewed over with wreck stuff," near the position at which *Ocean Monarch* was last reported. Captain Page and the rest of the crew, last seen on the deck of *Ocean Monarch*, were presumed to have been lost.

*Built for general freighting business. Was much the largest ship of this character heretofore built. She was built in the year 1856. Draft of water when launched, with all masts on end and most of the yards on deck, 30 ton anchor and chains, iron tank 6-1/4 tons and hawsers on board, was forward 9 ft. 7 in., aft 10 ft. 5 in. First cargo, chiefly of wheat, corn and rice, draft was 23 feet. The ship carried over the bar, as then existing, at the mouth of the Mississippi River on each of two voyages, over 7,000 bales of compressed cotton on a draft of about 20 feet.*

### UNCOWAH
### Hull 108

The medium clipper *Uncowah* was built for Wakeman, Dimon & Co., of New York, in 1856 for service in the East India trade. She measured 169 feet by 35.8 feet by 22 feet. A vessel of 988 tons, she was launched at 10 o'clock on the morning of 15 October 1856. Her first register document was dated 27 November 1856 and she cleared New York for San Francisco the following day. Her commander was Captain N. Kirby, Jr., who drove her to her destination in 116 days on her first voyage. Her 1859 passage took 143 days. On 2 February 1865, *Uncowah* cleared New York for Shanghai under Captain

Rudolph on what was to be her last voyage for Wakeman, Gookin & Company, that firm having succeeded Wakeman, Dimon & Company a short time before. Later that year, she was sold to N. Larea, of Lima, Peru, to carry coolies across the Pacific. After more than five years in this trade, which supplied labor for the Peruvian guano workings on the Chincha Islands, she was lost is a particularly horrible accident. An unidentified newspaper reported, "Again we are called upon to chronicle another horror in the traffic in coolies." *Uncowah* sailed from Macao for Callao in the autumn of 1870, and was reportedly set afire by the coolies when she was near an island variously identified as Nipune or Neptune. It was soon found that it would be impossible to save the ship, and, the newspaper continued, "The captain, officers and crew were compelled to take to the boats and leave their living freight as food for the devouring element." Those in the boats were picked up by a passing ship, along with 112 of the coolies. No fewer than 425 others perished in the flames or drowned.

### WILLIAM H. WEBB
### Hull 109

Mr. Webb's namesake vessel was a sidewheel towboat launched on 6 September 1856 for the firm of Chambers & Heiser. *William H. Webb* was a vessel of 655 tons, and measured 190 feet by 30 feet 2 inches by 12 feet. A double beam engine, having two cylinders 44 inches in diameter with a stroke of 10 feet, was installed. She was first enrolled on 18 November 1856, and was used primarily in offshore assignments out of the port of New York, under the command of Captain Charles Hazzard, a veteran towboatman who had previously been master of the ocean towboat *Leviathan*. Later *William H. Webb* was acquired by Charles Morgan's Southern Steamship Company for towing service at New Orleans.

During the Civil War, she had a varied career. In May of 1861 she was given a privateer's commission, but never engaged in that activity. After a period as a transport, she was acquired by the Confederate Army and converted to a ram for use on the Mississippi and Red Rivers. With the cottonclad ram *Queen of the West* (formerly a Union Army ram, sunk and later raised by the Confederates) she captured the Union ironclad *USS Indianola* (an unusual vessel in that she was propelled by both paddle wheels and propellers) on 24 February 1864, forcing that vessel aground near the mouth of the Red River, a strategic location in Admiral Porter's river campaign. The

Antonio Jacobsen's 1892 painting of the extreme clipper *Young America*, built in 1853 for George B. Daniels. She survived until 1886, when she went missing while under the Austrian flag. Jacobsen painted several thousand ship portraits between 1870 and 1920.

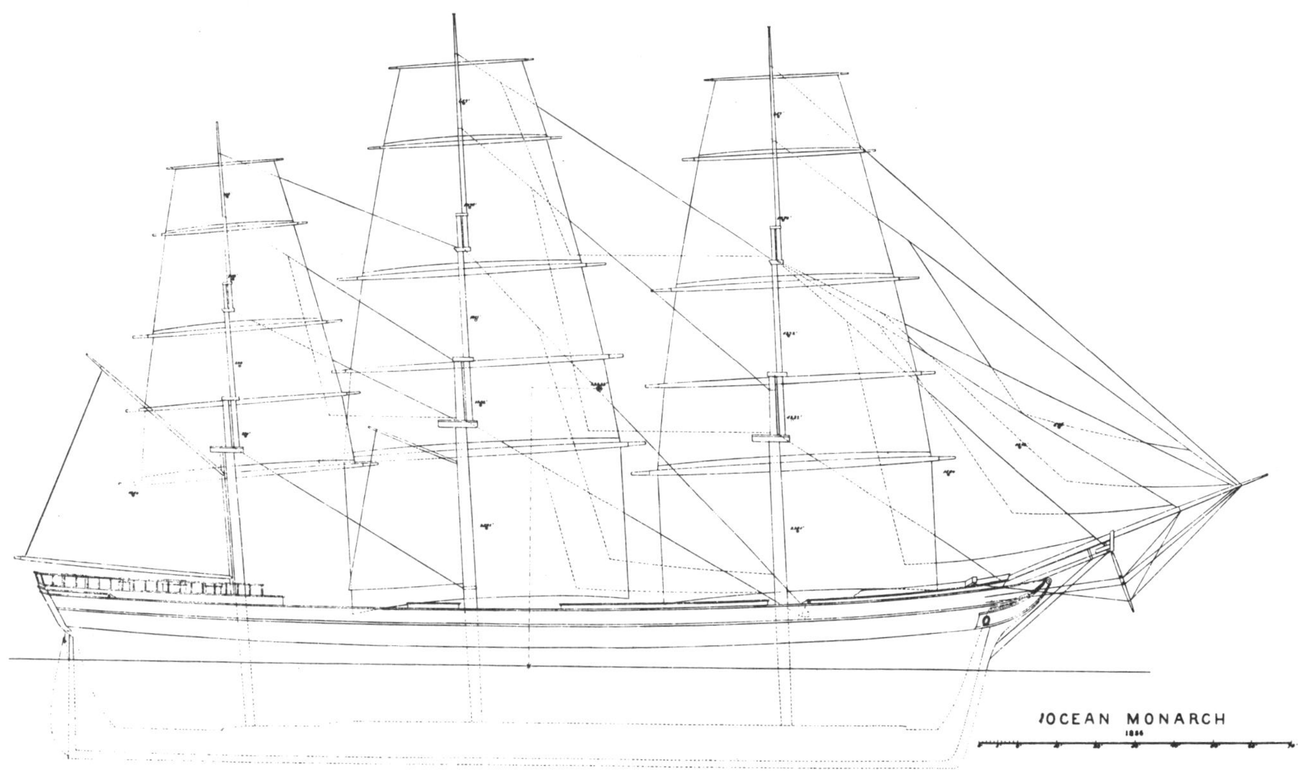



The differences in the way packet ships and clippers were rigged is apparent when this sail plan of *Ocean Monarch*, the largest packet when she was built in 1856, is compared with those of *Young America* and *Challenge* elsewhere in this volume. *Ocean Monarch*'s cavernous holds could accommodate up to 2700 tons of cargo—then a scarcely believable amount. Note the spencer, a rarely seen or used fore-and-aft sail, rigged on the mainmast.

Confederates attempted to salvage *Indianola*, but were frightened by the appearance of a coal barge disguised as a monitor with fake paddle boxes, an equally fake turret, smoke-spewing dummy stacks and logs posing as guns. The Union Navy had set the barge adrift above *Indianola*, and when she hove into the view of the salvors, they blew up the magazines of the grounded ironclad to prevent her recapture. After the transfer of *William H. Webb* to the Confederate Navy early in 1865, she was ordered to sea in April of that year as a commerce raider. Leaving the Red River on the 23rd of April, she ran the Union blockade and passed New Orleans successfully, but when about 24 miles below the city *USS Richmond* was sighted, and *CSS William H. Webb* (also referred to as *CSS Webb*) was run ashore and burned. The incident occurred 13 days after Lee's surrender at Appomattox Court House signalled the end of the war. What Mr. Webb thought of having his name on a Confederate vessel of war is not recorded.

Other vessels have been named for Mr. Webb over the years. In 1920, a cargo vessel from the Merchant Shipbuilding Corporation's yard at Chester, Pennsylvania, (a reincarnation of John Roach's famous, old Delaware River Iron Shipbuilding and Engineering Company) carried his name; sister vessels recalled shipbuilders Donald McKay, Henry Steers, John Stevens, John Englis and John Roach. During World War II, a Liberty ship built by the Bethlehem-Fairfield yard in 1943 honored Mr. Webb.

*Double beam engines of high power. Built in 1856. Draft of water when launched without machinery on board, was forward 3 ft. 9-1/2 in., aft 4 ft. 2 in. This vessel was originally built for a tug boat in New York Harbor. Afterwards employed at New Orleans as such. During the rebellion of the Southern States of the United States, she was fitted by the rebels as a Ram and employed in their defences [sic] on the Mississippi River, and proved formidable and an exceedingly fast vessel.*

## BLACK HAWK
### Hull 110

*Black Hawk* was a medium clipper ship built in 1857 for the mercantile firm of Bucklin & Crane, of New York. By the time she was built, the day of the extreme clipper had passed, and the evolution of the type as driven by commercial needs led to a more conservative, but still fast, cargo carrier. The hull form of *Black Hawk* represented the best of that later period—the medium clipper—upon which the design of sailing vessels of the following decades would be based. Her dimensions were 178 feet by 38.5 feet by 23.9 feet, and she had a tonnage of 1175. The vessel was launched in December of 1856. Her first register document was issued on 25 March 1857 and she departed on her first passage to San Francisco on the 2nd of April under Captain B. P. Bowers. The voyage was made in 119 days. Two of her later passages to San Francisco during the 1850s took 142 and 107 days.

Although clippers are popularly associated with tea and silk, many were used from time to time to carry more mundane commodities. Grain, salt, coal and other bulk cargoes frequently occupied holds that might have held tea and silk on a previous voyage. *Black Hawk* carried guano from Jarvis Island (a remote and little known dot in the Pacific Ocean about 1300 miles south of Hawaii) to New York in the summer of 1859, and made the voyage in an unexceptional 93 days. Westbound cargoes to San Francisco in the 1850s and 1860s were in the main composed of the multitude of everyday things that must have been in short supply in that new and growing community: from cast iron stoves to calico; from miners' tools to paint and wallpaper. Despite the handsome appearance of nearly all of the clipper ships and the aura of romanticism that has surrounded them for over a century, they were built to carry cargo—at a profit to the owners—and the opportunity to turn a profit was what filled their holds.

*Black Hawk* was sold to George Howes & Company in the mid-1860s, and she continued to trade for them as she had for Bucklin & Crane, primarily between New York and San Francisco, carrying general merchandise westward and grain back to the East coast. *Black Hawk* made no less than 20 round-trip Cape Horn passages, her last arrival from San Francisco having taken place on 22 January 1881, after a passage of 131 days. Howes then sold the vessel to D. H. Waetjen & Company, of Bremen, for service between the German ports of Bremen and Hamburg and New York. She cleared New York under her new colors on the 10th of March, and was a regular caller at her old home port for the next eight years, initially under Captain J. F. Haesloop and later Captain Wachsmuth. During that period, New York was a major oil refining center, with refineries at Constable Hook and Long Island City. (The country's first long-distance petroleum pipeline was built in 1881 to supply feedstock to these two facilities from the oil fields of Western New York state). Retaining her original name while under the German flag, *Black Hawk* Joined the fleet of sailing vessels that carried American oil to European markets. Her cargoes at this time consisted of refined petroleum in barrels (consisting mainly of kerosene for use in lamps) eastbound and empty petroleum barrels westbound. On occasion, iron, cement or salt made up a part of the ship's westbound manifest to lower her center of gravity by providing additional weight in her lower hold. A full cargo of empty barrels made the ship tender under sail. In 1889, *Black Hawk* was sold to Norwegian owners who changed her name to *Christiania*. She was reported to have been damaged by fire and condemned in February of 1891.

*Built in 1857 for the California and East India trade. Was a good carrier, proved a fast sailer and good sea boat and very successful vessel. A exemplification in her model that great deadrise is not necessary to secure high speed. Draft of water when launched, with mast [sic] standing, yards [sic] in deck, one anchor and three chains (21 tons), iron water tank (4 tons), hausers [sic] and rigging on board was, forward 8 ft. 7 in., aft 9 ft. 10 in. Deep loaded 21 ft. 6 in.*

## ROGER A. HEIRN
### Hull 111

*Roger A. Heirn* was built in 1857 for Post, Smith & Company as an Antwerp packet. She was a ship-rigged vessel of 1088 tons, measuring 172 feet by 36 feet by 23 feet, and was a duplicate of the packet *John H. Elliott*, built for the same owners by the Webb yard the year before. *Roger A. Heirn* was launched on 9 April 1857 and was completed about the first of June. Her first register document was issued on the 2nd of June, and she soon departed for Antwerp for the Hurlbut Line, for which Post, Smith & Company were then the New York agents. Captain James Stewart was in command on the vessel's first, and all later voyages. In 1858, she was on Tapscott's Line of Liverpool packets for a while, and two years later ran on the Stanton & Thompson line to New Orleans. *Roger A. Heirn* then reverted to European routes—first to London and, in the late spring of 1862, to Liverpool. She traded to that port carrying cargo and emigrants through the late summer of 1864. On 6 January 1865 she cleared New York for Matamoros, Mexico, to load Confederate cotton, a trade described earlier in the history of the bark *Alice Tainter*. For this service, *Roger A. Heirn* was transferred to British registry, thereby avoiding the thorny issue of Union vessels trading with the Confederacy. By this time her original owners, Post, Smith & Company, had been succeeded first by J. & N. Smith & Company, and then by Smith & Dunning. As a New York-based commission merchant, the firm was an important part of the transportation system that brought Texas cotton to the North and to Great Britain throughout the Civil War.

*Roger A. Heirn* was back in New York on 6 August 1865 with merchandise and 406 steerage passengers from Liverpool. She was next reported at Mobile, where she loaded 2644 bales of cotton for Liverpool. She proceeded to sea on the 16th of December with a fresh breeze from the east. As she drew abeam of Sand Island, just inside the bar at the mouth of Mobile Bay, the breeze shifted to the southeast. Captain Summerville decided that it would be unsafe to try to cross the bar with a head wind and anchored. During the night the wind increased to gale force. The vessel lost an anchor with its cable and drifted onto the West Bank at Sand Island. Her masts were cut away, and, on the 17th there was water in her holds to a depth of nearly 15 feet. She was soon afterwards declared a total loss.

### *TRIESTE*
### Hull 112

The bark *Trieste* was constructed in 1857 on speculation and sold after launching to the firm of W. A. Sale & Company. She measured 135 feet by 28 feet 11 inches by 12 feet 9 inches, and had a tonnage of 549. She was launched without a name on 27 June 1857, but her first register document was not issued until 12 November 1857, indicating that Mr. Webb may have had some difficulty in finding a buyer. *Trieste* was used by W. A. Sale in trading to Mediterranean ports, most of the westbound cargoes consisting of fresh fruit in season. During the Civil War, she was still owned by W. A. Sale and was trading as a tramp. The details of her later career are not known with certainty except that she made at least one voyage to China, arriving at New York with a cargo of tea on 5 January 1863. In November of that year, an Austrian bark named *Trieste* landed a cargo of Brazilian coffee at New York after a 39-day passage from Rio de Janeiro. It is possible the W. A. Sale had Austrian connections, and that, for them, the Austrian flag was a wartime flag of convenience, in the same way that British or Bermudian registry was used by other shipowners. Foreign registry was an attractive expedient during the war, for it permitted a vessel to trade freely while avoiding the risk of capture or destruction by Confederate raiders such as *CSS Alabama* or *Florida*. Then as now, a vessel built in the United States and placed under a foreign flag lost her coastwise privileges if she was returned to United States registry. The legislation that today affords protection to domestic shipping in the United Staates is referred to as the "Jones Act" (named for Republican Senator Wesley L. Jones, of Washington, its principal sponsor. The "Jones Act" is actually only Section 27 of the Merchant Marine Act of 1920,) but the roots of this protection may be traced back to the Coastwise Act of 1789.

*Built in 1857 for the Mediterranean trade. Proved a good sea boat and carried well.*

### *MOSES TAYLOR*
### Hull 113

The handsome sidewheel steamship *Moses Taylor* was constructed for the United States Mail Steamship Company of Marshall O. Roberts, for service between New York and Aspinwall. She was a vessel of 1370 tons, measuring 246 feet by 34 feet by 17 feet. Her propulsion plant consisted of two beam engines 50 inches in diameter by 10 feet stroke that had previously been in United States Mail's steamship *El Dorado*. The machinery was arranged in such a way that the engines normally turned both paddle wheels together, but, by disconnecting a shaft coupling, each engine was connected independently to one wheel. *Moses Taylor* was launched on 1 August 1857 and completed at the end of the year. Her first register document was issued on 5 January 1858, and on that day she departed on the first of 18 voyages to Aspinwall. On her first return trip she carried $1,565,799 in gold. There followed six voyages to Havana and New Orleans for Roberts, after which she was sent to the Pacific coast to operate the western segment of a line in competition with Pacific Mail. In 1866, the line was sold to the North American Steamship Company, owned by a group headed by Mr. Webb. Sailings were continued until the North

American company collapsed six months later. *Moses Taylor* and the rest of the North American fleet (at that time consisting of the steamships *Nebraska*, *Dakota* and *Nevada*) then came under the sole ownership of Mr. Webb, who organized the United States, New Zealand and Australia Steamship Company. In April of 1871, *Moses Taylor* was placed on the route between San Francisco and Honolulu, at which port the passengers and mails were transferred to one of the company's larger ships to continue their travels to Auckland and Sydney. On her first voyage, a furnace collapsed in one boiler, scalding to death the entire engine room crew—six men— on watch at the time. The unfortunate engineers and firemen were buried at sea and the vessel proceeded to Honolulu on one boiler. *Moses Taylor* was too small to handle all of the cargo received from the larger eastbound steamships during peak periods, and it was necessary for the line to charter tonnage from time to time to supplement her limited capacity. The service between San Francisco and Honolulu (and the through service to Australia) lasted until March of 1873, just two years after the line commenced running. Pacific Mail later purchased *Moses Taylor*, removed the engines and boilers and used her hull as a stores hulk at Acapulco. She was disposed of by Pacific Mail some time after 1877. "Rollin' Mose" (so called because of her propensity to roll heavily at sea) was unique among steam vessels built by Mr. Webb in that she was the only one that was ever solely owned and operated by him.

### *RESOLUTE*
### Hull 114

When Mr. Webb built the ship *Resolute* for Williams & Guion in 1857, the golden age of the American packet was nearing its end. These vessels were still carrying thousands of emigrants on their westbound voyages, but steamships had captured the cabin class trade and, after the Civil War, the days of the packet were numbered. Many of the old lines went out of business at that time. Some, like Williams & Guion, converted to steam; others gave up completely. *Resolute* was a three-decked vessel of 1412 tons, measuring 190 feet by 42 feet 6 inches by 28 feet 3 inches. She was launched on 5 September 1857, and completed early in October. With Captain T. W. Freeman (late of the Webb-built packet *Silas Wright*, lost in a gale at Liverpool the previous November) in command, she cleared New York for Liverpool on her first voyage the 6th of October.

After Williams & Guion gave up their packets in the late 1860s, Stephen B. Guion, American-born but based in Liverpool, formed the Liverpool & Great Western Steamship Company (the famous Guion Line), which enjoyed many years of success on the North Atlantic during which they achieved a place in history with their outstanding steamships *Arizona*, *Alaska* and *Oregon*. This trio, built in 1879, 1881 and 1883 respectively, held the eastbound Blue Riband of the North Atlantic from May of 1879 to May of 1884, and westbound for a somewhat shorter period. But the fortunes of the line declined, and, following a series of misfortunes which included a succession of poor trading years, the death of Guion, the forced sale of *Oregon* to Cunard in the summer of 1884 and drastic quarantine restrictions by the United States following a serious cholera outbreak in 1892, the line gave up. *Arizona* made the last Guion Line voyage from Liverpool to Nerw York in May of 1894.

*Resolute* went tramping after the Williams & Guion packet line ceased running. She was sold to Captain Jonathan C. Nickels, of Searsport, Maine, in 1871, and under his ownership and command continued to work her way around the oceans for a number of years. She was then sold to S. J. Melchers, of Schiedam, The Netherlands, in July of 1884, and transferred to Dutch registry. This was to prove a short-lived chapter in her story, for she was abandoned in the North Atlantic in March of 1886 while bound from Philadelphia to London. During a bout with heavy weather, she had lost her rudder and several sails, and, when Captain Hemmes and her crew were taken off by the British bark *Annie Burrill*, to be landed at Le Havre on the 14th of April, *Resolute* was left in a sinking condition with 13 feet of water in her hold.

### *HARRIET LANE*
### Hull 115

The famous *Harriet Lane* was constructed for the United States Revenue Cutter Service, the predecessor of the Coast Guard, in 1858 as the first steam-propelled vessel in that service. She was named for the niece of James Buchanan, who acted as the country's First Lady during the administration of the bachelor president. The contract for the cutter was awarded by the Revenue Cutter Service on 13 July 1857, following a design competition in which Mr. Webb's proposal was chosen from a field of eighteen. The board appointed to select the vessel's design included Samuel H. Pook, Henry Hunt and Charles W. Copeland. The three men were associated with the United States Navy in that service's pioneering days in steam. Pook was an eminent naval architect, while Hunt and Copeland were marine engineers. In the 1830s, Copeland had been associated with the West Point Foundry Association, of Cold Spring, New York, one of the earliest builders of marine steam engines in the United States.

The vessel measured 180 feet by 30 feet by 12 feet 6 inches and had a tonnage of 674. An double inclined engine with two cylinders 45 inches in diameter and a piston stroke of seven feet was built by the Allaire Works. *Harriet Lane* was launched at 11:30 on the morning of 19 November 1857 and delivered to the Revenue Cutter Service in the early spring of 1858. She was temporarily attached to the Navy late in 1858 to lead a mission to Paraguay. Having completed that task, she reverted to her revenue cutter role and, in September of 1860 carried the Prince of Wales, Edward Albert to Mount Vernon. The prince was the first member of the British Royal Family to visit the United States. She was transferred again to the Navy on 17 September 1861 for Civil War duty. Her most notable assignment during the war was as flagship of Commander D. D. Porter's Mortar Flotilla, which led in the spring and summer of 1862 to Union control of the lower Mississippi River and the port of New Orleans. October of 1862 found *Harriet Lane* at Galveston, where, with the aid of a number of naval vessels (including two former New York harbor ferries outfitted as gunboats), she bombarded and captured this important harbor. On 1 January 1863, she was herself captured and Galveston returned to Confederate control after a fierce fight with the rebel gunboats *Bayou City* and *Neptune*. She served the Confederate Army for a short time and was then sold to T. W. House, renamed *Lavinia*, and converted to a blockade runner. She made one run to Havana with cotton in April of 1864 and was interned. After the war, she was returned to the United States.

Early in 1869, she was rebuilt at East Boston as a bark for Nehemiah Gibson and others. At that time she was renamed *Elliot Ritchie* and her home port was changed to Boston. She then traded quietly until the spring of 1884. On the 22nd of March of that year she departed from Brunswick, Georgia, for Buenos Aires with Captain Perkins in command and a cargo of lumber stowed in her hold. While off the Brazilian coast near Pernambuco, she sprang a leak and became waterlogged. Her crew of ten men was taken off by the British ship *Galgate* and the disabled vessel set afire in an attempt to prevent her

becoming a menace to navigation. On the 13th of May, the crew was transferred to the British iron steamship *Pleiades*, bound from Santos to the United Kingdom via Rio de Janeiro and Pernambuco, and they were landed at the last named port the following day. Weeks later, about 3,000 pieces of pine lumber drifted ashore south of Pernambuco—the last testament of the long-lived *Harriet Lane*.

*Built for the United States Treasury Department for Revenue Cutter Service, being the first steamer used by that Government for that purpose. An unusual competition was manifested for the contract to build this vessel, eighteen different models and plans having been presented, that of Mr. W. H. Webb having been awarded the preference by the Board of Experts. Battery — one long 32-pounder, 4 Dahlgren 24-pounder Howitzers, one boat howitzer. She was transferred to the United States Navy, participated in the expedition to Paraguay, fired the first gun off Charleston in defence of the U. S. Government when Fort Sumpter [sic] was taken by the Rebels; was finally captured at Galveston Harbor (Captain Wainwright commanding killed) by the Rebels who converted her into a blockade runner. With coal, water, provision for crew of seventy-three officers and men and battery all ready for a voyage [draft] was 8 ft. 10 in. She proved an excellent sea boat, having moderate speed, was very popular and gave marked satisfaction. The Harriet Lane had a national reputation.*

### *JAPANESE*
### Hull 116

*Japanese* (sometimes spelled "Japanis" because of variations in the transliteration of Cyrillic characters) was a 1530-ton steam corvette built in 1858 for the Imperial Russian Government, the third of four vessels constructed by Mr. Webb for the service of the Tsar. *Japanese* was an armed screw steamer for use on the Amur River, the upper reaches of which today form part of the boundary between Russian and Chinese territory. Russia had acquired certain lands along the Amur from China in 1858, and the vessel was ordered to provide protection along the new border. Her hull dimensions were 217 feet by 36 feet 8 inches by 17 feet 6 inches. She was launched on the morning of 16 November 1857, three days before the launching of the revenue cutter *Harriet Lane*. A double oscillating engine, with cylinders 51 inches in diameter and a piston stroke of 36 inches, provided the power for her single propeller. Steam was supplied by two boilers. She was delivered early in 1858. No details of her Russian naval career are known.

*Built in 1858 for the Imperial Russian Government. 10 side guns, 32 pounders, bow and stern guns, 2 oscillating engines, each 51 inches in diameter of cylinders and 3 feet stroke, 2 flue boilers, propeller 12-1/2 feet in diameter. Draft when launched, with anchors and chain and 328,000 pounds of machinery on board, forward 5 ft. 6 in., aft 7 ft. 1 in., drag of keel being 1 ft. 6 in. With boilers, iron bulkheads, spars and rigging, 3 bow and 1 stern anchor and chain (150 fathoms), forward 6 ft. 10 in., aft 7 ft. 9 in., with 125 tons of coal, water and stores on board, forward 11 ft. 6 in., aft 12 ft. 9 in., with extra coal and stores on board for long voyage to Kamtschatka, was 14 ft. 6 in. Proved a good sea boat and fast sailer.*

### MARTINHO DE MELLO
### Hull 117

The bark *Martinho de Mello* is somewhat of a vessel of mystery. She was built by Mr. Webb's brother Eckford at the Greenpoint shipyard of Webb & Bell under sub-contract in 1858. At this time, the artisans at the Webb yard were devoting their energies to building the ironclad frigate *General Admiral* for the Imperial Russian Government, and skilled manpower was not available to permit work on vessels other than the immense frigate. *Martinho de Mello* was built at a cost of $40,000 for the Portuguese Government, but the purpose of the vessel is not clear. She measured 138 feet 2 inches by 31 feet 6 inches by 18 feet 6 inches, and had a tonnage of 622. Her launching took place on 2 October 1858, and she was completed and delivered early in the next month. An American register document was issued for the vessel, indicating that she flew the American flag on her delivery voyage to Lisbon, Portugal. This voyage commenced on the 8th of November, when she departed from New York for Lisbon under the command of a Captain Giraud. No information has been found that describes her career after her maiden voyage to Portugal.

### YORKTOWN
### Hull 118

*Yorktown* was a sidewheel steamship built for the New York & Virginia Steamship Company's line from New York to Richmond. The company pioneered that service in 1851, and was a predecessor of the famed Old Dominion Line. She was a vessel of 1403 tons, measuring 250 feet by 34 feet by 17 feet, and was propelled by a double beam engine from the Morgan Iron Works. Each engine had a cylinder 50 inches in diameter by 10 feet stroke of piston. *Yorktown* was launched in May of 1859 and entered service on the 15th of November, when she departed for Havana under charter to Hargous & Com-

pany. She took her place on the Richmond route a short time later, joining the line's other vessels, *Jamestown* and *Roanoke*. When Virginia seceded from the Union on 17 April 1861, *Yorktown* was seized and turned over to the Confederate Navy. She emerged as *CSS Patrick Henry*, mounting one 10-inch and six 8-inch guns, and was attached to the James River Squadron. During the battle at Hampton Roads between *USS Monitor* and *CSS Virginia* on 8-9 March 1862, *Patrick Henry* played an important supporting role. Later she became the schoolship for the Confederate States Naval Academy at Drewry's Bluff, a short distance down the James River from Richmond. On 3 April 1865, as Richmond was being evacuated and the war was winding down, she was burned to prevent her capture by the Union forces. Six days later, General Robert E. Lee capitulated to General Ulysses S. Grant at Appomattox Court House, 92 miles west of Richmond, and the war was essentially over. Hostilities did not end until the 2nd of June, when the Confederate garrison at Galveston lowered its flag.

*Built in 1859 for the trade [from] New York to Richmond, requiring very light draft of water. Draft when launched, forward 6 ft. 0 in., aft 6 ft. 6 in. Machinery, anchors, cables and all equipment on board, forward 8 ft. 0 in., aft 8 ft. 6 in., with boilers filled with water, forward 9 ft. 6 in., aft 8 ft. 6 in., with 140 tons of coal on board, forward 9 ft. 5 in., aft 9 ft. 10 in., draft deep loaded, 13 ft. 0 in. Was fast and proved an excellent sea boat, a very serviceable and successful vessel.*

### HARVEST QUEEN
### Hull 119

It is difficult today to visualize a three-masted square-rigged sailing vessel, all of 114 feet in length, built for carrying general cargo on the high seas. A craft of this description and size would more likely have been schooner-rigged and would be found on coastwise routes. But *Harvest Queen* was such a vessel. She was launched in June of 1860 and completed late in July for C. B. Fessenden. Her first registry document was issued on the 25th of that month. She was a 315-ton bark and measured 114 feet by 26 feet 6 inches by 11 feet 3 inches, dimensions that are typical today of a harbor tug. Details of the career of *Harvest Queen* are unclear, except that she operated out of New York during the early part of the Civil War. After the war, her home port was changed from New York to San Francisco, whence she traded under the ownership and command of Captain R. C. Johnson. She was dropped from United States registry in the mid-1870s.

*Built in 1860 for general freighting business. Proved a good sea boat and satisfactory vessel. Draft very deep loaded 10 ft. 2 in.*

## *MISSISSIPPI*
### Hull 120

The sidewheel steamer *Mississippi* was ordered in 1860 by Samuel L. Mitchill for the New York & Savannah Steam Navigation Company, but, upon the outbreak of the Civil War and the cessation of nearly all coastwise shipping between northern and southern ports, she never ran for that company. Instead, she saw service as a naval vessel during the war and ran to Brazil for a decade after peace came in 1865. She was a vessel of 2026 tons, measuring 250 feet by 38 feet by 22 feet 6 inches. Her propulsion plant consisted of a beam engine from the Morgan Iron Works, having a cylinder 80 inches in diameter and a piston stroke of 11 feet. She was launched as *Mississippi* in January of 1861, and a trial trip was run on the 22nd of May. On 18 July 1861, the completed vessel was purchased by the United States Navy for $200,000 and converted for purposes of war. Four 32-pounders and a 12-pound rifle were fitted and she was commissioned as *USS Connecticut* on the 23rd of August. Her wartime activities included service transporting men and supplies, convoy duty and a year with the North Atlantic Blockading Squadron. On two successive days, 9 and 10 May 1864, she captured the British steamships *Minnie* and *Greyhound*, yielding valuable cargoes of cotton, tobacco, turpentine and gold, as well as the celebrated Confederate spy, Belle Boyd.

Her wartime service completed, *Connecticut* was sold at auction on 21 September 1865 to Daniel B. Allen. Allen and his partner, Cornelius K. Garrison, had a short time before formed the United States and Brazil Mail Steamship Company, to provide a monthly mail service from New York to Rio de Janeiro, with northbound calls at Bahia, Pernambuco, Para and St. Thomas. *Connecticut*, now renamed *South America*, departed from New York for Rio de Janeiro on her first voyage the Brazil Line on 29 November 1865. She made regular passages every three months. Upon her return to New York from her 39th voyage on 20 October 1875, *South America* was laid up, and the line was wound up a month later. She was towed to Boston to be broken up in September of 1878.

*Built for passenger business New York to Savannah — bought when new by the United States Gov't at the outbreak of the Rebellion, and converted into a war cruiser mounting several guns. Was a very handsome vessel, proved an excellent sea boat, very fast and was a great favorite — had very light draft of water — draft when launched (only one anchor chain on board) was forward 5 feet 10 in., aft 6 feet 7 in. This vessel was later named the South America and ran as a mail packet New York to Brazil and was the favorite vessel in the service.*

## *ALEXANDER MARSHALL*
### Hull 121

*Alexander Marshall* was the last but one in the long line of sailing packets built by Mr. Webb for C. H. Marshall & Company for service on the Black Ball Line between New York and Liverpool. Launched in November of 1860, she was to be the last sailing vessel to come from the Webb yard for more than eight years. Rigged as a ship, she was of 1177 tons and measured 179 feet by 37 feet 10 inches by 27 feet 6 inches. Her first register document was issued on 20 December 1860, and her maiden voyage commenced the same day. The vessel's first master has been reported erroneously by some sources as having been Edward C. Marshall, a brother of Charles H. Marshall, but she was actually in charge of Charles A. Marshall, a son of Alexander C. Marshall, for whom the packet was named. Alexander, whose death had occurred earlier in the year, was another brother of Charles H., and had been master of several early Black Ball Line packets before retiring from the sea in the 1840s.

*Alexander Marshall* was to remain in service under the company's ownership almost until they gave up the operation of packets in 1881. Her last scheduled voyage on the Black Ball Line ended at New York in March of 1879. She then made a tramping voyage to London and commenced a second such trip on 29 January 1880 with Captain William A. Gardner in command. Captain Gardner had relieved Charles A. Marshall as the vessel's master in the summer of 1869, when the latter was put in charge of the newly-built packet *Charles H. Marshall*, the last vessel built by Mr. Webb. The cargo of *Alexander Marshall* consisted of 457,979 gallons of refined petroleum and 14,561 gallons of lubricating oil, all in wooden barrels. About two weeks out, she encountered gale winds and mountainous seas, which completely dismasted her and left her leaking badly. The crew manned the pumps and threw barrels of oil over the side in an attempt to keep the vessel afloat by lightening her. On the night of the 17th of February, after the pumps had run continuously for five or six days, the lights of the Danish steamship *Thingvalla* were sighted. The Danish vessel saw *Alexander Marshall*'s distress signals and hove to until morning. A

The steamship *Mississippi*, ordered by Samuel L. Mitchill for his New York & Savannah Steam Navigation Company, never ran for that line, but served during the Civil War as *USS Connecticut*—a more appropriate name for a Union naval vessel. She is shown here in a heavily retouched photograph in her naval guise. (Courtesy of The Mariners' Museum, Newport News, Virginia)

lifeboat was sent to the stricken ship in heavy seas and picked up most of the crew, who were forced to jump from the stern of their ship. Several other crew members managed to launch one of the stricken vessel's boats and reached the Danish steamer. *Thingvalla* landed the exhausted men at Yarmouth, on the Isle of Wight, on the 23rd. On the 13th of April, the Belgian steamship *Plantyn* passed several pieces of bulwark and a nameboard marked with the unfortunate vessel's name at a position over 500 miles from where the rescue had taken place.

### *CONSTITUTION*
### Hull 122

The wooden sidewheel steamship *Constitution* of 1861 was the first vessel of the second generation of the Pacific Mail Steamship Company fleet, the company's first new-

building since 1854 and their first from the Webb yard since the unfortunate *San Francisco* was delivered late in 1853. She was also the largest steamship built at New York since the Collins Line's *Adriatic* came out of George Steers' yard in 1856. *Constitution* was a vessel of 3315 tons, measuring 340 feet by 45 feet by 31 feet 6 inches. Her beam engine from the Novelty Iron Works had a cylinder 105 inches in diameter with a 12-foot piston stroke, perhaps the largest such engine ever installed in a steamship up to that time. Most of the later Pacific Mail steamships would have machinery of the same size and type. Unlike the other Pacific Mail steamers built by Mr. Webb, she had two stacks, with two boilers forward and two aft of her paddle wheels. The vessel was launched on the morning of 25 May 1861 and completed about 16 November 1861 when her first register was issued. She was chartered to the United States Government for several months as a troop transport, and finally sailed from New York to take up her duties on Pacific Mail's coastal service from Panama to San Francisco on 29 June 1862. She also ran on the company's Pacific Northwest route.

On 5 October 1877, while about 50 miles from San Francisco on a southbound voyage from Victoria, British Columbia, *Constitution* was found to be afire in the vicinity of her after boilers. She proceeded through the Golden Gate and her after hold was pumped full of water, sinking the vessel to her guards, putting her hard aground on the Mission Flats and extinguishing the fire. Over the next two weeks she was raised, the cargo and baggage removed and a survey made of the damage. The cost of repairs was nearly the value of the ship, and she was cast aside on the mud flats, or, as the contemporary marine press so colorfully stated after the survey, "she will be put in the bone yard." In November of 1878, *Constitution* and her fleetmate *Colorado* were sold to Thomas P. A. Whitelaw for demolition.

## RE D'ITALIA
### Hull 123

The single-screw ironclad frigates *Re d'Italia* and *Re Don Luigi di Portogallo* were built for the Italian Navy in 1864. The construction of these vessels was a direct result of the partial confederation of Italian states in 1860, in which a strong navy was perceived as a necessity. The reputation enjoyed by Mr. Webb from building *General Admiral* for the Russians, led the Italians to his shipyard and the signing of a contract for the two powerful warships. Much of the technology of the Russian frigate went into the Italian vessels. *Re d'Italia* had a displacement of 6150 tons and her hull measurements were 282 feet 3 inches by 54 feet 2 inches by 33 feet 7 inches. Her rig was that of a bark. The same propulsion plant was fitted as in *General Admiral*. Six boilers supplied steam to two horizontal back-acting engines with cylinders 84 inches in diameter and a piston stroke of 45 inches. These engines turned a single propeller 19 feet in diameter. As was the practice on war vessels that were fitted for both steam and sail, the propeller was arranged so that it could be raised from the water when under sail. Her single telescoping smokestack could be lowered when using sail power. *Re d'Italia* was laid down on 21 November 1861 and launched on 18 April 1863. A trial trip was held on 18-19 February 1864, after which the vessel returned to the harbor and was moored at Ford's Wharf, Brooklyn, for last minute adjustments to her machinery and to load

stores for her transatlantic delivery voyage. She left New York on the 8th of March and arrived at Naples on the 2nd of April after a passage of 25 days. Her commissioning took place on 14 September 1864. In the political turmoil that troubled Europe in the mid-19th century, Italy found itself allied with the Prussians in the Austro-Prussian War of 1866. With the future of Venice at stake, Admiral Tegethoff and the Austrian fleet engaged a numerically superior Italian naval force of ironclads and lesser vessels on 20 July 1866 at the Island of Lissa (now Vis) on the present-day Yugoslavian coast. Admiral Persano, the commander of the ill-trained Italian fleet, made a number of tactical blunders that resulted in the ramming of *Re d'Italia* by the Austrian ironclad *Erzherzog Ferdinand Max* and her subsequent loss. The less-than-illustrious career of *Re d'Italia* had lasted two years and three months.

*Iron-clad 32 broad-side guns built for the Royal Italian Gov't., draft when launched, forward 11 ft. 1-1/2 in., aft 15 ft. 3 in. and was with all iron cuirass fitted, engines and machinery complete including duplicates, anchors, chains, hausers [sic], boilers & Water Tanks filled, forward 18 ft. 6 in., aft 20 ft., with 700 tons of coal in bunkers, forward 22 ft. 5 in., aft 21 ft. 6 in., with extreme after hold and after shell rooms filled with coal 1010 tons in all[,] boilers half filled, forward 21 ft. 10 in., aft 22 ft. 9 in., with everything including three months stores and provisions for 300 men, all the boilers filled and 8 guns on board was forward 22 ft 6 in., aft 23 ft. 8 in.—weight of hull only 2570 tons, made voyage in winter with heavy weather from New York to Naples 5300 miles in twenty days being the first iron clad vessel to cross the Atlantic Ocean without an escort.*

## RE DON LUIGI DI PORTOGALLO
### Hull 124

A sister ship to *Re d'Italia*, the ironclad frigate *Re Don Luigi di Portogallo* was named for King Louis I of Portugal, who had succeeded to the throne of that country in 1861. Her dimensions, 279 feet 9 inches by 54 feet 2 inches by 33 feet 7 inches, appear to be slightly different from those of her sister, but they were essentially identical with the same propulsion plant, rig and armament. *Re Don Luigi di Portogallo* was laid down alongside *Re d'Italia* in December of 1861 and launched on 29 August 1863. On the morning of 3 August 1864, she went to sea for a trial trip, returning to New York the same afternoon. Her delivery voyage commenced several days later, the passage from New York to Naples taking 20 days and 22 hours. Although naval reference sources state that she was commissioned on 23 August 1864, the date corresponds approximately to that of her arrival in Italy. The vessel's career was similar to that of her sister, with little to note until the Battle of Lissa on 20 July 1866. As a part of the Italian fleet of Admiral Persano, in which his vessels were scattered and without an overall plan of battle, *Re Don Luigi di Portogallo* and her sister may not have had the opportunity to prove themselves as warriors. The Austrian fleet, under Admiral Tegethoff, consisted of seven ironclads and one elderly frigate, but the tactics used by Persano were so poor that he was roundly defeated. Persano was later brought to trial on charges of incapacity, negligence, disobedience, cowardice and treason, but was convicted of only the first two. *Re Don Luigi di Portogallo* was rammed by the Austrian *Kaiser Max*, but it was not a mortal wound. Her later career was quiet, and she was stricken from the Italian Navy on 31 March 1875. The Battle of Lissa demonstrated that the ironclad warship, well protected from damage by shot, was vulnerable to the age-old act of ramming. Despite the fact that the ram bow persisted in the navies of the world until the 20th century, ramming rapidly became obsolete as a tactical maneuver as the fire power of warships and their ability to destroy targets from a distance improved.

### *GOLDEN CITY*
### Hull 125

Mr. Webb built the sidewheel steamship *Golden City* for the Pacific Mail Steamship Company early in 1863. Similar to *Constitution*, built at the Webb yard two years before, but with all of her boilers forward of the paddle wheels, she had a tonnage of 3373 and measured 343 feet by 45 feet 1 inch by 30 feet. Propulsion power was provided by a beam engine having a 105-inch cylinder and a piston stroke of 12 feet. Her hull construction was typical of that of large wooden vessels of the day, with solid floors, watertight bulkheads and double-diagonal iron cross bracing. Frames and bottom planking were of white oak; top timbers of hackmatack; ceiling, beams and side planking of pitch pine; and deck planking of white pine. Iron and "yellow metal" were used for inside fastenings, while those exposed to salt water and salt-laden air were copper and galvanized iron respectively. *Golden City* was launched on 24 January 1863, and was towed to the Novelty Iron Works, at the foot of East 12th Street (a short distance north of the Webb yard) for the installation of her machinery. She was completed in August of that year (receiving her first register document on the

11th,) and departed shortly thereafter for the Pacific by way of the Straits of Magellan. After an uneventful delivery voyage, she took her place on the company's Panama-San Francisco route. The career of *Golden City* on the Panama route was quiet until her 53rd voyage. She left San Francisco on 18 February 1870 with 400 passengers and a valuable cargo which included gold from the California fields and tea transshipped from one of the Pacific Mail's transpacific steamships. All went smoothly until the morning of the 22nd, when she stranded in calm seas and a dense fog at Cape San Lazaro, about 180 miles north of the tip of Baja California. Passengers, baggage and the $791,000 in gold that was a part of her cargo were brought safely ashore and a distress signal was displayed. By the 24th, *Golden City* had begun to break up in the worsening sea, and the northbound Pacific Mail steamer *Colorado* spotted the signal, but could not land her small boats because of the surf. While Captain Comstock and several crew members remained behind to guard the specie, the passengers and the rest of the crew walked overland to a sheltered place where it was possible to embark on *Colorado*. They landed at San Francisco on the 1st of March. The *New York Times* reported on 3 March 1870 that,

> The passengers had no confidence in Captain Comstock and paid no attention to his orders. They had an indignation meeting aboard the Colorado, and passed resolutions to the effect that, in their opinion, the wreck of the Golden City was due to the incapability or gross carelessness of the commanding officer, and that much unnecessary suffering on shore was to be traced to the same source.

Although the passengers endured some hardship while at Cape San Lazaro, their resolutions seem to have been rather hard on Captain Comstock. Navigation on the Baja California coast was not an easy undertaking in the mid-19th century. Frequent dense fogs and a bare minimum of aids to navigation made the area a graveyard of ships. One entertaining, but potentially nasty incident occurred prior to the overland march. It was reported that, as the ship broke up, casks of alcoholic beverage floated ashore and were promptly broken open by some of the shipwrecked party. In the wake of the drunkenness and rowdyism that ensued, persons unknown of a more temperate bent smashed casks as they reached the beach. Could it have been that at least a part of the antagonism directed towards Captain Comstock was somehow related to the destruction of these casks?

*Built in 1862 [sic] for the San Francisco and Japan trade,*

*Golden City* is shown at anchor in San Francisco Bay in this undated photograph. Her seven-year career, from 1863 to 1870, was spent on the Pacific Mail's coastwise express route between San Francisco and Panama. (Collection of the late Cedric Ridgely-Nevitt)

*passengers and freight. Proved an excellent sea boat, large carrier and a very successful vessel. Draft of water when launched, forward 7 ft. 0 in., aft 8 ft. 6 in. When leaving New York for Panama, deep loaded, forward 17 ft. 4 in., aft 18 ft. 4 in. Made the passage from Panama to San Francisco in the year 1865 (3,260 knots [sic]), with full passenger list and 2,000 tons of freight on board, consuming only 485 tons of coal, average 36-1/2 tons per day. Draft forward 17 ft. 6 in., aft 18 ft. 3 in.*

### SACRAMENTO
### Hull 126

The hull of *Sacramento*, a sidewheel steamship ordered from Mr. Webb by the Pacific Mail Steamship Company shortly after *Golden City*, was built under subcontract by his brother, Eckford Webb, of the firm of Webb & Bell. Slightly smaller than *Golden City*, she measured 304 feet by 42 feet 6 inches by 29 feet 2 inches and her tonnage was 2647. The Novelty Iron Works supplied a beam engine which had a cylinder 103 inches in diameter (two inches less than those on Pacific Mail's line-haul steamers) and a piston stroke of 12 feet. *Sacramento* was launched in May of 1863 and not delivered until June of 1864. After she received her register document on the 10th of June she proceeded to the west coast and entered the owner's service between Panama and San Francisco. Both her life and her loss were remarkably similar to those of

*Golden City*. After eight uneventful years plying the dangerous coastal route, she stranded on 5 December 1872 on a reef near San Geronimo, Baja California, about 190 miles south of San Diego. The accident, the sixth to befall a Pacific Mail steamship over a period of a few months, occurred in dense fog and calm seas, as in the case of her late running mate. All hands came ashore safely, and the second officer sailed a small boat to San Diego to bring news of the wreck. He arrived on the 9th, and on the following day, the company's steamship *Montana* was sent to the scene. All passengers and most of the crew were embarked on *Montana*, which landed them at San Diego on the 12th. The captain and 25 crew members stood by at the wreck site, intending to save as much of the cargo as possible, but *Sacramento* broke up on the reef as the weather worsened, and little was salvaged other than the $335,648 in specie that had been in her strong room.

*Built in 1863 for the Pacific Mail S. S. Co. for the trade between San Francisco and Japan. Proved a good sea boat and very serviceable vessel.*

### COLORADO
### Hull 127

*Colorado* was built for Pacific Mail Steamship Company's coastwise express service between Panama and San

Francisco. She was a vessel of 3357 tons, and measured 340 feet by 45 feet by 31 feet 6 inches. *Colorado* was launched on 21 May 1864, and spent nearly a year in fitting out. She was completed in March of 1865 and her first register document issued on the 29th of that month. She departed on the 1st of April for the west coast by way of Rio de Janeiro, the Straits of Magellan and Callao, arriving in Panama on the 12th of June. *Colorado* was chosen to inaugurate Pacific Mail's transpacific route, pending the delivery of the magnificent steamers *Great Republic*, *China*, *Japan* and *America*. Of these, only the second was constructed by Mr. Webb; Henry Steers built the others at his Greenpoint yard. To prepare *Colorado* for her new service, she was fitted with a mizzen mast. The line's coastwise liners were two-masted; the trans-oceanic vessels had three. Additional coal bunkers were also provided for the long-haul route.

Pacific Mail's first transpacific voyage commenced on 1 January 1867 with *Colorado*'s departure from San Francisco for Yokohama and Hong Kong. When she made her first call at Hong Kong on 31 January 1867, *USS Hartford*, then on Asiatic station, saluted her with appropriate ceremony. She made a total of 18 transpacific voyages from 1867 onwards, with coastwise voyages to Panama interspersed in 1868-69 and again in 1875-76. On 19 May 1876, she arrived at San Francisco for the last time and was laid up. After a relatively short and uneventful career, *Colorado* (along with Webb's *Constitution*) was sold to Thomas P. A. Whitelaw in November of 1878 to be broken up. Her demolition was not completed until mid-1881 when she was burned for her metal fastenings. Along with the Pacific Mail's other wooden sidewheel steamers, this majestic vessel was retired before her time as a result of the appearance of modern American-built iron screw steamships.

### *DUNDERBERG*
### Hull 128

Despite the fact that Mr. Webb's celebrated clipper ships were outstanding vessels of their type, the ironclad steam ram *Dunderberg* may have been his most significant contribution to American shipbuilding. Officially termed an "ocean-going ironclad frigate ram," she was ordered by the United States Navy on 3 July 1862, in the wake of the battle at Hampton Roads between *USS Monitor* and *CSS Virginia* (ex-*USS Merrimack*,) at a contract price of $1,250,000. Few shipbuilders in the country would dare tackle a vessel of her great size. She measured 377 feet 4 inches by 72 feet 10 inches by 22 feet 7-1/2 inches, and

was of unusually heavy wooden construction. Her displacement was 7,000 tons, including 1,000 tons of armor and an equal amount of coal. A two-cylinder back-acting engine of 5,000 horsepower, built by John Roach's Etna Iron Works, gave her a sea speed of 15 knots. The cylinders measured 100 inches in diameter and the piston stroke was 45 inches. Six main and two auxiliary boilers supplied steam. Her propeller was 21 feet in diameter and weighed over 15 tons. In appearance, she resembled *CSS Virginia*, with her low hull and sloping-sided iron casemate that provided protection for her heavy battery, engines and boilers. The battery consisted of four 15-inch Rodman guns and fourteen 11-inch Dahlgren guns. Originally, she was to have been fitted with two Timby revolving turrets atop the casemate, but these were deleted prior to construction. As she was being built, many local sages and armchair shipbuilders predicted that she would never be launched, while others confidently prophesized that when she was launched she would go straight to the bottom and never reappear. But *Dunderberg* was successfully launched at 9:03 on the morning of 22 July 1865 with more than 900 persons on board and before a crowd estimated to number 20,000, that filled the shipyard, the neighboring piers and many spectator vessels. Fitting out the vessel with machinery, armor and armament occupied many months.

The Civil War having ended two months before the launching, the Navy had little need for the vessel. Attempts were made to sell her to the Chilean and Peruvian governments, but the officialdom of Washington would not permit Mr. Webb to proceed with the sale. At Mr. Webb's request, *Dunderberg* was at last released to him by an Act of Congress of 2 March 1867, and he refunded to the Government the sum of $1,092,887.73 in progress payments. She was immediately sold to the French Government for $2,500,000, and, as *Rochambeau*, was to have become a formidable unit of the French Navy. However, because of the political upheaval in France that accompanied the fall of Napoleon III and the start of the Third Republic, *Rochambeau* was in commission for only a short time in 1870, and was disposed of in 1872. The revolving turret that was the most visible feature of *USS Monitor* had been chosen by the United States Navy for its large fleet of wartime ironclad gunboats (which were termed "monitors" after the Ericsson craft,) and later this novel method of mounting heavy armament was used with success in nearly every type of major combatant vessel. The other features of *Dunderberg*, an elongated ram bow and topsides with exaggerated tumble-home, profoundly influenced French naval design in the 1870s.

ADMIT THE BEARER,

ON BOARD THE

# DUNDERBERG,

ON LAUNCHING DAY.

## July 22d, at Nine o'clock, A. M.

*Please present this Ticket at Entrance Gangway.*

W. H. WEBB.

When the ram *Dunderberg* was launched on 22 July 1865, the crowds were expected to be so large that printed tickets were required for guests to gain access to the shipyard. As predicted, many thousands took advantage of every available vantage point to watch the spectacle. This ticket was found in Mr. Webb's scrapbook, a part of Webb Institute's collection of William H. Webb's personal papers.

*Dunderberg* is depicted as a state-of-the-art example of the technology of the 1860s. As she was being fitted out, the *New York Times* echoed the popular impression of the vessel as an indomitable weapon of war on 14 July 1866, when it stated,

> The Dunderberg will be the largest and most complete vessel of her kind in the world, and it is thought will be able to overcome anything with which she comes in contact.

But her design had its genesis prior to that of John Ericsson's revolutionary *USS Monitor*. *Dunderberg* was heavily built and heavily armed, but she never had the opportunity to display her prowess in battle, and so we shall never know how successful she might have been. Had "Thunder Mountain" been at Hampton Roads rather than *Monitor*, and the battle fought between two casemated vessels, or had she been built with the originally proposed Timby revolving turrets in addition to her other armament, would the naval history of the Civil War have read the same? Or would naval warfare have evolved as it did in the late 19th century? The comparative vital statistics of *Dunderberg* and *CSS Virginia*, *USS Monitor*'s opponent at Hampton Roads, suggest that the famous battle might have resulted in a decisive victory for the Union Navy rather than the stalemate that it was.

> *Method or plan adopted for launching Ram Dunderberg and other long and sharp vessels, the launching ways being laid on a circular line, so as to secure an early immersion of the after end of the vessel and thus obtain buoyancy for that part sooner than if the ways were laid on a straight line and so prevent hogging.*

[Note: Although *Dunderberg* is considered to be one of the most important vessels built by William H. Webb, the "Book of Plates" does not contain any of his notes or observations except in relation to the declivity of the vessel's launching ways quoted above. Launching ways "laid on a circular line" have been used many times since the launch of *Dunderberg* for the reason stated by Mr. Webb.]

### *HENRY CHAUNCEY*
### Hull 129

*Henry Chauncey*, named for an early backer of the Pacific Mail Steamship Company, was built for service on the company's coastwise express line between Panama and San Francisco. She was one of three vessels ordered in 1864 (the others being *Montana* and *Arizona*) to replace some of Pacific Mail's older ships. *Henry Chauncey* and

*Montana* were contracted for by Mr. Webb, but only the former was built by him; he subcontracted *Montana* to his brother, Eckford Webb, and *Arizona* was built by Henry Steers. *Henry Chauncey* was a sidewheel steamship of 2656 tons, measuring 319 feet 5 inches by 43 feet by 28 feet. She was fitted with the standard Pacific Mail beam engine, measuring 105 inches in diameter by 12 feet stroke. She was launched in October of 1864 and departed from New York on her maiden voyage on 1 November 1865. *Henry Chauncey* never made it to the Pacific but spent her entire career running between New York and Aspinwall, whence travelers and freight crossed the Isthmus to awaiting steamers at Panama. Most of her passages were uneventful, but she broke her port paddle-wheel shaft shortly after leaving Aspinwall on 2 June 1870, and limped back to that place on one wheel. On the 14th, two days after her estimated arrival time, she was listed as overdue at New York, but arrived there on the 28th. She had left Aspinwall on the 7th, and it is assumed that she made the voyage using sails, her starboard wheel and hard left rudder. Meanwhile, the reserve steamship *Ocean Queen* was hurriedly placed in commission to make *Henry Chauncey*'s scheduled departure on the 21st. After an extensive repair, the ship returned to her route on 20 December 1870.

In August of 1871, *Henry Chauncey* was an innocent participant in a hoax involving, in the words of the *New York Herald*, "base trafickers in false alarms in the neighborhood of Wall Street." The vessel had departed from New York for Aspinwall on 15 August 1871. On the 18th, several New York newspapers published a report that the vessel was afire and beached at Bodie Island, North Carolina, basing their stories on copies of a letter from one James Brown, of Norfolk, addressed to the Sun Mutual Insurance Company, reporting the ship's destruction. Pacific Mail stock dropped three and one-half percent that day and an additional two percent the day following. Meanwhile, the insurance company denied knowing a "James Brown" or having received the letter, and the Associated Press at Norfolk remarked, "Nothing is known here; not even a rumor." In addition, the wrecking steamer *RESOLUTE,* which "James Brown" had claimed was standing by the stricken vessel, was sighted on the morning of the reported accident passing Fortress Monroe inbound with a waterlogged schooner in tow. It was soon determined that the "base trafickers in false alarms" had been engaging in what the *Herald* alliteratively called "scandalous stock speculating." Pacific Mail posted a $5,000 reward for the arrest and conviction of the perpetrators and *Henry Chauncey* arrived, safely and on schedule, at Aspinwall. On 9 December 1875, *Henry Chauncey* commenced the last voyage between New York and Aspinwall by one of the company's sidewheelers. When she returned to New York she was laid up, her place on the line taken by an iron-hulled screw steamship. In July of 1877 she was sold for scrap, and the era of the wooden sidewheel steamer in the Pacific Mail fleet finally came to an end.

*Built in 1864 for Mail Service, Panama and San Francisco. Draft of water when launched, with anchors and chains, most of cabin joiners work, forward 6 ft. 10 in., aft 8 ft. 6 in. Draft on leaving Panama on first voyage with freight (measurement and weight) 900 tons; stores 285 tons; baggage 200 tons; coal 666 tons; passengers 930; crew 139, was forward 17 ft. 6 in., aft 16 ft. 11 in. Proved a fast, serviceable and popular ship.*

### *MONTANA*
### Hull 130

In 1864, the Pacific Mail Steamship Company contracted for the construction of three large wooden-hulled sidewheel steamships, *Henry Chauncey, Montana* and *Arizona*. Mr. Webb was to build the first two vessels and Henry Steers, the third. Because of the backlog of work at the Webb yard, *Henry Chauncey* alone came the ways at the foot of Sixth Street. *Montana* was built across the East River in Greenpoint by his brother Eckford Webb, of the firm of Webb & Bell, in whose shipyard she was launched on 28 February 1865. She was a sister of *Henry Chauncey*, having a tonnage of 2676 and measuring 320 feet by 43 feet 7 inches by 27 feet 4 inches. The Novelty Iron Works built her beam engine, which had a cylinder measuring 105 inches in diameter with a piston stroke of 12 feet. The vessel did not receive her first register document at New York until 20 July 1866, nearly 17 months after her launching. She left shortly afterwards left for her Pacific duties. Constructed for the Pacific Mail's express line from Panama to San Francisco, she ran on that route until the summer of 1869. By that time the 'Golden Spike' of the Pacific railroad had been driven at Promontory Point, Utah—the Central Pacific, from the west, and the Union Pacific, from the east, having met at that place on the 10th of May. Traffic across the Isthmus and up the coast on Pacific Mail steamers decreased to the point where maintaining the express passenger service caused a serious drain upon the company's finances. Thereafter, *Montana* was used mainly on their local service on the coast of Mexico and California. In February of 1872 she was chartered by Webb's Australian Line to take the mails from San Francisco to Honolulu and return with part of the cargo of the eastbound steamship from Australia. *Moses Taylor*, which ran Webb's regular service between San Francisco and Honolulu, was of insufficient size to handle the large eastbound shipments, which consisted mainly of wool. In common with the other side-wheelers in the Pacific Mail's fleet, *Montana* was retired in 1877, when her place taken by iron-hulled screw steamers. Her machinery was then removed and her hull sold for scrap.

## *BRISTOL*
## Hull 131

*Bristol* and her identical sister *Providence* were the only large steamboats for inland waters ever built by Mr. Webb at a time when there were scores of such vessels running out of New York alone. They were ordered in 1866 by the Merchants Steamship Company, a recent consolidation of the Neptune Line from Providence and the Stonington Line, for a new service between New York and Bristol, Rhode Island, whence passengers would continue to Boston by train. Within a 12-month period, the Merchants company suffered the loss of three of its steamers, and, after suspending operations, was succeeded by the Narragansett Steamship Company in 1867. This company purchased the two uncompleted steamboats, which measured 360 feet by 47 feet 2 inches by 16 feet. Their tonnage was 2962, and their accommodations were commodious and elegant, the grand saloon being 275 feet long, 28 feet wide and 21 feet high. They were powered by 2,800-horsepower beam engines built by John Roach's Etna Iron Works. The engine's single cylinder, 110 inches in diameter with a piston stroke of 12 feet, was the largest of the type constructed up to that time, and was not surpassed in size until the engine for the Fall River LIne's *Pilgrim*, with the same cylinder diameter and a 14-foot stroke, was built in 1883. *Bristol* (which, according to some sources, was to have been called *Puritan* by the Merchants company) was launched on 4 April 1866. She and *Providence* were placed on their route on Monday, 17 June 1867. The owning company was subjected to a complex sequence of corporate changes over the years, but *Bristol* and *Providence* continued to serve travellers between New York and New England on what may be termed the Fall River Line for many seasons.

Life on the route through Long Island Sound was anything but placid. Late in 1876, *Bristol*'s passengers and crew were subjected to a harrowing experience. She left New York for Newport, Rhode Island, on the afternoon of the 9th of December. At about 11 o'clock, opposite New Haven, her rudder chain broke. Termporary repairs were made, but the chain parted four more times. Attempts were made to steer her with a jury rig at the tiller, but the ropes soon broke. The wind was high and the seas unusually rough that night, and *Bristol* drifted at their mercy for nine hours while the vessel was steered with the jury gear and the engine backed frequently to keep her headed into the weather. Eventually *Bristol* arrived at Newport, where her weary and discommoded passengers disembarked and repairs were made to the steering gear. In other incidents, she ran into and sank a bark in 1869, collided with the schooner *Fred Warren* in 1872, sank the bark *Bessie Rogers* the same year, grounded near Newport in 1874, brushed against the notorious rocks of Hell Gate in August of 1875, collided with an unidentified schooner two months later, and ran aground once more at Newport in 1877. Nearly all of these events occurred in dense fog. That *Bristol* and her fleet mates were able to maintain the line's service at all under such conditions—with only the skills of their pilots to guide them—is quite remarkable. It is currently fashionable to speak of occupational stress; that of the steamboat pilot, who was obliged to rely almost entirely upon his memory and a keen sense of hearing while his vessel was enshrouded in fog, may be ranked with the most stressful of today's occupations.

*Bristol* was destroyed in a spectacular fire at her Newport pier on 30 December 1888. She arrived from New York at about three o'clock that morning on the last trip of the season. At about six o'clock, flames were seen and the Newport Fire Department summoned, but after an hour, the fire was out of control. Fortunately, there was little or no wind, for the line's steamers *City of New Bedford* and *Pilgrim* were moored nearby; these vessels and the pier received little, if any, damage. With the exception of her paddle-wheel boxes, the superstructure of *Bristol* was completely consumed, but her hull timbers were so impregnated with salt (as were the wheel boxes) that they would not burn. What remained of her hull later sank and, on 25 January 1889 was raised and moved to a nearby location for removal of her engine.

*Were similar in model, engines, and all other respects; built for the route, New York to Newport [sic], via Long Island Sound. Launched in the year 1866. Furnished accommodations for 1,200 [sic] night passengers, and room for a large quantity of deck freight. Had unparalleled speed of twenty miles per hour continuous the whole route, and gave very great satisfaction, and for many years the most popular vessels ever built for the service.*

### *PROVIDENCE*
### Hull 132

The sidewheel steamer *Providence* was a sister to *Bristol*. She was ordered in 1866 by the Merchants Steamship Company and completed during the following summer for the Narragansett Steamship Company for service between Bristol, Rhode Island, and New York. She was of the same dimensions as her sister, 360 feet by 47 feet 2 inches by 16 feet, and of the same tonnage, 2962. Her beam engine, built by the Etna Iron Works, of New York, had a single cylinder 110 inches in diameter and 12 feet stroke. Speaking of these two remarkable steamboats, the *Fall River Line Journal* stated, on 31 May 1909:

> "The 'Bristol' and 'Providence' in their construction were so far advanced of the type of steamboats heretofore built that they were looked upon as marvels and their fame was world-wide. They were larger, more powerful, and carried more passengers than any of their rivals . . . An additional deck upon these boats permitted a gallery tier of rooms. When they were building they attracted much comment because of the great height from the keel to the dome deck, and some folks declared that they would be top heavy. They were lighted by gas, and later on steam heating and steam steering gear were installed."

*Providence* (which was to have been named *Pilgrim* by Merchants, in the opinion of some sources) was launched in August of 1866. She and *Bristol* entered service on Monday, 17 June 1867, with one vessel departing from New York and the other from Bristol. *Providence* appears to have led anything but a quiet life. She collided with the schooner *S. D. Hart* in dense fog in 1874, the bark *Ocean Gem* in 1876, an unidentified schooner in 1877, the New York pilot boat *Pet* in 1885, a coal barge the following year, the steam yacht *Adelaide* in 1888, and the schooner *Benjamin A. Van Brunt* in 1892. In addition, she was struck by an immense wave in 1875 which smashed her port paddle wheel box and flooded her main deck, broke her shaft in 1880 and suffered a near-fatal grounding near Bristol in 1887. Otherwise, *Providence* was more fortunate than her sister and survived longer than any of Mr. Webb's steamships. She was towed to Boston in the autumn of 1901 and there dismantled. Her hull was later burned for its metal fastenings. Her last years had been spent in semi-retirement, *Providence* having been succeeded on the line by the famous, more modern iron or steel sidewheelers—*Pilgrim*, *Puritan*,

*Plymouth* and *Priscilla*—for which the Fall River Line is best remembered. However, *Bristol* and *Providence* provided the ultimate in speed and comfort during their halcyon days on the line.

### *CHINA*
### Hull 133

If the clipper *Young America* was Mr. Webb's masterpiece in sail, then the magnificent sidewheeler *China* was the pinnacle of his merchant steam vessel designs. With a tonnage of 3836 and measuring an impressive 360 feet by 48 feet by 31 feet 9 inches, the three-masted, three-decked, bark-rigged *China* was one of the largest wooden sidewheelers ever built for ocean service. Her overall length of 376 feet exceeded that of the Collins Line's *Adriatic*, the largest wooden steamship when built in 1856, by 31 feet, although that vessel's tonnage was no less than 4145. *China*, one of four vessels specially built to serve Pacific Mail's transpacific route, was surpassed in tonnage by her nearly-identical sister *Great Republic*, built by Henry Steers at Greenpoint and admeasured at 3881 tons. The remaining two vessels, also built by Henry Steers, were *Japan* and *America*, slightly longer and having about 20 inches greater beam than the first pair, measured 4351 and 4454 tons respectively. *America* had the greatest tonnage of any wooden steamship.

*China* was ordered in 1865 by the Pacific Mail Steamship Company and launched as *Celestial Empire* at 9:40 on the morning of 8 December 1866. Her original name honored China (her sister ship, *Great Republic*, launched a few weeks before, celebrated the United States). The launching was witnessed by an estimated 2000 and 3000 persons, who braved a driving rain to watch the vessel enter the waters of the East River. No less than 500 persons, invited aboard for the launching, "cheered lustily," in the words of the *New York Sunday News*, as the enormous wooden hull slid gracefully down the ways. The *New York Times* reported somewhat less colorfully that they had "cheered loudly." The vessel, fitted with her three masts, was decorated with flags and bunting for the event. An American flag flew at her stern and a union jack forward. The blue and white Pacific Mail house flag was at the foremast head; a red and white banner proclaiming her name flew on the mainmast; and another American flag was at the mizzen. Once she was waterborne, several tugs towed her the short distance to the wharf of the Novelty Iron Works, where her machinery was to be installed. The morning's events were exciting

This idyllic scene shows the Fall River Line's steamer *Providence* steaming serenely and majestically as she leaves her Eastern terminus for New York on what must have been a warm summer evening. Note the tromp l'oeil decoration of her paddlebox (repeated on her sister *Bristol*) that depicts the interior of a domed chamber.

to all, but especially to a ten-year-old boy who, spellbound from watching the immense structure of *China* enter her native element, fell from a pier; he was quickly rescued by an alert newspaper reporter.

*China* was propelled by a beam engine having a cylinder 105 inches in diameter with a piston stroke of 12 feet, for which steam was supplied at a pressure of 30 pounds per square inch by four return-tube boilers. Each boiler— 24 feet 3 inches long, 13 feet 6 inches deep and 12 feet high—had six furnaces and 594 three-inch tubes. The engine turned a pair of paddle wheels, each of which was

40 feet in diameter and equipped with 34 floats measuring 12 feet in length and two feet wide, at a leisurely 10 revolutions per minute. The power of this machine, 1500 horsepower, was sufficient for the vessel to make a speed of between 10 and 11 knots—Pacific Mail always emphasized reliability over speed on their transpacific route. Her hull was strongly built. Diagonal iron strapping was fitted to the outside of white oak frames; yellow pine planking was then laid and caulked, followed by another set of diagonal strapping and a second layer of planking— yellow pine above the water line and white oak below.

The China Cabin, a structure removed from Pacific Mail's steamship *China* when she was broken up in 1886, has been lovingly restored at Belvedere, California. The China Cabin is the only extant piece of a Webb-built vessel. Here it is shown from the landward side. (Photograph courtesy of Eugene Schorsch).

The resulting hull was extremely strong, but expensive. The vessel was stated to have cost about $1,000,000.

After completing a trial trip on 4 June 1867 as *Celestial Empire*, she was delivered as *China* to the Pacific Mail Steamship Company later that month, having received her first register document on the 27th. She cleared New York for San Francisco on her delivery voyage via the Straits of Magellan on 1 July 1867. Captain George H. Bradbury was in command.

She arrived at San Francisco on the 18th of September, after an uneventful voyage that included stops en route for coal and a revenue call at Panama. *China* entered the transpacific service of Pacific Mail on 14 October 1867, when she sailed for Yokohama and Hong Kong under the command of Captain E. W. Smith. Her initial voyage, which was typical of many made by the company's side-wheelers, took her first to Yokohama, where she arrived on the 6th of November. Hong Kong was reached on the 14th, and the return leg of the voyage commenced on the 26th. Another call was made at Yokohama, whence she departed for San Francisco on the 6th of December. The California port was reached on the 31st.

*China* completed 31 voyages to the Far East on the company's transpacific service (with trips to Panama interspersed from time to time) before being laid aside in favor of iron-hulled screw steamers, the first of which appeared in the Pacific Mail fleet in 1873. *China* made the company's last Pacific crossing by a sidewheeler during the summer of 1879, and later made occasional trips from San Francisco to Panama or to ports in the Pacific Northwest. She was sold to the Oregon Railway & Navigation Company in July of 1883, served as a receiving ship for smallpox victims in 1884 and, in 1885, was sold for scrap. Towed to Belvedere, near Tiburon, California, for the removal of her machinery, her hull was burned the following year at nearby California City to recover the tons of valuable metal fastenings that were a part of every wooden hull. Mr. Webb's double diagonal iron strapping made this an even more profitable venture.

The only known extant fragment of a vessel built by Mr. Webb is the "China Cabin" at Belvedere, California. This elegant mid-victorian structure, described as a social cabin but probably the enclosed weather-deck entry to the grand staircase that led to her main deck saloon, was removed from the vessel when she was dismantled and used for many years as a private residence and, later, a clubhouse. It has been lovingly restored; its joiner work and etched-glass windows allow today's visitor to experience one small facet of mid-19th century shipboard life. The cabin was originally located aft of her paddle wheels at the foot of her mainmast.

An apocryphal tale that the Pacific Mail sidewheelers had short lives because green timbers had been used in their construction has appeared and reappeared over the years, but the story is simply untrue. Mr. Webb certainly would not have jeopardized his reputation as a shipbuilder of the highest integrity by resorting to such a practice. It is true that these wooden-hulled vessels became obsolete soon after their construction, but the underlying reason was the introduction by Pacific Mail of modern iron-hulled, propeller-driven steamships on both the Pacific and Atlantic in the early 1870s. These new ships were built on the Delaware River, where iron construction was enthusiastically embraced by the shipyards. The builders at New York did not do so, and the center of American shipbuilding soon moved from the East River to the Delaware.

As a result, *China* was nearly the last ocean-going wooden steamship built at New York. Four wooden-hulled screw steamers were subsequently built by John Englis & Sons at Greenpoint for F. Alexandre & Sons' line to Mexico, but the age of ocean-going steamships constructed of white oak, live oak, chestnut and locust came to its inevitable end in 1874, when *City of Vera Cruz* the very last of her type, was handed over to the Alexandre Line. The line's next vessel had an iron hull.

The austere exterior of the China Cabin belies the joy that awaits a visitor inside. Decorated in white and gold, with gold-capitaled brown faux columns and etched-glass clerestory windows, this jewel was typical of the way that steamship and steamboat interiors were decorated during the nineteenth century. Compare these views with that of the saloon of *Bristol* and *Providence* elsewhere in this volume. (Photographs courtesy of Eugene Schorsch).

The Englis yard, unable or unwilling to convert to iron and steel shipbuilding, built wooden-hulled inland vessels for another twenty years and then became a ship joinery sub-contractor, specializing in the installation of wooden superstructures on steel-hulled steamboats built by others.

*Built for the Pacific Mail Steam Ship Company for regular mail and passenger service between San Francisco, Japan and China; had accommodations for 1,200 passengers, was a very beautiful model, a remarkable good sea-boat, and proved a very superior vessel in every respect, with high speed on a very moderate consumption of coal; gave great satisfaction and was the model steamer of the Company for that service. Launching draft with two anchors, 5,000 lbs. each, and two chain cables 120 fathoms each on board, and without any machinery was forward 8 feet 5 in., and aft 9 feet 10 in. Draft with coal 2,107 tons, cargo and extra machinery 105 tons = 2,212 tons (2,240 lbs. each) was 20 feet 4 in. This vessel has double outside planking from keel to wale height.*

### JAMES A. BORLAND
### Hull 134

The bark *James A. Borland* was built in 1869 as a general cargo carrier for the New York firm of S. W. Lewis & Company. She was the first vessel that had been built at Mr. Webb's yard in two years, reflecting the sharp downturn in shipbuilding that followed the Civil War. She measured 143 feet by 32 feet by 18 feet 10 inches, and had a tonnage of 637. Launched at 3:30 on the afternoon of 23 December 1868, *James A. Borland* sailed on her first voyage during the following month under Captain Obed Baker, Jr., who was a part owner of the vessel. Although she was built to be a general cargo carrier—a tramp—it was reported by the press at the time of her launching that "no pains have been spared to make her a first class vessel in every respect." After an extensive rebuild in 1884, she was sold to Lewis S. Davis, of New York, for whom she ran until about 1887 or 1888. At that time, she was sold to Louis Sloss & Company, of San Francisco, under whose ownership she lasted until 1896, her last days probably spent carrying lumber along the Pacific coast. She was wrecked on Tugidak Island, Alaska, on 1 September 1896, shortly after she had departed from Karluk, on Kodiak Island, for an unreported destination. Vessel and cargo were a total loss, but the crew was saved. In an extraordinary coincidence, "Lewis" (or "Louis") appears in the names of each of the three successive owners of *James A. Borland.*

*Built for the general freighting business, proved an excellent sea boat, was a fair carrier, good sailer and desirable*

*class of vessel for the business for which she was intended, gave great satisfaction, was the last vessel barque rigged built in the Port of New York. Draft of water when launched with all masts standing, yards on deck, one anchor and two chains on board forward 6 ft. 4 in., aft 8 ft. 2 in.*

### CHARLES H. MARSHALL
### Hull 135

What better way to honor the most eminent of the packet ship operators than to name the very last of the breed in his honor? The association of Mr. Webb with Charles H. Marshall (who died at New York in September of 1865) began with the packet *Montezuma* in 1843 and led to the construction of a further twelve vessels of the type, plus a steamship and two ocean towboats. *Charles H. Marshall* was a three-decked ship-rigged packet, named for the founder of C. H. Marshall & Company, the proprietors of the Black Ball Line, and designed for trading to the Far East as well as service as a Liverpool packet. Her dimensions were 193 feet by 39 feet 2 inches by 28 feet 6 inches, and her tonnage was 1683. She was launched at 9:30 on the morning of 26 May 1869, at which time she was christened by Miss Chrissie Metzger, the daughter of the shipyard foreman, Christian Metzger. Little is known of the sponsors of most of Mr. Webb's vessels. In this case, however, we know that, as Miss Metzger broke the traditional bottle of champagne over the vessel's bow, she was reported to have stated, in language that now seems archaic but entirely fitting:

"I christen thee 'Charles H. Marshall,' may'st thou be as successful as thy predecessors on the Black Ball Line to whatever part of the world thou may'st go."

*Charles H. Marshall* cleared New York for Liverpool on her first voyage on the 14th of July, with Captain Charles A. Marshall, a son of Alexander C. Marshall, in command. The vessel remained active until the end of the Black Ball Line and well beyond, her latter days under the hand of Captain Charles Hutchinson, whose career as a Black Ball Line shipmaster began in the 1850s. After the line gave up regular passages to Liverpool in 1879, they dispatched their vessels to other ports, including Bremen, Antwerp and London, with occasional trips to Liverpool. But *Charles H. Marshall* ended her trips to Liverpool long before the line folded. Her last scheduled voyage as a Black Ball liner ended on 12 April 1878, when she arrived at New York from Liverpool after a passage of 33 days. Her next assignment was a voyage to San Francisco for the New York mercantile house of

Simonson & Howes, commencing on 8 June 1878. She did not return to New York until 27 August 1879, via the Far East and Liverpool. Next came four voyages to European ports, the last one, which ended with the arrival of the vessel on 15 September 1881, apparently marking the end of C. H. Marshall & Company's days as a ship operator. After the firm ceased running its vessels, *Charles H. Marshall* was acquired by Captain Hutchinson and chartered to the mercantile firm of Snow & Burgess, for whom she ran on the North Atlantic for a few years, making occasional passages to London or Antwerp as cargo became available. She was sold to A. Melin, of Stavanger, Norway, in late 1887 and renamed *Souverain*. The vessel never returned to New York, and caught fire and was scuttled while loading coal at Penarth Docks, near Cardiff, Wales, on 20 March 1891. Her remains were later sold and hulked.

*Charles H. Marshall*, a maritime anachronism when she was constructed, was not only the last vessel built by Mr. Webb, but was the last square-rigged vessel built at New York—city and state. The shipyards of Maine would carry on the tradition of constructing large sailing vessels (including a handful built of steel) for another thirty years, as the steamship encroached even further and the supply of shipbuilding timber—which at one time had appeared to be endless—dwindled. By the time *Charles H. Marshall* was launched in May of 1869, regularly scheduled packets on the North Atlantic had nearly ceased to exist. Of the many lines that prospered before the Civil War, only the Black Ball Line to Liverpool and the Red Swallowtail Line to London continued to sail on a scheduled basis. Black Ball lasted until 1879 and Red Swallowtail, owned by Grinnell, Minturn & Company, struggled on until 1881. Both lines wound up their days carrying cargo only, the emigrant trade having been taken over by steamships. Thomas Dunham's Nephew & Company was the last of the former emigrant carriers to survive. That firm's final voyage, by the ex-Black Ball liner *Hamilton Fish*, ended in December of 1894. Dunham's remaining packets were then rebuilt as schooner barges to carry coal along the east coast. *Hamilton Fish*, built in 1856 at Waldoboro, Maine, as *Wm. F. Storer*, was the only vessel acquired by the Black Ball Line after 1843 that had not been built by Mr. Webb. She ended her days as a coal barge under the ownership of Lewis Luckenbach, the founder of the Luckenbach Steamship Company. Lewis Luckenbach's nephew, J. Lewis Luckenbach, was a fondly remembered trustee and benefactor of Webb Institute of Naval Architecture. *Hamilton Fish*, the last survivor of the Black Ball Line, burned and sank on the night of 6 December 1906 off Barnegat, New Jersey, while bound from Newport News to Providence under tow of the tug *Edgar F. Luckenbach*.

It is perhaps fitting that the building that housed the offices of C. H. Marshall & Company at 38 Burling Slip (now John Street) is still standing as a little-recognized reminder of the Black Ball Line and, indirectly, of William H. Webb and his last vessel. Appropriately, it forms a part of a restaurant that, in name at least, commemorates the golden age of American sail—"The Yankee Clipper."

*Three decker. Built in 1869 for the European and East Indian trade. Launched May 26th, 1869, last vessel built by W. H. Webb and last square-rigged vessel built in New York City or State. Was a large carrier and a good sea boat. Draft of water when launched with masts in end, yards on deck, one anchor, and two chain cables (90 fathoms each) on board was forward 10 ft. 3 in., aft 10 ft. 5 in. Deep draft 24 ft.*

## ACKNOWLEDGEMENTS

In writing of 19th-century marine history, one is heavily dependent upon extant source materials that originally were part of the day-to-day operations of the shipbuilders, ship owners, commission merchants and underwriters of the period under review. What in 1860 were routine working documents containing routine information—such as ledgers, crew lists, cargo manifests or ship's logbooks—today hold the keys that unlock many doors to some particularly knotty aspects of the historian's subject. Two organizations were most helpful in making contemporary records available for our use.

The Marine Library of the Atlantic Mutual Companies (of whose predecessor company, Atlantic Mutual Insurance Company, Mr. Webb was a trustee from 1865 to 1899) was a valuable reference source. The "Vessel Disaster Books" in this library—over three hundred bound volumes of clippings and notes describing partial and total losses of vessels from 1852 onwards—represented a veritable Golconda in the search for clues to the loss of many of the vessels. We wish to thank Atlantic Mutual for its kindness in permitting us to use the library. We commend Atlantic Mutual for its keen sense of history, usually so lacking in the modern corporate world, in maintaining this priceless collection.

Of great assistance in tracing the owners of the ships and in ferreting out the technical details of the vessels was the nearly complete collection of early record books of the American Shipmasters' Association (the predecessor of the American Bureau of Shipping) owned by the latter organization. We are indebted to ABS for allowing us to consult these volumes.

Also to be thanked are Webb Institute of Naval Architecture and the library of the South Street Seaport Museum, whose extensive collections were consulted from time to time during the preparation of this work. The treasure trove found in a scrapbook kept by Mr. Webb from the late 1840s through the 1870s was of inestimable value. Despite a lack of dates (a common failing in many scrapbooks of the type) it is one of the gems of what little we have of Webb memorabilia.

## BIBLIOGRAPHICAL NOTES

Professor Dunbaugh's difficulty in uncovering the details of Mr. Webb's personal life was not duplicated in writing of many of his ships—in particular the clippers, where scores of persons—some entirely competent, others less so—have written of the vessels and their careers. It is easy to compile an extensive bibliography that addresses the clippers, but many authors have embellished their stories with romantic codswallop to the point that they seem to have become the wave-borne chariots of their god-like masters. That most of these vessels were handsome is readily granted, but it must not be forgotten that these were hard-working craft built for the express purpose of making a profit for their owners. A list of the authors who recognized this is a short one indeed.

The lives of many of Mr. Webb's lesser ships tax the ingenuity of the researcher. Some of the vessels, unheralded by even the most meticulous of marine historians, seem to have left little or no wake once they left the shipyard.

The list of published works below includes the major reliable sources consulted in the preparation of the ship biographies which form the latter part of this volume. Each contributed some small part to the whole, and each author represented below typically addressed only one aspect of the overall task.

Of particular importance were the major works of Carl C. Cutler, Professor Cedric Ridgely-Nevitt and Howard I. Chapelle. Cutler is well-known for his highly readable volumes relating to clipper ships and packets, in which he was able to steer well clear of the romanticism and inaccuracy that have marred the efforts of most authors. Mr. Webb's ships play an important role in Cutler's writings.

Professor Ridgely-Nevitt's deep interest was the American ocean steamship. An important and lasting contribution to marine history is his authoritative work that details the development of American steamships on the Atlantic up to about 1870. Again, Mr. Webb's accomplishments appear prominently throughout the text.

Chapelle wrote extensively of the American sailing ship, but his most valuable contribution remains his erudite and insightful analysis of the evolution of the fast sailer, or, in his words, the "search for speed under sail." Needless to say, Mr. Webb's design philosophies are dissected alongside those of McKay, Griffith, Pook and the other prominent naval architects of the day.

Many hours were spent in examining the ship arrival and departure columns of the *New York Times* for the period from 1851 to the mid-1880s, as well as those of the *New York Maritime Register* and the *New York Shipping and Commercial List* for this period and earlier. These provided much data relating to the movements of many of Mr. Webb's ships, and to the fortunes and misfortunes of the companies, ships and men that made up the 19th century. Placed in perspective, the industry's troubles today seem in essence to be little different from—and no more serious than—those that recurred with alarming regularity a century or more ago, or, to quote the malapropism attributed to baseball's Yogi Berra, "It's deja vu all over again."


Albion, Robert Greenhalgh, *The Rise of New York Port [1815-1860]*, New York, N.Y., 1939 (Second Edition 1970)

Albion, Robert Greenhalgh, *Square-Riggers on Schedule: The New York Sailing Packets to England, France and the Cotton Ports*, Princeton, New Jersey, 1938 (Reprinted by Archon Books, Hamden, Connecticut, 1965)

Bonsor, N. R. P., *North Atlantic Seaway*, Volumes I and II, Brookside Publications, Jersey, Channel Islands, 1975 and 1978

Chappelle, Howard I., *The History of American Sailing Ships*, W. W. Norton & Co., New York, N.Y., 1935

Chappelle, Howard I., *The Search for Speed Under Sail*, W. W. Norton & Co., New York, N.Y., 1967

Covell, William King, *A Short History of the Fall River Line*, Newport, R.I., 1947

Crighton, Richard E., *The Wreck of San Francisco*, The American Neptune, Vol. XLV No. 1, Winter 1985

Cutler, Carl C., *Greyhounds of the Sea*, Halcyon House, New York, N.Y., 1930

Cutler, Carl C., *Five Hundred Sailing Records of American Built Ships*, The Marine Historical Association, Inc., Mystic, Connecticut, 1952

Cutler, Carl C., *Queens of the Western Ocean*, United States Naval Institute, Annapolis, Md., 1961

Fairburn, William Armstrong, *Merchant Sail*, Fairburn Marine Educational Foundation, Inc., Center Lovell, Maine, 1945-55

Heyl, Erik, *Early American Steamers*, Buffalo, New York, 1953-1969

Holdcamper, Forrest R., Compiler, *List of American-Flag Merchant Vessels That Received Certificates of Enrollment or Registry at the Port of New York 1789-1867*, The National Archives, Washington, D.C., 1968

Howe, Octavius T., M.D., and Frederick C. Matthews,

*American Clipper Ships 1833-1858*, Maritime Research Society, Salem, Massachusetts, 1926

Matthews, Frederick C., *American Merchant Ships, 1850-1900*, Maritime Research Society, Salem, Massachusetts, 1930

Nordhoff, Charles, *Life on the Ocean*, Cincinnati, Ohio, 1874. (Facsimile edition, Jerome Ozer Publishing, Inc., 1970.)

Ridgely-Nevitt, Cedric, *The United States Mail Steamer George Law*, The American Neptune, Vol. IV No. 4, 1944

Ridgely-Nevitt, Cedric, *American Steamships on the Atlantic*, University of Delaware Press, Newark, Del., 1981

Sibley, Marilyn McAdams, *Charles Stillman: A Case Study of Entrepreneurship on the Rio Grande, 1861-1865*, Southwestern Historical Quarterly, Volume LXXVII No. 2, October 1973

Stackpole, Edouard A., *The Wreck of the Steamer San Francisco*, The Marine Historical Association, Inc., Mystic, Connecticut, 1954

Whipple, A. B. C., *The Challenge*, William Morrow & Co., Inc., New York, 1987

Yanaway, Philip E., *The United States Revenue Cutter Harriet Lane*, The American Neptune, Vol. XXXVI No. 3, July 1976

*American Lloyd's Register of American and Foreign Shipping*, E. & G. W. Blunt, New York, N.Y., 1863

*Dictionary of American Naval Fighting Ships*, Volumes I through VIII, Naval Historical Center, Department of the Navy, Washington, D.C., 1959-81

*List of Merchant Vessels of the United States*, United States Treasury Department, Bureau of Navigation, Washington, D.C., Various dates, 1884-1901

*Merchant Steam Vessels of the United States 1790-1868*, The Steamship Historical Society of America, Inc., Staten Island, N.Y., 1975

*Photographic Portraits of American Ocean Steamships 1850-1870*, The Steamship Historical Society of America, Inc., Providence, R.I., 1986

*Plans of Wooden Vessels Selected as Types from One Hundred and Fifty of Various Kinds and Descriptions, from a Fishing Smack to the Largest Clipper Ships and Vessels of War, Both Sail and Steam, Built by Wm. H.*

*Webb, in the City of New York, from the Year 1840 to the Year 1869,* (The "Book of Plates"), William H. Webb, New York, Undated [About 1895]

*Preliminary List of Merchant Vessels of the United States,* United States Treasury Department, Director of the Bureau of Statistics, Washington, D.C., 1868

*Record of American and Foreign Shipping,* American Shipmasters' Association, New York, N.Y., Various dates, 1869-1898

*The New York Times,* Marine Intelligence columns, Various dates, 1851 through 1887

*New York Maritime Register,* Various dates, 1876 through 1887

*New York Shipping and Commercial List,* Marine Intelligence columns, Various dates, 1840 through 1867

# Index of Ships' Names

Entries in capital letters indicate original names of vessels built by William H. Webb.